Policy for Agricultural Research

About the Book and Editors

The contributors to this volume, based on the Agriculture Research Seminars held annually at the University of Minnesota, examine the role of government, multinationals, and the emerging private sector (in both domestic and international contexts) in determining agricultural research policy.

Vernon W. Ruttan is Regents Professor at the University of Minnesota. *Carl E. Pray* is associate professor of agricultural economics, Rutgers University.

Policy for Agricultural Research

edited by
Vernon W. Ruttan
and Carl E. Pray

Westview Press / Boulder and London

Westview Special Studies in Agriculture Science and Policy

This Westview softcover edition is printed on acid-free paper and bound in softcovers that carry the highest rating of the National Association of State Textbook Administrators, in consultation with the Association of American Publishers and the Book Manufacturers' Institute.

Published in 1987 in the United States of America by Westview Press, Inc.; Frederick A. Praeger, Publisher; 5500 Central Avenue, Boulder, Colorado 80301

Library of Congress Cataloging-in-Publication Data
Policy for agricultural research.
(Westview special studies in agriculture science and policy)
1. Agriculture--Research. 2. Agriculture and state. I. Ruttan, Vernon W. II. Pray, Carl E. III. Series.
S540.A2P65 1987 630'.72 87-14277
ISBN 0-8133-7369-7

Composition for this book was provided by the editors.
This book was produced without formal editing by the publisher.

Printed and bound in the United States of America

The paper used in this publication meets the requirements of the American National Standard for Permanence of Paper for Printed Library Materials Z39.48-1984

6 5 4 3 2 1

CONTENTS

LIST OF TABLES

LIST OF FIGURES

PREFACE

The chapters in this book represent selections from a series of Agricultural Research Policy Seminars held at the University of Minnesota in 1983, 1984, 1985, and 1986. The seminars were sponsored by the University of Minnesota Agricultural Experiment Station, the International Service for National Agricultural Research (ISNAR), and the United Nations Food and Agriculture Organization (FAO).

Each year the seminar attracted thirty-five to forty senior agricultural research policymakers and administrators from the United States and abroad. In the 1983-1986 seminars, the book by V. W. Ruttan, Agricultural Research Policy, was used as a core reference for seminar participants.

The papers included in this book were selected with a view to using the book as a core reference for future national and international seminars and workshops focusing on agricultural research policy.

The authors are indebted to Mary Strait for editorial assistance and to Janet Cardelli and Linda Schwartz for translating the drafts onto the word processor for transmission to Westview Press.

Vernon W. Ruttan
Carl E. Pray

PART 1

GLOBAL PERSPECTIVE

INTRODUCTION

Vernon W. Ruttan

We are, in the closing years of the twentieth century, completing one of the most remarkable transitions in the history of agriculture. Before the beginning of this century almost all increases in food production were obtained by bringing new land into production. There were only a few exceptions--in limited areas of East Asia, the Middle East, and Western Europe. By the first decade of the next century almost all increases in world food production must come from higher yields--from higher output per hectare.

In most areas of the world the transition from a resource-based to a science-based system of agriculture is occurring within a single century. In a few countries this transition began in the nineteenth century. In most of the presently developed countries it did not begin until the first half of this century, and most of the countries of the developing world have been caught up in this transition only since mid-century.

In the first paper in this section, Judd, Boyce and Evenson document the exceptionally rapid growth of agricultural research capacity in the developing world between the late 1950s and the early 1980s. At the beginning of this period the developing countries were typically highly extension-intensive. Extension workers and extension budgets exceeded research workers and research budgets by several multiples. Although research expenditures and budgets have grown more rapidly, particularly since 1970, the developing countries are still relatively extension-intensive.

The Judd-Boyce-Evenson paper also sheds new light on a controversy about the impact of the International

Agricultural Research Centers (IARCs), sponsored by the Consultative Group on International Agricultural Research (CGIAR), on the development of national agricultural research systems. A number of observers, including Evenson himself, had speculated that the investment in the IARCs tended to displace investment in national agricultural research systems. Statistical analysis indicates, however, that investment in the CGIAR centers has typically had a positive impact on investment in national agricultural research systems. The research conducted by the IARCs has, in effect, reduced the national centers' costs of advancing technology. National governments have responded by increasing their own research investments to take advantage of more profitable research opportunities. In addition to the favorable effect of the international centers on national research investment, Judd, Boyce, and Evenson document the substantial direct impact of the international centers on crop productivity and production.

The paper by Jock R. Anderson and Robert W. Herdt documents in greater detail the impact of the international centers on crop production and productivity. Their paper draws on a massive study conducted by the CGIAR in 1984 and 1985. The initial impacts were greatest in the case of wheat and rice. By the early 1980s there were beginning to be substantial impacts on the production of potatoes, field beans, cassava, maize, sorghum, and forage crops. The system has played a major role in germplasm collection and enhancement, farming systems innovations, and institutional arrangements in the areas of commodity policy and rural development. The centers have also been an important source of training for agricultural research personnel in the developing world.

In the last paper in this section I discuss some of the challenges that face both the international and national agricultural research systems as we move toward the turn of the century. The international system remains incomplete. A number of nonassociated centers supported by the same group of donors that fund the CGIAR system have grown up outside the system. Some of them address resource maintenance, development, and management issues in addition to commodity production. Their funding is less secure and the system of monitoring their performance is less adequate than in the case of the CGIAR centers. A global system that effectively links the IARCs, the national systems in the developing countries, the national systems of the developed market economies, and the centrally planned economies has yet to emerge.

A major challenge over the next several decades, for both national governments and assistance agencies, is to strengthen the national research systems. In spite of the progress that has been made, there are probably not more than half a dozen national systems in developing countries that are effectively institutionalized--that have achieved the professional, managerial, and fiscal viability necessary to perform at a level consistent with the needs of their countries.

Achieving viability in the national systems of the smaller countries poses particularly difficult problems. These systems will, of necessity, remain dependent on other national systems, regional networks, and the international system for advanced training, for advances in basic knowledge, for improvements in research methods, and for much of their technology development.

I also argue, in this paper, that new ways must be sought by the international community to provide financial assistance to developing national systems. Over the longer run, a move toward a formula approach in which donor support is based on national performance would seem desirable.

1 Investment in Agricultural Research and Extension*

M. Ann Judd, James K. Boyce, and Robert E. Evenson

The capacity to increase the supply of food and other agricultural products is of obvious importance to developing countries, particularly those faced with rapidly growing populations. Historically, growth in agricultural supply has passed through stages. As long as it was possible to bring new areas under cultivation at low cost, increased agricultural supply was achieved primarily through the expansion of cultivated area. As low-cost land conversion possibilities became exhausted, higher-cost sources of growth were exploited, notably investments in irrigation and drainage. Investments in agricultural research and extension systems have also produced growth in agricultural supply, as documented by a large number of economic studies. These studies, however, have revealed strong interactions of improved agricultural technology with soil and climate factors, which impede the diffusion of technology across broad regions. Agricultural experiment stations and associated extension services must, therefore, be located to serve specific subregions if their growth-producing potential is to be realized fully (Kislev and Evenson 1975; Evenson, Waggoner, and Ruttan 1979).

Agricultural research systems are complex institutions. The formidable public good problems encountered in most branches of scientific research are particularly acute in the case of agriculture. Patent laws protect mechanical and chemical inventions more easily than biological inventions. Hence crop and livestock research--plant breeding, phytopathology, entomology, agronomy, soil science, animal nutrition, and so on--are today primarily public-sector activities. The amount and type of agricultural research provided by the public sector in a given country is influenced by a number of factors, including the supply of research personnel and public funds, the perceived opportunities for productive research, and the political strength of those who stand to gain from it whether as agricultural producers, input suppliers, or consumers. Today the industrialized countries retain their historical dominance in worldwide agricultural research investment, but recent decades have witnessed substantial growth in agricultural research in developing countries. In addition to the national systems, many of which have grown rapidly, sixteen International Agricultural Research Centers (IARCs) have been established beginning with the forerunner of CIMMYT (the International Wheat and Maize Improvement Center) in Mexico in the 1940s. International aid donors have provided considerable funding for national agricultural research systems as well as the IARCs (Oram and Bindlish 1981).

If agricultural technology were highly transferable between countries, a strategy of heavy investment in extension could be expected to yield high returns. Extension programs could "screen" internationally available technology for effectiveness and extend this technology to farmers. A reading of early development literature and a perusal of government and aid agency budgets indicate that this was the chief agricultural development strategy during the 1950s and 1960s. In general, however, this strategy produced very little growth. Agricultural technology simply is not very transferable.

As countries recognized this fact, they began to expand agricultural research programs, and thereby to facilitate a somewhat more sophisticated indirect transfer of technology. A country could benefit from the research programs of neighboring countries and international centers by <u>adapting</u> their research findings to its own geo-climatic conditions. Depending on the degree of complementarity between its own research and that of its neighbors, a country may have an incentive to "free ride" on its neighbor's research. That is, if a neighbor is undertaking an active research program, the country may find it feasible to invest in only a minimal adaptive research program. In the absence of good neighbors, the country would have to spend more on research to achieve the same growth objective.

In this paper we first present data on trends in national agricultural research and extension programs between 1959 and 1980. We then present the findings of a statistical analysis of the determinants of investments in agricultural research and extension programs (Judd, Boyce, and Evenson, 1983).[1] In the last section we discuss policy implications that these findings have for individual countries, for donors, and for the International Agricultural Research Center (IARC) system.

A SURVEY OF PUBLIC SECTOR AGRICULTURAL RESEARCH AND EXTENSION

The tables in this section summarize data on research and extension expenditures and manpower as well as measures of research and extension intensity. The data are summarized by geographic region for three time periods: 1959, 1970, and 1980. The data are drawn from an earlier work by the authors that contains constructed time-series data for approximately 106 countries for eight time periods between 1959 and 1980 (Judd, Boyce, and Evenson 1983).

Compiling data of this kind that will be comparable on an international basis is subject to several problems. One of the most difficult is that of currency conversion, especially for countries that have experienced very high rates of inflation (e.g., Brazil and Argentina). Another problem is that of "quality" variation in scientists. Standards for what constitutes a research scientist vary from country to country. In addition, in many countries forestry and fisheries research is funded through the same agency or agencies as agricultural research. Wherever possible, expenditures on forestry and fisheries research were eliminated, but totals for some countries may still include these categories. The reader, therefore, must interpret the data bearing these problems in mind.

EXPENDITURE AND MANPOWER DATA

Tables 1.1 and 1.2 provide data on agricultural research and extension expenditures and manpower by geographic region. Total public-sector agricultural research spending in 1980 was approximately US$7.4 billion; public-sector extension spending was approximately US$3.4 billion. These data do not include spending by the IARCs of approximately US$140 million in 1980. National research systems have increased real spending by a multiple of 3.68 since 1959 (and by a multiple of 1.4 since 1970). Scientist man-years have increased by a multiple of 3.14 since 1959, reflecting the rise in expenditures per

TABLE 1.1
Agricultural Research Expenditures and Manpower

Region/Subregion	Expenditures (Constant 1980 US$ Thousands)			Manpower (SMY)		
	1959	1970	1980	1959	1970	1980
Western Europe:	247,984	918,634	1,489,588	6,251	12,547	19,540
Northern Europe	94,718	230,135	409,527	1,818	4,409	8,027
Central Europe	141,054	563,334	871,233	2,888	5,721	8,827
Southern Europe	39,212	125,165	208,828	1,545	2,417	2,686
Eastern Europe and Soviet Union:	568,284	1,282,212	1,492,783	17,701	43,709	51,614
Eastern Europe	195,896	436,094	553,400	5,701	16,009	20,220
Soviet Union	372,388	846,118	939,383	12,000	27,700	31,394
North America and Oceania:	760,466	1,485,043	1,722,390	8,449	11,688	13,607
North America	688,889	1,221,006	1,335,584	6,690	8,575	10,305
Oceania	91,577	264,037	386,806	1,759	3,113	3,302
Latin America:	79,556	216,018	462,631	1,425	4,880	8,534
Temperate South America	31,088	57,119	80,247	364	1,022	1,527
Tropical South America	34,792	128,958	269,443	570	2,698	4,840
Caribbean and Central America	13,676	29,941	112,941	491	1,160	2,167

Africa:	119,149	251,572	424,757	1,919	3,849	8,088
North Africa	20,789	49,703	62,037	590	1,122	2,340
West Africa	44,333	91,899	205,737	412	952	2,466
East Africa	12,740	49,218	75,156	221	684	1,632
Southern Africa	41,287	60,752	81,827	696	1,091	1,650
Asia:	261,114	1,205,116	1,797,894	11,418	31,837	46,656
West Asia	24,427	70,676	125,465	457	1,606	2,329
South Asia	32,024	72,573	190,931	1,433	2,569	5,691
Southeast Asia	9,028	37,405	103,249	441	1,692	4,102
East Asia	141,469	521,971	734,694	7,837	13,720	17,262
China	54,166	502,491	643,555	1,250	12,250	17,272
World total	2,063,553	5,358,595	7,390,043	47,163	108,510	148,039

Sources: J. K. Boyce and R. E. Evenson, National and International Agricultural Research and Extension Programs (New York: Agricultural Development Council, 1975); and M. Ann Judd, James K. Boyce, and Robert E. Evenson, "Investing in Agricultural Supply," Discussion Paper no. 442 (New Haven, Conn.: Yale University, Economic Growth Center, 1983).

TABLE 1.2
Agricultural Extension Expenditures and Manpower

Region/Subregion	Expenditures (Constant 1980 US$ Thousands)			Manpower (Workers)		
	1959	1970	1980	1959	1970	1980
Western Europe:	234,016	457,675	514,305	15,988	24,388	27,881
Northern Europe	112,983	187,144	201,366	4,793	5,638	6,241
Central Europe	103,082	199,191	236,834	7,865	13,046	14,421
Southern Europe	17,950	71,340	76,105	3,330	5,704	7,219
Eastern Europe and Soviet Union:	367,329	562,935	750,301	29,000	43,000	55,000
Eastern Europe	126,624	191,460	278,149	9,340	15,749	21,546
Soviet Union	240,705	371,475	472,152	19,660	27,251	33,454
North America and Oceania:	383,358	601,950	760,155	13,500	15,113	14,966
North America	332,892	511,833	634,201	11,500	12,550	12,235
Oceania	50,466	90,067	125,954	2,080	2,563	2,731
Latin America	61,451	205,971	396,944	3,353	10,782	22,835
Temperate South America	5,741	44,242	44,379	205	1,056	1,292
Tropical South America	47,296	136,943	294,654	2,369	7,591	16,038
Caribbean and Central Ameria	8,414	24,786	57,911	779	2,135	5,505

Africa:	237,883	481,096	514,671	28,700	58,700	79,875
North Africa	84,634	176,498	172,910	7,500	14,750	22,453
West Africa	53,600	181,324	204,982	9,000	22,000	29,478
East Africa	39,496	86,096	106,030	9,000	18,750	24,211
Southern Africa	60,153	37,178	30,749	3,200	3,200	3,733
Asia:	143,876	412,937	507,113	86,900	142,500	148,780
West Asia	28,211	97,315	119,780	7,000	18,800	16,535
South Asia	56,422	87,727	82,194	57,000	74,000	80,958
Southeast Asia	19,747	55,441	63,959	9,500	30,500	33,987
East Asia	39,496	172,454	241,180	13,400	19,200	17,300
China	N.A.	N.A.	N.A.	N.A.	N.A.	N.A.
World total	1,427,913	2,722,564	3,443,489	177,521	294,483	349,337

Sources: See table 1.1

scientist over the period. The comparable multiples for extension spending and manpower are 2.50 and 2.05.

From table 1.1 it is evident that the industrialized regions of the world--Western Europe, Eastern Europe/Soviet Union, and North America/Oceania--continue to spend the most on agricultural research. In addition, China appears to have become, by 1970, one of the "big spenders" in this area.[2] However, if one were to calculate expenditure shares, these would reveal the declining importance of both North America/Oceania and Eastern Europe/Soviet Union relative to the rest of the world and the increasing importance of Western Europe, Latin America, and Asia. When China is included, Asia's share of world expenditures almost doubles between 1959 and 1980 (rising from 12.7% to 24.3%); without China it increases from 10.3% to 17.1%. Latin America's share rises from 3.9% to 6.3%, and Western Europe's share increases from 13.3% to 20.2%. At the same time, the share of Eastern Europe/Soviet Union drops from 27.5% to 20.2%, whereas that of North America/Oceania declines sharply from 36.9% to 23.3%. Africa's share of total expenditures remained virtually unchanged during the period, indicating that the region has generally not experienced an expansion in its research program comparable to that of Latin America and Asia.

The pattern revealed by the scientist man-year (SMY) figures is somewhat different from those that characterize the expenditure figures. Eastern Europe/Soviet Union and Asia accounted for over 60% of the world's SMYs in both 1959 and 1980 (this is true even if one excludes China). The share of Eastern Europe/Soviet Union declined somewhat between 1959 and 1980 (from 37.5% to 34.9%), but it remains several times larger than that of either Western Europe (13.2% in 1980) or North America/Oceania (9.2% in 1980), a reflection of the relative capital intensity of the research programs in these latter regions.[3] Although Africa's share of world SMYs only increased from 4.1% to 5.5% during this period, SMYs in 1980 were over four times what they had been in 1959. Many African countries have gone through a postcolonial adjustment period during which highly paid British and French civil servants have been replaced by somewhat lower paid national scientists. Therefore, these countries have been able to increase the number of SMYs devoted to agricultural research at a faster rate than they have expanded expenditures.

The increase in the number of SMYs in Latin America follows the same pattern as the increase in expenditures. Tropical South America (mainly Brazil) and the

Caribbean/Central America (mainly Mexico) have experienced the greatest increases. South Asia and Southeast Asia have both increased SMYs more slowly than expenditures.[4]

Computing shares for extension expenditures would reveal that, in 1980, the international distribution of this spending was remarkably even--Western Europe accounted for 15.6% of worldwide expenditures, Eastern Europe/Soviet Union 21.0%, North America/Oceania 22.4%, Latin America 12.7%, Africa 14.8%, and Asia 13.5%. Although the industrialized regions were spending slightly more than other regions, differences in expenditure levels are much less significant than in the case of research. This points out quite clearly that the industrialized regions have placed more emphasis on research than on extension, while the developing regions have tended to do the opposite. The most dramatic change between 1959 and 1980 was the expansion of extension expenditures in Latin America (Latin America's share increased from 4.3% in 1959 to 12.7% in 1980). Africa's share decreased slightly during the period, but the region continues to spend more on extension than either Latin America or Asia (excluding China).

The data on extension workers show a very different distribution pattern. In 1959, almost half of the world's extension workers were in Asia, and this share was still substantial in 1980, 45.1%. If extension data were available for China, Asia's share would, of course, be much larger. The number of extension workers in Latin America increased dramatically between 1959 and 1980--from 3,353 to almost 23,000. Within Latin America, the countries of Tropical South America account for 70% of all extension workers. Worldwide, Latin America's share of extension workers increased from 1.9% to 7.2%.

MEASURES OF EXPENDITURE AND MANPOWER INTENSITIES

In table 1.3, expenditures on research and extension are given as a percentage of the value of agricultural product. This measure has been calculated for five country groups as well as for the geographic regions. Countries were assigned to a group based on the classification used by the World Bank in its World Development Report, 1980 (World Bank 1980). The only change made in the classification was to divide the middle-income category into two parts--middle-income developing and semi-industrialized--using a per capita income level of $1,050 as the dividing line between the two classes.[5]

TABLE 1.3
Research and Extension Expenditures as Percentage of the Value of Agricultural Product: Public Sector

	Agricultural Research Expenditures			Agricultural Extension Expenditures		
Subregion/Country Group	1959	1970	1980	1959	1970	1980
Northern Europe	.55	1.05	1.60	.65	.85	.84
Central Europe	.39	1.20	1.54	.29	.42	.45
Southern Europe	.24	.61	.74	.11	.35	.28
Eastern Europe	.50	.81	.78	.32	.36	.40
Soviet Union	.43	.73	.70	.28	.32	.35
Oceania	.99	2.24	2.83	.42	.76	.98
North America	.84	1.27	1.09	.42	.53	.56
Temperate South America	.39	.64	.70	.07	.50	.43
Tropical South America	.25	.67	.98	.34	.71	1.19
Caribbean and Central America	.15	.22	.63	.09	.18	.33
North Africa	.31	.62	.59	1.27	2.21	1.71
West Africa	.37	.61	1.19	.58	1.24	1.28
East Africa	.19	.53	.81	.67	.88	1.16
Southern Africa	1.13	1.10	1.23	1.64	.67	.46
West Asia	.18	.37	.47	.25	.57	.51
South Asia	.12	.19	.43	.20	.23	.20
Southeast Asia	.10	.28	.52	.24	.37	.36
East Asia	.69	2.01	2.44	.19	.67	.85
China	.09	.68	.56	N.A.	N.A.	N.A.
Country group:[a]						
Low-income developing	.15	.27	.50	.30	.43	.44
Middle-income developing	.29	.57	.81	.60	1.01	.92
Semi-industrialized	.29	.54	.73	.29	.51	.59
Industrialized	.68	1.37	1.50	.38	.57	.62
Planned	.33	.73	.66	...	...	...
Planned, excluding China	.45	.75	.73	.29	.33	.36

Sources: See table 1.1; and U.S. Department of Agriculture, Indices of Agricultural Production (various issues).

[a]For definition of country groups, see n. 5.

All regions increased the percentage of agricultural product invested in agricultural research between 1959 and 1980. Eastern Europe, the Soviet Union, and North America, however, experienced declines in the ratios during the 1970s. This measure is as low as it is for North America primarily because the United States has been spending only about 1% of the value of its agricultural product on research (0.98% in 1980 compared with, e.g., 3.34% in

Japan, 3.03% in Australia, 2.26% in New Zealand, and 2.22% in Canada). Some caution is required in drawing conclusions about the situation in Eastern Europe and the Soviet Union, since the recent data for these countries are estimates, but it appears that the relative lack of expansion that currently characterizes the U.S. system would describe the systems of Eastern Europe and the Soviet Union as well.

While almost all regions have increased the percentage of agricultural product invested in research, the basic relationship between the industrialized and developing countries has not changed very much since 1959. It is still the case that the more affluent a country is, the more it is apt to spend on research relative to the value of its agricultural product. The relationship between middle-income developing and semi-industrialized countries has changed somewhat, with the middle-income countries achieving roughly the same ratios as the semi-industrialized countries by 1980.

The data for extension show generally that the lower-income countries spend a larger share of the value of agricultural product on extension than on research, and higher-income countries spend a smaller share. Five regions--North, West, and East Africa; Tropical South America; and West Asia--are investing more in extension than in research. These differences in investment levels can perhaps be explained by factors such as the number of farms per unit of output and the schooling or literacy levels of farmers. In general, the agricultural systems of developing countries are made up of many small farms, whereas those of developed countries comprise relatively few large farms. Education and literacy levels are also higher in developed countries. It can be argued, therefore, that lower-income countries and regions are spending a higher share of the value of agricultural product on extension simply because there are more farmers to reach than in developed countries and their educational needs are greater. Unfortunately, the data bases used in this paper do not include either numbers of farms or measures of literacy, so this point cannot be examined in depth.

Table 1.3 does show, however, that by 1980 low-income developing countries as a whole were actually spending a slightly higher percentage of the value of their agricultural product on research than on extension. This indicates a major change in priorities since 1959, when the percentage for extension expenditures was twice as high as that for research expenditures.

Table 1.4 presents calculations of SMYs and extension workers per US$10 million (constant 1980 dollars) of agricultural product by geographic subregion and country group. In general, the industrialized countries engage more scientists and fewer (many fewer) extension workers for every dollar of agricultural product than do the developing countries. The number of SMYs per dollar of product is highest for the planned and middle-income developing economies; the number of extension workers per dollar of product is highest for low-income developing countries. The SMYs relative to the value of agricultural product have been increasing over time in almost all regions and for almost all country groups. Extension workers relative to agricultural product have changed little or decreased somewhat in Western Europe and North America/Oceania and have tended to increase in the regions of Latin America, Asia, and Africa. Most regions of Asia and Africa have very high levels of extension workers per dollar of product.

Although table 1.3 indicated that the developing countries may be beginning to shift their priorities from extension to research, table 1.4 shows that the number of extension workers relative to the value of agricultural product has increased steadily since 1959 for the developing countries.

EXPENDITURES PER MANPOWER UNIT

The data on expenditures relative to manpower reported in table 1.5 show immediately that, without exception, expenditures per SMY are substantially higher than those per extension worker. Not only are salaries for research scientists higher than those for extension workers, but the level of support (laboratory facilities, technicians, etc.) required to maintain an effective research program is many times that required to maintain an effective extension program.

The levels of expenditures per SMY in 1980 were highest in North America/Oceania and lowest in Asia and Eastern Europe/Soviet Union. This was also the case in 1959. Expenditures per SMY have been decreasing in Africa, indicating that the cost of a research scientist has been declining, probably as the result of a shift from expatriate to indigenous personnel. It is interesting to

TABLE 1.4
Research and Extension Manpower Relative to the Value of Agricultural Product

Subregion/Country Group	SMYs Per US$10 Million (Constant 1980) Agricultural Product			Extension Workers Per US$10 Million (Constant 1980) Agricultural Product		
	1959	1970	1980	1959	1970	1980
Northern Europe	1.05	2.01	3.14	2.76	2.56	2.61
Central Europe	.80	1.21	1.56	2.19	2.77	2.73
Southern Europe	.93	1.17	.96	2.00	2.76	2.69
Eastern Europe	1.44	2.97	2.84	2.36	2.88	3.13
Soviet Union	1.38	2.37	2.34	2.26	2.33	2.50
Oceania	1.91	2.64	2.43	2.26	2.17	2.11
North America	.84	.89	.84	1.44	1.31	1.08
Temperate South America	.46	1.15	1.32	.26	1.19	1.26
Tropical South America	.41	1.41	1.77	1.71	3.95	6.46
Caribbean and Central America	.53	.86	1.20	.82	1.53	3.12
North Africa	.91	1.44	4.24	18.83	28.45	22.23
West Africa	.33	.61	1.42	7.61	14.01	18.08
East Africa	.32	.77	1.76	16.28	22.41	26.64
Southern Africa	1.90	1.96	2.47	8.73	5.94	5.62
West Asia	.33	.84	.88	4.39	7.25	6.54
South Asia	.50	.65	1.29	20.83	19.51	19.53
Southeast Asia	.47	1.28	2.07	9.81	13.07	19.72
East Asia	3.80	5.29	5.72	6.57	7.05	6.13
China	.22	1.66	1.49	N.A.	N.A.	N.A.
Country group:						
Low-income developing	.43	.67	1.40	18.14	18.61	20.43
Middle-income developing	.69	1.31	2.40	8.89	14.68	15.98
Semi-industrialized	.70	1.21	1.36	2.80	4.95	5.21
Industrialized	1.24	1.71	1.85	2.37	2.31	2.12
Planned	1.02	2.27	2.13	...	...	...
Planned, excluding China	1.40	2.54	2.50	2.29	2.49	2.63

Source: See tables 1.1 and 1.3.

note that while there are significant differences in expenditures per SMY between the industrialized countries, planned economies and other countries, there are no significant differences between the low- and middle-income developing and semi-industrialized countries.

Extension expenditures per extension worker in 1980 were also highest in North America/Oceania. They were lowest in Asia and Africa. The breakdown by country group

TABLE 1.5
Agricultural Research/Extension Expenditures per SMY/Extension Worker (constant 1980 US$ thousands)

Region/Subregion/Country Group	Research Expenditures per SMY			Extension Expenditures per Extension Worker		
	1959	1970	1980	1959	1970	1980
Western Europe:	44	73	76	15	19	18
Northern Europe	52	52	51	24	33	32
Central Europe	49	98	99	13	15	16
Southern Europe	25	52	78	5	13	11
Eastern Europe and Soviet Union:	32	29	29	13	13	14
Eastern Europe	34	27	27	14	12	13
Soviet Union	31	31	30	12	14	14
North America and Oceania:	90	127	127	28	40	51
North America	100	142	130	29	41	52
Oceania	52	85	117	24	35	46
Latin America:	56	44	54	18	19	18
Temperate South America	85	56	53	28	42	34
Tropical South America	61	48	56	20	18	18
Caribbean and Central America	28	26	52	11	12	11
Africa:	62	65	53	8	8	6
North Africa	35	44	27	11	12	8
West Africa	108	97	83	6	8	7
East Africa	58	72	46	4	5	4
Southern Africa	59	56	50	19	12	8
Asia:	23	38	39	2	3	3
West Asia	53	44	54	4	5	7
South Asia	22	28	34	1	1	1
Southeast Asia	20	22	25	2	2	2
East Asia	18	38	43	3	9	14
China	43	41	37	N.A.	N.A.	N.A.
Country group:						
Low-income developing	34	40	47	2	2	2
Middle-income developing	42	44	47	7	7	6
Semi-industrialized	41	45	46	10	10	11
Industrialized	55	80	93	16	25	29
Planned	33	32	31	...	...	...
Planned, excluding China	31	25	30	13	13	14

Sources: See table 1.1.

shows a strong correlation between development level and expenditures per extension worker. It should also be noted that expenditures per extension worker did not change for low- and middle-income developing and semi-industrialized countries between 1959 and 1980. In fact, only the industrialized countries have increased expenditures per worker.

It is worth noting that the differences in expenditures per manpower unit summarized in table 1.5 go a long way toward explaining the investment patterns observed in the earlier tables. Most developing countries have been able to train extension workers at low cost and to staff extension programs at very low costs per extension worker. This is particularly true in Asia, notably in South and Southeast Asia. The same cannot be said for researchers. The capacity to train researchers at an advanced level was practically nonexistent in most developing countries in 1959. Most developing countries faced very high training costs for scientists (usually trained abroad) and very high costs of operating research programs. The African countries in particular faced high costs. This situation has changed somewhat since 1959, of course, but researchers are still costly for most developing countries. India, the Philippines, Brazil, Mexico, and a few other countries have developed the capacity to produce doctorates in the agriculture sciences, but this capacity is quite limited. Thus even in 1980 the costs of operating research programs continued to be high for the developing world.

INVESTMENT BY COMMODITY

Data on research investment by commodity are extremely difficult to obtain. The agricultural ministries and research councils of many countries can provide aggregate data on expenditures and manpower for research programs, but they are not able to allocate expenditures or manpower to commodities. In view of the relevance of such data for policy, a methodology was developed for estimating the commodity orientation of research for twenty-six large developing countries. These countries account for more than 90% of the research undertaken in developing and semi-industrialized countries, excluding China.

The methodology entails the following steps. First, through computer search of the Commonwealth Agricultural Bureau (CAB) Abstracts, a count was obtained of the scientific articles and books published by commodity orientation and by the country for the periods 1972-75 and 1976-79. The CAB abstracts provide very extensive coverage of the world's agricultural science publications, and the search process provided reasonable data on publications by commodity. Research expenditures per publication vary, however, from commodity to commodity. The second step, therefore, was to compare the data on publications and on research expenditures by commodity for Brazil, where previous work has provided good expenditure data by commodity (Evenson 1982). These data enabled publications to be "standardized" to obtain equivalent spending units.[6] Third, research expenditure budgets were allocated to commodities according to proportions of equivalent spending units. Estimated research expenditures by commodity were thus obtained for each of the twenty-six countries. Note that publications data were utilized only to allocate expenditures among commodities.

Table 1.6 summarizes these data in terms of research expenditures as a percentage of the value of the commodity. Several observations may be readily made. First, expenditures on livestock research are generally quite high. Second, several commodities--specifically cassava, sweet potatoes, and coconuts--receive little research attention anywhere in the world. Other commodities--field beans, groundnuts, cotton, sugar, potatoes, and even rice and maize--receive very modest research attention in Asia,as does maize in Latin America. Export crops--citrus, coffee, and cocoa--are generally given heavy emphasis.[7] Third, the international research centers are not large enough in terms of expenditures to alter this picture, except in potatoes and possibly field beans. Many observers fail to realize that the IARCs are not very large relative to national systems.[8] Furthermore, the IARC investment is not allocated so that it improves the general distribution of research attention by commodity. The IARCs have not concentrated their research efforts on the neglected commodities.

A MODEL OF THE DETERMINANTS OF RESEARCH AND EXTENSION INVESTMENT

Several previous studies have investigated the factors influencing the levels of investment in research and extension. Some have attempted to derive empirical specifications from relatively simple models of government behavior (Guttman 1978; Otsuka 1979). These models are not

TABLE 1.6
Research as Percentage of the Value of Product, by Commodity: Average 1972-79 Period

Commodity	Region: Africa	Region: Asia	Region: Latin America	All Countries	International Centers
Wheat	1.30	.32	1.04	.51	.02
Rice	1.05	.21	.41	.25	.02
Maize	.44	.21	.18	.23	.03
Cotton	.23	.17	.23	.21	...
Sugar	1.06	.13	.48	.27	...
Soybeans	23.59	2.33	.68	1.06	...
Cassava	.09	.06	.19	.11	.02
Field beans	1.65	.08	.60	.32	.04
Citrus	.88	.51	.57	.52	...
Cocoa	2.75	14.17	1.57	1.69	...
Potatoes	.21	.19	.43	.29	.08
Sweet potatoes	.06	.08	.19	.07	...
Vegetables	1.56	.41	1.13	.73	...
Bananas	.27	.20	.64	.27	...
Coffee	3.12	1.25	.92	1.18	...
Groundnuts	.57	.12	.60	.25	.005
Coconuts	.07	.03	.10	.04	...
Beef	1.82	.65	.67	1.36	.02
Pork	2.56	.39	.60	1.25	.02
Poultry	1.99	.32	1.12	1.64	...
Other livestock	1.81	.89	.42	.71	...

Sources: M. Ann Judd, James K. Boyce, and Robert E. Evenson, "Investing in Agricultural Supply," Discussion Paper No. 442 (New Haven, Conn.: Yale University, Economic Growth Center, 1983); and U.S. Department of Agriculture, Indices of Agricultural Production (various issues).

well suited to the present purpose. There are two reasons for this. First, they presuppose a relatively simple form of optimal planning on the part of governments. The actual behavior of governments reflects response to different interest groups (Rose-Ackerman and Evenson 1985). More important, however, prior models have generally been applied to the investment decisions of a single country. The data in our study are international and can be used to compare the experiences of a number of countries in order to draw inferences regarding factors influencing investment behavior. The present analysis also investigates the extent to which one country's investment is affected by the investment of other countries.

The specification of the model was arrived at in a stepwise fashion. First, the variables that would be relevant in a simple social planning model were identified. Second, this perspective was modified further to consider special features of internationally traded commodities and differences in factor-product price ratios. The model was modified further to consider externalities in international technology markets by means of transfer-related variables designed to capture free-riding behavior by countries. Finally, two variables that measure political constraints were developed. Investment equations for research expenditures and for extension expenditures are required to be estimated with two bodies of data. The first set of data is the commodity data for twenty-six countries and two time periods summarized in table 1.6. These data are suited to the estimation of the research investment function only. They are subject to "errors of attribution" to commodities, but since these errors are in the dependent variable, they may not be too serious from an econometric perspective. The commodity detail in the independent variables, however, provides valuable added information. The main set of data, for estimation of both aggregate research and extension investment functions, is the expenditures series summarized in tables 1.1 and 1.2. This data set encompasses 106 countries (grouped into several classes) for eight 3-year time periods (1959-80).

Table 1.7 provides definitions of the variables actually used and presents means for the data sets. Independent variables are treated as exogenous variables in the analysis. In some cases they are lagged to reduce simultaneity biases.[9] In others they are expressed in ratio form to eliminate errors of measurement. The model treats both research and extension spending decisions as jointly determined by the set of independent variables. The dependent variables could have been expressed in manpower quantity units, that is, SMYs and extension workers, rather than in expenditure units. However, the expenditure measures were judged to be more reliable and less subject to definitional problems than were the manpower quantity units.

A relatively simple planning perspective was employed in developing a model of the determinants of research and extension investment. This model assumes an economic system where the resource base and existing distribution of income-producing assets are taken as given by a planner who seeks, in any given period, to maximize GNP or some other measure of aggregate income for the given resource base of the economy. In reaching this objective, the planner would consider the technology stock relevant to the country to be part of the resource base. Resources required to purchase or imitate technology that originated abroad would be allocated along with all other resources in order to achieve the income objective. Thus, even in a static economy, expenditures for the purchase, copying, and extension of technology would occur. The planner attempts to maximize the present value of aggregate income or consumption over a number of periods subject to a social rate of time preference, the existing technology in the economy, and the "technology of producing technology" through investments in research. The planner's solution would be one in which the internal rate of return to all types of investment projects would be equalized. For purposes of this analysis, this means that, for a given agricultural supply growth objective, a planner would attempt to find the lowest cost investment combination to achieve the growth.

In general, growth can come about through: (1) additions to the arable land stock by clearing, draining, and irrigating land; (2) improvement of the existing arable land stock by draining, irrigating, leveling, terracing, etc.; (3) additions to the labor force in agriculture; (4) additions to the stock of animal and machinery capital; (5) utilization of more fertilizer and chemical inputs; (6) elimination of inefficiencies in resource allocation caused, for example, by price distortions or land tenure patterns; (7) additions to farmers' human capital, through education and training programs; (8) investment in extension programs to diffuse existing but unused technology to farmers; and (9) development of improved location-specific agricultural technology through research investment.

Each of these alternatives has its own cost configuration for producing growth, and these costs vary over time and between countries. Adding to the land stock, for example, is a low-cost source of growth when a land frontier exists. As such opportunities are exhausted, more costly activities are required. Drainage of swamps and investment in irrigation can be quite costly, particularly when the "easy" projects, from an engineering standpoint, have been exhausted. Growth can also be achieved by additions to the agricultural labor force, but the planner is probably interested in obtaining more product per

TABLE 1.7

Variables Dictionary and Means: Investment Analysis

Variable	Commodity Data Set 26 Developing Countries: Average 1972-75 1976-79[a]	General Data Set[b]				
		Low-Income Developing Countries	Middle-Income Developing Countries	Semi-Industrialized Countries	Industrialized Countries	Planned Economies
Dependent:						
RESEXP (expenditures on agricultural research)	1.008	8.51	8.19	17.81	144.5	189.02
EXTEXP (expenditures on agricultural extension)	N.A.	10.49	11.97	15.85	63.03	87.38
Independent:						
1. Economic:						
PRODUCTION (value of production)	226.75	2754.42	1417.14	3152.50	11,515.83	26,918.29
CROPSHARE (share of crops in total agricultural product)	N.A.	.88	.84	.73	.43	.67
ARABLE EXPANSION (ratio of arable land currently to 6 years previous)	1.09	1.05	1.06	1.03	.99	.99
DIVERSITY (inverse of sum of squared shares of total production in commodity/geo-climate combinations)	.408	.497	.445	.420	.325	.334
RES/EXT PRICE (ratio of expenditures per SMY to expenditures per extension worker)	10.18	17.25	7.36	6.40	3.98	2.39
2. Economic-political						
XPORT (value of exports)	25.47	472.08	602.78	1048.61	4307.61	1434.28
MPORT (value of imports)	16.24	277.73	233.91	602.32	5676.26	2546.77
FERT/RICE PRICE (ratio of urea price to rice price, prior period)	2.77	N.A.	N.A.	N.A.	N.A.	N.A.

Variable						
3. Transfer-related						
GCNSYM (SMYs devoted to research in similar geo-climate regions in other countries)	N.A.	6133.49	6327.59	7843.73	17,787.34	21,347.45
GCNRES (research expenditures by commodity in similar geo-climate regions in other countries)	8.54	N.A.	N.A.	N.A.	N.A.	N.A.
GCNIARCSP (expenditures by IARCs in similar geo-climate regions)	N.A.	24.13	17.62	10.07	N.A.	N.A.
IARCSP (expenditures by IARCs in commodity)	.953	N.A.	N.A.	N.A.	N.A.	N.A.
DOM (dummy = 1 if IARC located in country)	.0183	.16	.23	.05	N.A.	N.A.
4. Political						
ECONAG (% of economically active labor force in agriculture)	53.18	80.70	58.99	36.99	13.20	37.09
URBANIZATION (% of population living in urban areas)	34.53	9.91	30.53	48.84	65.22	43.63

SOURCES: FAO, Production Yearbook (various issues); FAO, Trade Yearbook (various issues); U.S. Department of Agriculture, Indices of Agricultural Production (various issues); United Nations, Demographic Yearbook (various issues); World Bank, World Bank Development Report (various issues); see also table 1.1.

Note: For RESEXP, EXTEXP, PRODUCTION, XPORT, MPORT, GCNIARSCP, and IARCSP, entries are in 1980 US$ millions.

[a]Means expressed on a per-commodity basis.

[b]Means expressed as country averages.

capita. If animals, machines, fertilizer, and plant-protection chemicals are inexpensive, they can be low-cost means of obtaining more growth. Generally, however, modern inputs are low-cost sources of growth only when "compatible" biological technology is being produced.

The technology/human capital options become particularly viable options when other sources of growth become relatively high cost. Hence the economic variables ideally include measures of these relative costs. The economic variables included here are PRODUCTION, a variable measuring the value of total agricultural product, and, in the general data set, CROPSHARE, a variable that reflects differences in demand conditions and some natural resource conditions. The CROPSHARE variable may also control for the possibility that different support mechanisms for crop and livestock research may be in place. The variable ARABLE EXPANSION is designed to be a proxy for the cost of land expansion. It is defined as the ratio of arable land in the current period to arable land six years previous. If land expansion is high cost, this ratio should be low, and this should induce an expansion in research and extension spending.

The model also includes a variable measuring the diversity of agricultural production. The variable DIVERSITY interacts with the PRODUCTION variable. The a priori expectation is that the higher is DIVERSITY, the higher will be research spending <u>unless</u> there are substantial scale economies to combination-specific research. (One could visualize a situation in which a country would decide not to invest in a particular combination-specific research program because the fixed costs of doing so were high relative to the size of the production base.) The variable (RES/EXT PRICE) is defined as the ratio of expenditures per SMY to expenditures per extension worker, lagged three years. This variable is partly endogenous to the model, but, by expressing it in ratio form and by lagging it, much of the simultaneity should have been eliminated. The ratio form also eliminates exchange rate comparability problems. The nature of the data does not afford much opportunity for a more elaborate treatment. This variable has important policy implications, since it is subject to change through policy action. A rise in this price ratio would be expected to induce less spending on research and more on extension.

Imports (MPORT) and exports (XPORT) are classified as economic-political variables, since they allow for political responses to special interest groups. Most developing countries have pursued import substitution programs and have thus engaged in overvaluation of exchange rates. This penalizes the tradable sector, but governments might then compensate for this by investing in more research and extension in traded goods. In addition, since demand elasticities are higher for traded commodities, producer interest groups will have a stronger incentive to support investment in research and extension in traded commodities. Finally, the pattern of research and extension spending may continue to reflect the colonial legacy in many countries where most attention was given to commodities exported to the mother country.

Since all twenty-six countries in the commodity data set are rice producers, there is an additional economic-political variable in this set. The ratio of the urea price to the rice price in the prior period (FERT/RICE PRICE) is classified in the economic-political category because it reflects substantial market intervention by governments. High fertilizer-to-rice ratios reflect general discrimination on the part of governments against the agricultural sector, while low ratios reflect discrimination in its favor. The analysis here attempts to determine whether governments that discriminate via price policy display a similar tendency in research investment or whether they compensate for discriminatory price policies (e.g., by investing more in research when the price ratio is high).

The optimizing planner perspective can be extended to consider externalities in technology markets. The hypothetical planner will consider the costs of obtaining technology by simply imitating or copying technology produced abroad as an alternative to investing in the institutions to produce it domestically. This could lead to free riding in that a particular country might actually invest less if a geo-climate neighbor is investing at a high level, such that the cost of obtaining technology by copying or imitating is low relative to the cost of domestic production of new technology. Such free riding could lead to aggregate underinvestment as each country in a region attempts to free ride on others. It also could be a serious problem for the international aid investors in the IARCs: if IARC investment causes a reduction in national program investment, much of the impact of IARC investment may be lost.

Fortunately, another mechanism, the adaptive research mechanism, also enters into our calculations. Research programs build on past discoveries, which open up new research possibilities and thus are complementary to current research. Hence the past research discoveries in a neighboring country offer the option not only of imitation but also of adaptive discovery. This adaptive stimulus tends to offset the free-riding effect.

The following variables were specified in the general data set to measure these effects: GCNSMY (SMYs in geo-climate neighbors); GCNIARCSP (IARC spending in geo-climate neighbors); and DOMIARCSP (GCNIARCSP interacted with a dummy variable = 1 if an IARC is located in the country). The transfer-related variables used in the commodity data set are somewhat different: GCNRES (research expenditures, by commodity, in geo-climate neighbors); IARCSP (IARC spending in the commodity); and DOMIARCSP (IARCSP interacted with a dummy variable = 1 if an IARC is located in the country).

It is not possible to specify a priori whether net free riding or adaptive stimulus effects will be observed. Neighbors were defined to be countries in the same general geo-climatic region on the presumption that the bulk of the imitable or adaptable technology is produced in the same geo-climatic region. (Since a given country may have several geo-climatic zones, the transfer-related variables were defined in a proportional fashion.)

Two final variables, ECONAG and URBANIZATION, are designed to measure political interest group effects directly. They are obviously very crude measures of interest groups. The percentage of the population living in urban areas, URBANIZATION, serves as a proxy for industrial interests. The proportion of the labor force in agriculture, ECONAG, is a rough proxy for rural producer interests, although at low levels of per capita income it may actually be inversely related to political power, while the reverse may be true at higher income levels.

RESULTS OF THE STATISTICAL ANALYSIS

The results of the statistical analysis are presented in detail in our paper in Economic Development and Cultural Change (Judd, Boyce, and Evenson 1986). The more important results are summarized in this section.

Economic Variables

Greater diversity in commodity mix and geo-climatic characteristics generally results in an increase in research expenditures (except in the middle-income countries). This means that if, as production expands, the diversity of production does not expand proportionately, a larger country will be able to achieve a given growth rate objective with a lower ratio of research spending to production. The research costs required to meet a given growth target will be larger in a small, diversified economy such as Sri Lanka than in a small, homogeneous country such as Uruguay.

The costs of research and extension are important in determining research and extension expenditures. When countries are able to purchase research resources at low prices, they respond by purchasing more research resources and fewer extension resources.

Economic-political Variables

Research expenditures in the developing countries are sensitive to export possibilities. The higher the export propensity, the higher is research spending in the developing countries. The reverse is true for the industrialized countries. It also appears that a high export propensity shifts investment against extension in the low- and middle-income developing countries, but elsewhere the impact on extension is roughly similar to that estimated for research.

The import propensity effect on research is significant for cereal grains and pooled commodities in the commodity data set and strongest for middle-income developing countries in the general data set. In the low-income countries, higher imports similarly are associated with more extension investment, but in the semi-industrialized countries imports appear to cause a reduction in extension spending. These estimates are generally consistent with the view that most countries place a premium on foreign exchange savings (or earnings in the case of exports) and that they wish to avoid vulnerability to international price variation.

Countries that discriminate against agriculture by paying low product prices and charging high prices for inputs, such as fertilizer, also tend to have low expenditures on research. This supports the proposition that research investment, like price policies, reflects the relative political strength of the agricultural and nonagricultural sector.

Transfer-related Variables

The issue has frequently been brought up as to whether a country in which an international agricultural research center is located free rides on the IARC by reducing the investment in its own research. There is also a question of the extent to which countries depend on neighboring countries for research results. This analysis indicates that there is evidence of net free riding only in the industrialized countries, where research spending is lower when national spending by geo-climate neighbors is higher. There is no evidence that national research programs free ride on IARC spending. Indeed, the pooled commodity data set indicates a net adaptive stimulus of IARC spending. Moreover, the domestic location of an IARC appears to stimulate national research and extension spending. The net response to national research spending by geo-climate neighbors tends to be positive on research spending in developing countries, most notably in the semi-industrialized countries. It also has a positive effect on extension spending, suggesting that research by neighbors generally offers opportunities for both extension-based imitation and research-based adaptation.

Political Variables

The results generally indicate that in the developing countries, as the proportion of the labor force in agriculture rises, holding urban population constant, research spending declines, whereas it rises in the industrialized countries. The impact of the size of the agricultural labor force on extension spending is generally negative or negligible for all countries except the planned economies. This suggests a U-shaped relationship in which the political power of research supporters rises with declines in the proportion of the labor force in agriculture until the latter reaches a relatively low level.

The urbanization variable is really measuring concentration of the nonagricultural population, since it represents the proportion of the population living in large urban areas. There is some evidence in the commodity data that this concentration lowers research spending and, in the general data, that it increases both research and extension spending in industrialized countries.

IMPLICATIONS FOR POLICY

The investment in IARCs and much of the investment in national research programs in developing countries have been financed through international bilateral and multilateral aid, and it is likely that the majority of the capital budgets of national research institutions in the past three decades have been aid financed. The donors also provided technical assistance and visiting scientist services as part of their aid programs. This aid reflected the donors' belief that research and extension investment would produce low-cost agricultural growth.

There is little doubt that great progress has been made since 1959 in building research capacity in developing countries. This is reflected both in numbers of SMYs and in their levels of training. Most developing economies appear to have turned consistently to research and extension investments as land conversion and land development became more costly. The high cost of research resources and the low cost of extension resources encouraged a high degree of reliance on extension programs. A process of induced innovation, with a high degree of rational public-sector investment behavior, was clearly at work.

Studies of returns to agricultural research have consistently shown that research investments yield extraordinarily high returns to investments (Ruttan 1984). There are no indications that the expanded investment in research programs in developing countries has caused these returns to be lower. These studies indicate that optimal investment strategies will require further expansion and strengthening of agricultural research systems in developing countries.

It is often said that poor countries cannot "afford" as much research as rich countries. That is simply not true. Research and related extension programs are not consumer goods. They are investment goods and they produce growth in agricultural product just as investment in irrigation does. That product is in turn as valuable to a poor person or a poor country as to a rich person or a rich country. In fact, it is probably more valuable.

Indeed, under conditions where poor countries can actually purchase the services of research scientists at low prices relative to the prices associated with other growth-producing activities, rational planning would call for a higher degree of reliance on research than one would observe in countries where research scientists are costly. The data in this paper show quite clearly that many poor countries have responded to relatively low extension prices by investing heavily in extension. Most poor countries today have much higher public extension intensities (extension workers per unit of product) than any of the high-income countries ever had. The data also suggest that some high-income countries (notably the United States) are responding to high research prices by limiting spending.

Some poor countries face extremely high prices for scientific resources. If a country does not have a capacity to train scientists domestically, it faces high training costs in institutions abroad. If, in addition, it cannot provide an institutional setting conducive to keeping its costly scientists engaged in real research, its research costs can be very high indeed. In many countries today, a two-tier system of scientific manpower exists. The first tier consists of home-trained scientists educated to the bachelor's, master's, and, in some cases, Ph.D. levels. The second tier consists of those who have been sent abroad for scientific training and have returned. Agricultural research systems continually lose many of the best scientists from both tiers to the private sector, which pays higher salaries, and often to nonresearch jobs. Many of the best products of the system are also lost through international "brain-drain," often, somewhat ironically, to international agencies.

As a result, only a few developing countries are able to purchase low-cost SMYs. India, because of its size and the massive aid granted in the form of student fellowships in earlier years, has been able to realize relatively low-cost SMYs and has built a large and effective research system. But in recent years, India, too, has lost top agricultural scientists to other countries. Many other countries are in an unstable equilibrium in which research institutions are continually being raided and replacement of departing scientists is very costly.

The present study indicates that countries will respond to lower prices of national scientific resources. It would appear that after approximately two decades in which major attention has been given to building IARCs, the issue of training scientists at low cost in national programs now deserves much greater attention from aid donors. The development of a capacity to train scientists is fortunately highly complementary with the conduct of research. Graduate programs of quality must be research based. Aid programs to support graduate programs in agricultural sciences would not eliminate all the institutional barriers to optimal research investment levels, but they could ease a major one.

This study provides evidence that political factors have also affected the allocation of resources to research and to extension. There is some evidence that poor countries with large agricultural labor forces (and associated economic conditions) invest less in research and extension. In addition, traded commodities receive more research and extension attention than nontraded commodities. The data showing research intensity by commodity reflect these factors. Research attention to cassava and sweet potatoes is minimal in all regions, as is research on field beans in Asia. Clearly there is gross underinvestment in research on these commodities. This underinvestment is related to political influence. It remains true that the small farmers, staple food producers, and agricultural workers in much of the developing world have little influence on research and extension programs. This has produced a high degree of bias in agricultural research investment.

Investment in IARC systems has not improved the distribution of research across commodities appreciably. Boyce and Evenson argued in 1975 that the IARC system directed substantial international aid funding and scientific entrepreneurship away from national program development in the 1970s. They also suggested that in the long run the IARCs "may serve to facilitate international support for national systems" (Boyce and Evenson 1975, p. 64).

The present study provides the basis for a more positive perspective. It shows that, although the IARCs did direct funds and entrepreneurship away from national program support, national program development has generally been very substantial. There is no direct evidence whether the IARC system has now produced higher aid funds to national programs, but it would appear that the system is not presently a barrier to aid to these programs. More important, this study does not show significant net free riding on the IARC system by national programs. The commodity data suggest some positive adaptive stimulus. Thus it appears that the IARC system is achieving compatibility with national system development.

In summary, increased investment in agricultural research and extension throughout the world shows a fair degree of rational public-sector response to the goal of achieving an expansion in agricultural supply. Clearly some regions and some commodities have made little progress, however, and the specter of food shortages in many regions of the world is as real today as it was in 1959. Institutional barriers and political factors have affected overall investment levels as well as the allocation of investment among commodities. Nonetheless, the prognosis for growth in the capacity of the developing world to produce food to meet growing demand is encouraging.

NOTES

*This chapter is adapted from the paper by M. Ann Judd, James K. Boyce, and Robert E. Evenson, "Investing in Agricultural Supply: The Determinants of Agricultural Research and Extension Investment," Economic Development and Cultural Change 35(1)(October 1986): 77-113. T. W. Schultz, Dana Dalrymple, James Bonnen, Paul Waggoner, and Eliseu Alves provided constructive comments on an earlier version of this paper. In addition, the comments of anonymous reviewers inspired a fairly extensive revision of the paper and the investment model. This research was undertaken as part of the research program at the Economic Growth Center funded by the National Science Foundation under grant no. ISI-8018867.

1. A more detailed presentation of the methods and results of the statistical analysis of the determinants of investment is contained in M. Ann Judd, James K. Boyce, and Robert E. Evenson, "Investing in Agricultural Supply: The Determinants of Agricultural Research and Extension Investment," Economic Development and Cultural Change 35(1) (1986): 77-113.

2. It should be noted, however, that the data for China are subject to a relatively high degree of uncertainty owing to the paucity of source material.

3. Problems of definitional consistency in enumerating scientists may, however, account for part of the difference.

4. In compiling data on research spending, it was usually not possible to distinguish between capital and operating expenditures. Rapid periods of growth in national program expenditures often reflect increased capital spending that occurs when a country engages in program building. Although program-building expenditures may be used in part to increase the numbers of scientists engaged in research, they are more often used to improve laboratory facilities or raise salary levels of current research personnel. It is, therefore, not surprising that SMYs often increase at a rate slower than that of expenditures.

5. The five country groups are (1) industrialized countries--members of the OECD, except for Greece, Portugal, Spain, and Turkey; (2) planned economies--Eastern Europe, the Soviet Union, and China; (3) semi-industrialized countries--other countries with annual per capita income above $1,050; (4) middle-income developing countries--other countries with annual per capita income between $360 and $1,050; and (5) low-income developing countries--other countries with annual per capita income below $360 (see World Bank 1980, p. viii). The term developing countries is used to refer to the latter three groups collectively.

6. These were checked against U.S. data as well and found to be in close agreement.

7. Groundnuts, cassava, and coconuts are also significant export crops in some countries, however, and yet receive little research attention.

8. In terms of SMYs, the IARCs represent only 2 percent of the scientific manpower directed toward agricultural supply improvement in the developing and semi-industrialized countries.

REFERENCES

Boyce, James K., and Robert E. Evenson. National and International Agricultural Research and Extension Programs. New York: Agricultural Development Council, 1975.

Evenson, Robert E. "Observations on Brazilian Agricultural Research and Productivity." Revista de economia rural 20 (July-September 1982): 367-401.

Evenson, Robert E., Paul Waggoner, and Vernon W. Ruttan. "Economic Benefits from Research: An Example from Agriculture." Science 205 (September 1979): 1101-7.

Guttman, Joel. "Interest Groups and the Demand for Agricultural Research." Journal of Political Economy 86 (June 1978): 467-84.

Judd, M. Ann, James K. Boyce, and Robert E. Evenson. "Investing in Agricultural Supply." Discussion Paper No. 442. New Haven: Yale University, Economic Growth Center, 1983.

________. "Investing in Agricultural Supply: The Determinants of Investment in Agricultural Research and Extension." Economic Development and Cultural Change 35(1)(October 1986):77-113.

Kislev, Yoav, and Robert E. Evenson. "Investment in Agricultural Research and Extension: An International Survey." Economic Development and Cultural Change 23 (April 1975): 507-21.

Oram, Peter, and Vishva Bindlish. "Resource Allocations to Agricultural Research: Trends in the 1970s (A Review of Third World Systems)." Washington, D.C.: International Food Policy Research Institute; The Hague: International Service for National Agricultural Research, 1981.

Otsuka, Keijiro. "Public Research and Price Distortion: Rice Sector in Japan." Ph.D. diss. University of Chicago, 1979.

Rose-Ackerman, Susan, and Robert E. Evenson. "The Political Economy of Agricultural Research and Extension: Grants, Votes, and Reapportionment." American Journal of Agricultural Economics 67 (February 1985): 1-14.

Ruttan, Vernon W. Agricultural Research Policy, Minneapolis: University of Minnesota Press, 1984.

World Bank. World Development Report. Washington, D.C.: World Bank, 1980.

2
The Contribution of the CGIAR Centers to World Agricultural Research

Robert W. Herdt and Jock R. Anderson

Since the early successes of the Green Revolution, investment in agricultural research by national governments and international donors has grown rapidly. A small fraction of this investment has been directed to the international agricultural research centers sponsored by the Consultative Group on International Agricultural Research (CGIAR or "CG"), but with expectations of large payoffs. This paper briefly summarizes the results of a recent study of the impact of the CGIAR centers.[1]

THE CGIAR SYSTEM

The CGIAR was formed in 1971 to extend to other crops, regions, and ecosystems the successful U.S. Foundation-based research models developed for rice (IRRI) and wheat (CIMMYT). In 1971, when the CGIAR was set up, four centers existed. Since then, the CGIAR has facilitated the establishment of nine additional centers, resulting in the thirteen institutions briefly characterized on table 2.1. Subsets of the same forty or so donor governments and agencies that are members of the CG also support another dozen or so non-CG international and regional centers that conduct research on agricultural and related industries, but this paper is confined to examining the CG centers.

In 1983, several CG donors decided that it would be useful to take stock of what had been achieved by the CGIAR centers. The objective was to cast a deliberately broad net over the activities and measure what had been accomplished through the centers. But the international centers operate collaboratively with their partners in

TABLE 2.1
Centers Supported by the CGIAR, 1986

Acronym (Year established)	Center	Location	Research program, geographic focus	1986 Budget[a] ($ million)
IRRI (1960)	International Rice Research Institute	Los Banos, Philippines	Rice, global Rice-based cropping systems, Asia	21.6
CIMMYT (1966)	Centro Internacional de Mejoramiento Maiz y Trigo	Mexico City, Mexico	Maize, global Bread wheat, global Durum wheat, global Barley, global Triticale, global	21.9
IITA (1967)	International Institute of Tropical Agriculture	Ibadan, Nigeria	Farming systems, tropical Africa Maize, tropical Africa Rice, tropical Africa Sweet potato, yams, global Cassava, tropical Africa Cowpea, tropical Africa Soybean, tropical Africa	22.0
CIAT (1968)	Centro Internacional de Agricultura Tropical	Cali, Colombia	Cassava, global Field beans, global Rice, Latin America Tropical pastures, Latin America	21.4
CIP (1971)	Centro Internacional de la Papa	Lima, Peru	Potato, global	10.9

WARDA (1971)	West African Rice Development Association	Monrovia, Liberia	Rice, West Africa	2.1
ICRISAT (1972)	International Crops Research Institute for the Semi-Arid Tropics	Hyderabad, India	Chickpea, global Pigeonpea, global Pearl millet, global Sorghum, global Groundnut, global Farming systems, semi-arid tropics	21.2
ILRAD (1973)	International Laboratory for Research on Animal Diseases	Nairobi Kenya	Trypanosomiasis, global Theileriosis, global	10.4
IBPGR (1974)	International Board for Plant Genetic Resources	Rome, Italy	Plant genetic resources, global	4.5
ILCA (1974)	Intern. Livestock Center for Africa	Addis Ababa, Ethiopia	Ruminant livestock production systems, Africa	14.4
IFPRI (1975)	International Food Policy Research Institute	Washington, D.C., U.S.A.	Food policy, global	4.3
ICARDA (1976)	International Center for Agricultural Research in the Dry Areas	Aleppo, Syria	Farming systems, wheat, barley triticale, broad bean, lentil, chickpea, forage crops, dry areas of West Asia and North Africa	18.1
ISNAR (1980)	International Service for National Agricultural Research	The Hague, Netherlands	National agricultural research, global	3.8

Source: CGIAR Secretariat.

[a]CGIAR-supported core budget, net of capital, at the bottom of the bracket (December 12, 1985, estimate of the Secretariat).

development, the national agricultural research systems, which conduct location-specific adaptive research, disseminate the improved technologies (materials and methods), and promote institutions and policies that result from successful agricultural research. This raised an immediate difficulty of attribution. It was decided that imputation of credit for particular aspects of the centers' work was neither feasible nor useful. Accordingly, the stance taken was that measurement would be best addressed to the products of the centers' collaborative activities with national researchers rather than trying to identify separately the centers' contributions.

Another intention imposed at the outset of the study was to give emphasis to perceptions held about the CGIAR system by researchers in collaborating countries and institutions. The documentation of these perceptions would represent a unique record of what those researchers thought about the system. Within each country, personnel from a cross section of relevant research, administrative, and policymaking institutions were interviewed about the centers' modes of operation and any impressions that they had about the centers' success. Attempts were also made to document the more quantifiable aspects of center-related activities. This included the dissemination of new varieties associated with the centers, their other plant breeding and genetic resource conservation activities, the development of other agricultural technologies, training in agricultural research, institution-building efforts, and policy research. In view of the requirements for the above type of information, most of the study resources were directed at obtaining it. Little went to quantify productivity differentials associated with new center-associated technologies because time limitations precluded gathering farm-level and other disaggregated data, and because of the difficulties associated with separating the contribution of international research from those of domestic research, farmers' and public inputs, agricultural policies, extension, and general (nonagricultural) development.

INSTITUTIONAL ARRANGEMENTS FOR AGRICULTURAL RESEARCH

By the mid 1960s, it was increasingly clear that agricultural growth and productivity were not achieved simply by transferring technology to developing countries. Progress could only result from the generation and diffusion of technologies relevant to local ecologies and economies. The international centers were conceived as a mechanism that could organize research based on the stock of knowledge, scientific talent, and material (usually biological, especially plants) with the objective of producing technologies or technology components appropriate to the needs of the developing countries. These needs and circumstances were initially addressed through efforts of the IRRI and CIMMYT in developing widely adapted varieties of rice and wheat with a minimum of tailoring to local conditions.

As indicated in table 2.1, the evolution of the CGIAR system over time has seen the inclusion of more crops covering an increasingly wide range of environmental conditions. At the same time, it has become evident that there is usually a need to further modify technologies from international centers to make them suitable for local conditions.

IITA and CIAT were both operating before the organization of the CGIAR, concentrating on the agricultural production problems of high-rainfall tropical Africa and tropical Latin America, respectively. Each subsequently took on global responsibility (within the CG system) for crops of regional importance. CIP, the potato center, and WARDA, the center for rice in West Africa, were the first two additional centers to be supported by the newly organized CGIAR in 1971. The next was ICRISAT, with a charge to conduct research to improve agricultural technology for the semi-arid tropics, and the major food crops grown therein. Several years later, a center for the dry areas of West Asia and North Africa (ICARDA) was added. Two centers to help improve the productivity of livestock in Africa were opened in the mid 1970s, ILCA and ILRAD. Centers to assist countries with germplasm conservation (IBPGR), food policy research (IFPRI), and the development of national research systems (ISNAR) gave the system a broader diversity of coverage of ecological regions and subject matter.

The CG centers employed about 750 senior researchers in 1985, of whom most were working on crop improvement, defined to include plant breeding and the complementary research in plant pathology, virology, entomology, etc. The crop-oriented centers devote 25 to 40 percent of their senior staff resources to these crop improvement activities and to genetic resource conservation (table 2.2).

Other applied research activities, defined to include agronomy, economics, engineering, farming systems, and

TABLE 2.2
Approximate Proportion by Center of Internationally Recruited Senior Staff Members with Primary Responsibilities in Four Major Categories of Activities, 1984

Center	Strategic Research	Crop-Improvement Applied Research	Other Applied Research	Training, Research Support and Administration
CIAT	13	47	23	17
CIP	15	30	28	27
CIMMYT	14	45	24	17
IBPGR		83		17
ICARDA	11	33	10	45
ICRISAT	13	40	21	26
IFPRI			93	7
IITA	11	34	27	27
ILCA	6		59	34
ILRAD	83		6	11
IRRI	10	26	35	28
ISNAR			16	84
WARDA		10	34	56

animal nutrition, make up 15 to 30 percent of the senior staff resources of eight centers. IFPRI research is included in this category in table 2.2, as is ISNAR's research. However, the bulk of ISNAR's staff time is committed to activities that are more directly related to the second general objective of the centers--that of enhancing developing country capacity for research.

Training and other activities aimed at improving the capacity of developing country researchers are a major activity. A significant part of the staff time of the senior administrators is spent arranging to provide center products to national researchers, and most centers also have administrative or liaison staff designated to work with national programs. Training activities and research support activities like library work and communications are also significant contributors. The proportion of staff time devoted to these activities ranges from about 15 to 30 percent in most of the centers.

Virtually none of the centers' efforts are devoted to basic research conducted purely to build knowledge with no clear idea of how that might be used to enhance food production. A modest proportion of most centers' activities is devoted to strategic research, where the purpose is the solution of specific problems, but the product of which is far from being an immediately applicable technology. The bulk of ILRAD's researchers work at such activities, addressing the problems that need to be solved to generate technologies to control trypanosomiasis and theileriosis. The other centers typically devote from 10 to 15 percent of their staff resources to strategic research.

About 10 percent of the total staff are social scientists (distributed unevenly across the centers). IBPGR and ILRAD have no social scientists; IFPRI is comprised mainly of economists, and ISNAR has a few social scientists. Among the crop research centers, three organizational modes for social science research are evident: ICRISAT, ILCA, IRRI, CIMMYT, and CIP have specialized programs or departments dealing with the social sciences; IITA, ICARDA, and ILCA have social scientists in their farming systems programs; and CIAT and WARDA have social scientists in crop improvement programs.

As the semi-dwarf rice and wheat varieties spread, and as new centers were set up, it became increasingly apparent that technologies would have to be tailored to the climatic, soil, and plant disease conditions of specific localities. The realization that local adaptation to those conditions is a prerequisite of success led the centers to seek closer relationships with national researchers.

One manifestation of the closer relationships is the development of the centers' outreach and country programs whereby center researchers are stationed in countries other than at their center's headquarters. In 1984, about 225 staff from the centers were posted in country programs in all developing regions of the world, with about 140 in Africa. This had enabled centers to establish much closer working relations with their national counterparts. At the same time, the circumstances of many national programs have changed as the stock of human capital has grown. CGIAR training contributed to the growth, which was made more effective through networks of center alumni and growth in funding of national programs.

Donors provide significant funding to agricultural assistance worldwide, most of it directly to national governments, for a variety of purposes ranging from provision of food aid to sales of fertilizer and funds for low-interest farm credit. The CGIAR absorbs less than one-half of 1 percent of official development assistance, which was about one-twentieth of the value of food aid in 1980 (table 2.3). The CGIAR centers' budgets are about 5 percent of expenditures on agricultural research for the Third World, most of which is by the governments of the developing countries themselves.

PERCEPTIONS IN THE THIRD WORLD

Great importance was attached in the impact study to the views held by concerned individuals in the national organizations with which the centers collaborate. The sheer numbers of such individuals necessitated a case study approach, based on (a) the countries selected for study and, (b) a limited cross section of individuals who were available for comment. Knowledge of the centers and their research products varied greatly from person to person, reflecting, naturally, the experiences of the individuals.

The approach taken, while novel, was subject to the inherent limitations of the individuals involved, both those expressing their opinions and those collecting the data, and to partiality in terms of incomplete coverage, both between and within countries. The perceptions gathered are reported in detail in the country case studies published separately by the World Bank for the CGIAR Secretariat. Here we simply draw some generalizations about those perceptions.

Respondents who knew about a center's work on a particular crop in a given country, or had a more general understanding of the whole system, usually (but not always) expressed high regard for the center or the CG. Enthusiasm for different products varied according to individual knowledge, experience, discipline, and job responsibility.

While many responses were quite positive and supportive of the centers' work, they displayed considerable diversity, with some of the diversity related to the stage of development and capacity of the countries. Those with well developed research systems are able to use the centers' products effectively and to initiate requests for materials, training, publications, and other services. Countries with embryonic research systems are less likely

TABLE 2.3
Official Development Assistance (ODA), Worldwide, to Developing Countries, National Expenditures on Agricultural Research Worldwide, and CGIAR Expenditures

	1970	1980
	(billion, 1981 constant $)	
Official development assistance (ODA)[a]	21,300	36,210
Bilateral	18,450	28,650
Multilateral	2,860	7,560
Private voluntary associations	2,220	2,240
ODA commitments by purpose (1982)		
Total[b]		32,070
Allocable by sector		16,680
of which technical assistance		6,690
of which agriculture		1,000
Not allocable by sector		15,380
of which food aid		2,640
National agricultural research, worldwide	5,350	7,390
N. America, Oceania, W. Europe, Japan	2,400	3,210
USSR and E. Europe	1,280	1,490
Developing countries	1,670	2,680
CGIAR centers	20	140

Sources: Development assistance information from OECD (1983), agricultural research from Judd et al. (1984).

[a]From DAC countries (developed, western) plus OPEC + socialist + multilateral agencies.

[b]DAC and multilateral.

even to be able to test technologies effectively, to say nothing of interacting productively with center staff. Many countries with intermediate capabilities are able to interact to some extent with center researchers, but remain in the role of "receivers." The centers must thus tailor their approaches to different countries. The case studies document how, during the past twenty-five years, several research systems have grown from being rather dependent on the CG centers to being full partners in crop improvement, and how some are making independent contributions to other nations. Many, however, especially smaller countries, remain dependent on CGIAR centers for considerable assistance with varietal improvement research, training, and research methods.

THE FOCUS OF CGIAR CENTER RESEARCH

The major impact of the international centers to date has been in what are sometimes referred to as favorable environments--where rainfall is adequate to grow wheat or rice, or where irrigation is available. A review of adoption studies and data shows that the principal distinguishing feature of farmers who have not adopted modern wheat and rice varieties is the quality of their land and water resources rather than farm size, tenure, or social status.[2] This had led many observers to conclude that the CGIAR is concentrating its research attention on areas with favorable environments.

While it is true that, as yet, the CGIAR centers have little claim to having produced improved technologies for less favorable environments, this is not because of a lack of attention. Table 2.4 shows a breakdown of CGIAR research expenditure by commodity and region. Rice and wheat, which are the principal irrigated crops, receive 20 percent of CG research funding. Other cereals such as sorghum, millet, and maize, all grown in less favorable areas, receive 14 percent, roots and tubers 8 percent, food legumes 10 percent, and livestock 15 percent. Systems research, genetic conservation, and food policy research are all important. The balance goes to strengthening national capacities, to research support, and to management.

ICRISAT research is aimed entirely at the seasonally dry tropics of Africa, Asia, and Latin America; ICARDA research is concentrated on the dry cropping areas of the Middle East and North Africa. The high-rainfall tropics of Africa, where roots and tubers dominate, contain many environments that are unfavorable for food crop production in the sense that there are no production methods that are sustainable at a level significantly more productive than present systems. To date, the bush-fallow system is still usually perceived as best by farmers, but it will not long continue to support the rapidly increasing populations of tropical Africa. IITA is addressing this problem with a significant proportion of its resources.

Livestock production is the largest source of income for many subsistence farmers in the Sahel and other low-rainfall parts of Sub-Saharan Africa. ILCA is

TABLE 2.4
Approximate Annual Allocation of CGIAR Centers' Expenditures (in 1983 mil. dollars) to Major Activities and to Affected Continental Regions (1984-86 average, using budgeted levels for 1986)

Program area	Total	Africa	Asia	America	Middle East North Africa
Crop improvement					
Cereals	33.0[a]	8.8	15.6	5.3	3.3
Roots and tubers	8.2[b]	2.9	0.7	4.1	0.5
Food legumes	10.1[c]	2.4	2.5	2.7	2.5
Livestock					
Production systems	11.6	6.7	0	3.4	1.5
Disease control	4.1	4.1	0	0	0
Food Policy	2.4	1.0	0.8	0.3	0.3
Farming Systems	7.7	3.9	1.7	0	2.1
Genetic resources conservation	14.2	0.9	1.1	11.0	1.2
Research support	21.1	9.4	2.9	6.0	2.8
Strengthening national capacities					
Training and conferences	13.8	4.7	4.8	3.1	1.2
Information and communications	9.8	4.0	2.2	2.3	1.3
Technical assistance	2.6	0.5	0.7	0.5	0.9
Research management					
Administration and management	20.4	7.3	4.9	4.6	3.6
General operations	23.4	10.8	5.3	4.4	2.9
Total operations	182.4	67.4	43.2	37.7	24.1

Source: CGIAR Secretariat.

[a]Rice 41%, wheat 18%, maize 14%, sorghum 6%, millet 7%, other 14%.

[b]Cassava 23%, potatoes 48%, others 29%.

[c]Field beans 30%, groundnut 15%, other food legumes 55%.

concentrating on finding ways to increase livestock productivity in these areas. Similarly, CIAT is seeking methods to increase the productivity of the acid soil rangelands in the llanos of South America.

Thus, research inputs are distributed across a wide range of food crops and the ecologies of great importance for the developing world. The major successes have come in wheat and rice because research began on those crops and more was known about them. In addition, their effects have been more widely reported, but considerable resources are allocated to "less favorable areas."

For maximum productivity, research resources ought to be allocated to equate the ratio of the expected social marginal productivity of each research activity to its expected social cost across all alternative activities, and it is likely that present allocations differ from that ideal. It could be argued, however, that the CG's allocations may be closer to that optimum than those of many countries, despite the lack of apparent CG impact in "problem environments."

A majority of the world's farmers are female, but this fact has yet to be translated into agricultural infrastructure and priorities, especially in the area of extension. Such issues should be and sometimes are the subject of farming systems research work, an area in which the centers have played an important role. However, much remains to be done to sensitize research workers to the gender-specific impact of technological innovations.

MODERN CROP VARIETIES

Because of the many factors affecting agricultural production in the Third World, it is impossible to measure the impact of the CGIAR's contribution. One can, however, measure some instruments by which that impact is made. The varieties developed by or with the assistance of CG centers are a principal instrument in assisting agricultural researchers, so some effort was made to determine the utilization of varieties produced by centers or produced by countries from crosses made by centers and supplied to countries.

The CG system's major visible impact on world food production has come about from the development of semi-dwarf wheat and rice varieties. Developed and shipped around the world by CIMMYT and IRRI, beginning in 1965, semi-dwarf varieties of both of these crops were being grown on 125 million ha by 1983 (tables 2.5 and 2.6). Three distinct advantages were built into the new varieties. Their most dramatic feature is stature; the new varieties convert more of their dry matter to grain than to straw and are shorter and have sturdier stems than do conventional varieties, thus reducing the likelihood of lodging when higher rates of fertilizer are applied.

Second, the new cultivars are at least partially insensitive to day length and hence mature after a relatively fixed time compared with traditional varieties that flower only at a specific day length. This means that farmers can plant at any time of the year, double or triple cropping their land when other factors are not limiting. After the development of the first photoperiod-sensitive varieties, newer varieties were developed that were of shorter duration than the first semi-dwarfs, thereby permitting more intensification.

TABLE 2.5
Area under Semi-Dwarf Wheat, 1970 and 1983

Country	1970		1983	
	'000 ha	%	'000 ha	%
China	14.7	0.1	5126.0	17.0
India	6480.0	39.0	18550.0	80.0
Other Developing Asia	3458.5	40.1	7797.1	68.0
Afghanistan	232.0	10.5	400.0	13.0
Bangladesh			498.0	96.0
Nepal	98.3	49.2	377.6	92.0
Pakistan	3128.3	50.3	6521.5	88.0
Sub-Saharan Africa	69.8	5.0	556.3	52.0
Ethiopia	60.4	5.7	384.0	51.0
Kenya	7.9	5.3	83.8	72.0
Nigeria	1.0	33.3	10.0	71.0
Sudan		0.0	46.5	35.0
Tanzania		0.0	10.0	43.0
Zimbabwe	0.5	4.2	22.0	62.0
Latin America	794.5	10.8	8878.0	82.0
Argentina		0.0	6490.4	95.0
Bolivia	1.9	2.5	6.0	9.0
Brazil	56.1	3.1	826.5	43.0
Chile	61.2	8.3	329.7	70.0
Colombia	9.2	21.9	42.8	95.0
Ecuador		0.0	8.0	36.0
Guatemala	11.9	29.8	39.9	95.0
Mexico	651.9	88.1	942.5	95.0
Paraguay	2.1	6.6	6.0	8.0
Uruguay	0.2	0.0	186.2	62.0
Middle East/North Africa	1144.4	5.0	7690.3	33.0
Algeria	140.0	6.1	400.0	30.0
Egypt	0.0	0.0	306.2	53.0
Iran	63.0	1.3	891.7	14.0
Iraq	125.0	6.1	600.0	50.0
Libya	4.8	2.9	97.3	34.0
Morocco	90.0	4.6	721.6	36.0
Saudi Arabia		0.0	288.0	100.0
Syria	28.6	2.1	601.5	46.0
Tunisia	53.0	4.8	344.0	37.0
Turkey	640.0	7.4	3440.0	38.0
All Developing Countries	11962.0	14.0	48597.7	49.0

Source: Adapted from Dalrymple (1986b).

TABLE 2.6
Area under Semi-Dwarf Rice, 1970 and 1983

Country	1970 '000 ha	1970 %	1983 '000 ha	1983 %
China	26848.0	77.3	32265.2	95.0
India	5588.0	14.8	22180.0	54.1
Other Developing Asia	4281.5	10.0	19734.1	42.4
Bangladesh			2628.5	24.8
Burma	200.0	4.2	2370.1	50.4
Indonesia	1072.2	13.0	6626.9	72.8
Laos	53.6	6.0	9.7	1.4
Malaysia	164.6	23.3	254.8	36.4
Nepal	67.4	5.6	478.9	37.1
Pakistan	550.0	4.6	915.7	45.3
Philippines	1565.4	49.3	2757.0	83.5
S. Korea	2.7	0.2	418.6	34.1
Sri Lanka	73.6	11.2	749.7	81.0
Thailand	30.0	0.4	1200.0	12.8
Vietnam	502.0	20.1	1324.2	50.0
Sub-Saharan Africa	40.9	4.1	241.9	14.8
Cameroon			7.9	35.9
Ghana	36.8	89.8	35.0	43.8
Ivory Coast	2.1	0.7	32.7	7.1
Nigeria	1.0	0.4	60.0	10.0
Senegal	1.0	1.1	72.4	96.5
Sierra Leone			33.9	8.5
Latin America	252.4	4.2	1831.7	27.8
Argentina			27.3	33.7
Brazil			729.1	14.3
Colombia	41.0	17.4	364.3	91.8
Ecuador	15.7	10.5	40.3	53.1
Guatemala			3.5	29.2
Guyana			43.5	59.5
Haiti			11.0	22.0
Honduras	0.9	4.7	21.4	89.2
Mexico	123.3	66.6	154.2	83.4
Nicaragua	9.1	33.7	37.1	78.9
Panama	40.6	31.2	55.2	69.0
Paraguay	0.2	1.5	21.9	64.4
Peru	16.9	12.8	140.7	74.1
Surinam	4.7	13.1	48.7	69.6
Venezuela			133.5	79.9
Middle East/North Africa	2.1	0.3	80.7	11.0
Egypt	2.1	0.4	20.7	4.9
Iran			60.0	19.2
All Developing Countries	37012.9	30.1	76333.6	58.5

Source: Adapted from Dalrymple (1986a).

Third, the later versions of the new cultivars incorporate genetic resistance to many of the most important plant diseases and insect pests. These attributes were incorporated into the semi-dwarfs during the 1970s and 1980s, and farmers have been quick to recognize the advantages of these innovations. Today modern rice and wheat varieties with these characteristics make up the majority of the area planted to modern varieties (figure 2.1).

Success has also been recorded for other crops. Maize research has been slower to make an impact, but it is estimated that, by 1984, over six million ha of maize in the developing countries were planted to varieties derived from or related to CGIAR centers' maize research. This includes an estimated 25 percent of the lowland tropical maize in Africa.

Improved varieties of other crops related to or derived from the work of the centers are also being grown by farmers. In most cases these are of such recent vintage that there are few data on which to base impact studies. But, there is evidence that 40 to 60 percent of the area planted to field beans in Guatemala, Costa Rica, Cuba, and the important bean growing provinces of Argentina are planted to varieties selected from CIAT parents or breeding lines. Many of them are derived from "Dorado" varieties that are resistant to bean golden mosaic virus, a major disease problem in tropical America. In all, an estimated 18 countries in Latin America have named over 90 varieties of beans related to CG research through mid-1984. Six other countries have named similar varieties.

Twenty-three countries have named 63 potato varieties that they obtained through CIP's efforts to disseminate improved, disease-free germplasm. Other improved (albeit sometimes long-established) techniques, such as storing seed potatoes under diffused natural light, which increases their productivity, are also spreading through CIP's efforts with potato researchers in the developing countries.

Sixteen countries, including several in Africa, have released over 60 varieties of CG-center-related cassava. Six countries have released over 15 center-related sorghum cultivars; 6 varieties of center-related pearl millet have been released and another 15 are in the final stages of testing; 21 varieties of cowpeas have been released by 7 countries. The release of a variety does not guarantee that it will result in greater productivity, but it does indicate that it has been judged by knowledgeable

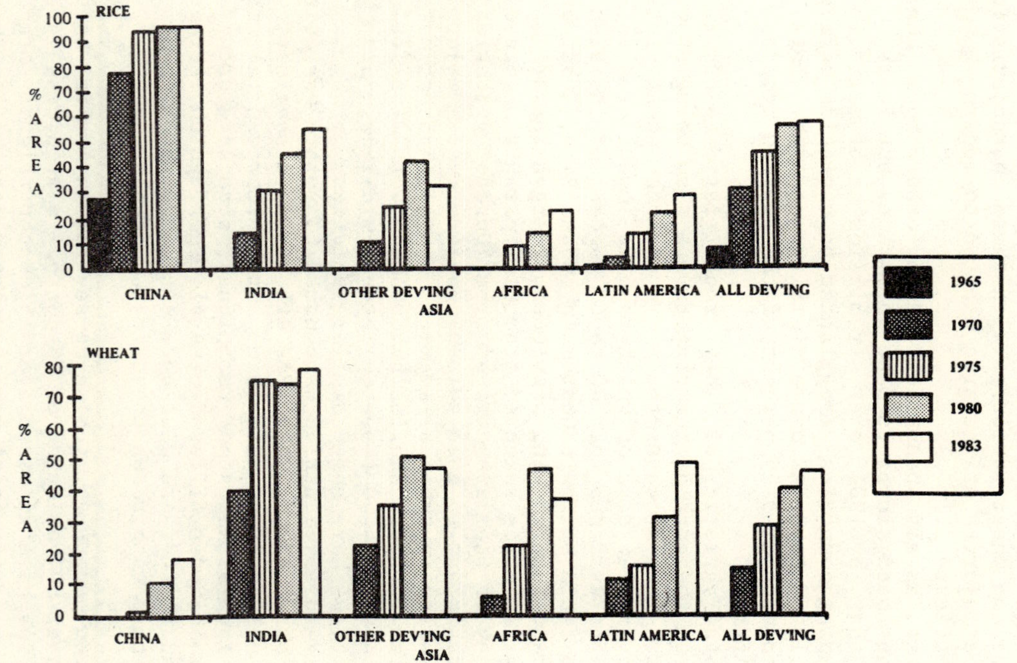

Figure 2.1. Adoption of semi-dwarf (HYV) rice and wheat varieties in China, India, and other developing regions, selected years 1965-1983

authorities as being an improvement over farmers' varieties.

DISTRIBUTION OF BENEFITS

The benefits of new technologies such as those embodied in modern crop varieties generate benefits because they permit farmers to produce crops at lower cost per unit of output. (Innovative ideas that entail higher costs per unit of output are not widely adopted without strong coercion.) Initially, farmers may retain the benefit of the lower costs but, if the new technology is widely spread, market supply increases and prices fall, so that consumers benefit. The final distribution of benefits between producers and consumers depends on the nature of the market for the product but, in general, the benefits are shared between producers who adopt the new technology and people who consume the product.

Early studies showed that large-scale farmers adopted modern varieties more readily than did landholders with small areas; later work has shown that small-scale farmers catch up within a few years, leaving the larger farmers, however, with the first years' gains (Lipton and Longhurst 1985). It has also been demonstrated that owner-farmers do not adopt new technologies more readily than do tenants, unless tenants are discriminated against in subsidized credit or input provision programs. Small-scale farmers may adopt technologies later than large either because they can avoid risk until they have seen that their wealthier neighbors have succeeded, or because they cannot purchase necessary inputs. Small-scale farmers ultimately adopt and utilize technology as intensively as do large-scale farmers and, because they have more family labor per hectare, small-scale farmers may produce higher yields. Studies such as the one for India from which the data in figures 2.2 and 2.3 were obtained show that there is no general link between modern-variety adoption or yield and farm size or owner-occupancy.

Figures 2.2 and 2.3 do illustrate dramatic differences between adoption in states like Haryana and Bihar. Areas that have not used modern varieties have performed comparatively poorly. Poor, non-adopting farmers and their employees suffer when burgeoning output depresses market prices. Yet, in non-modern-variety areas with poor soils, initial poverty is worse and less unequal so that the chances of fairly shared gains are better if modern

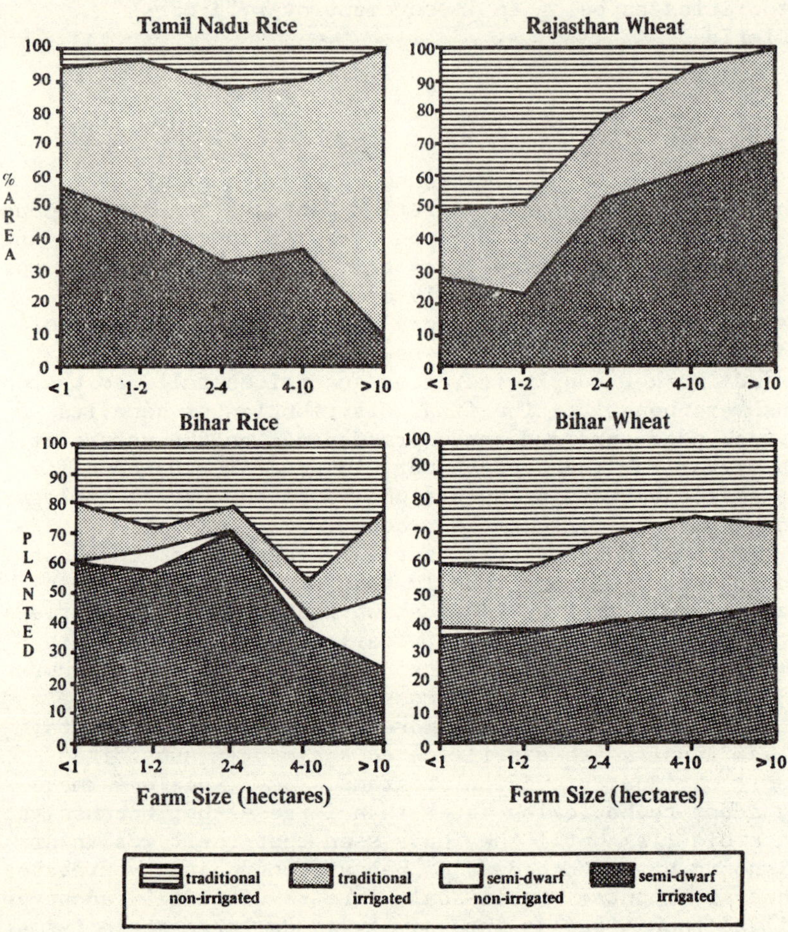

Figure 2.2. Percentage of area planted to semi-dwarf and traditional varieties on farms of five size groups in Tamil Nadu, Rajasthan, and Bihar, 1976

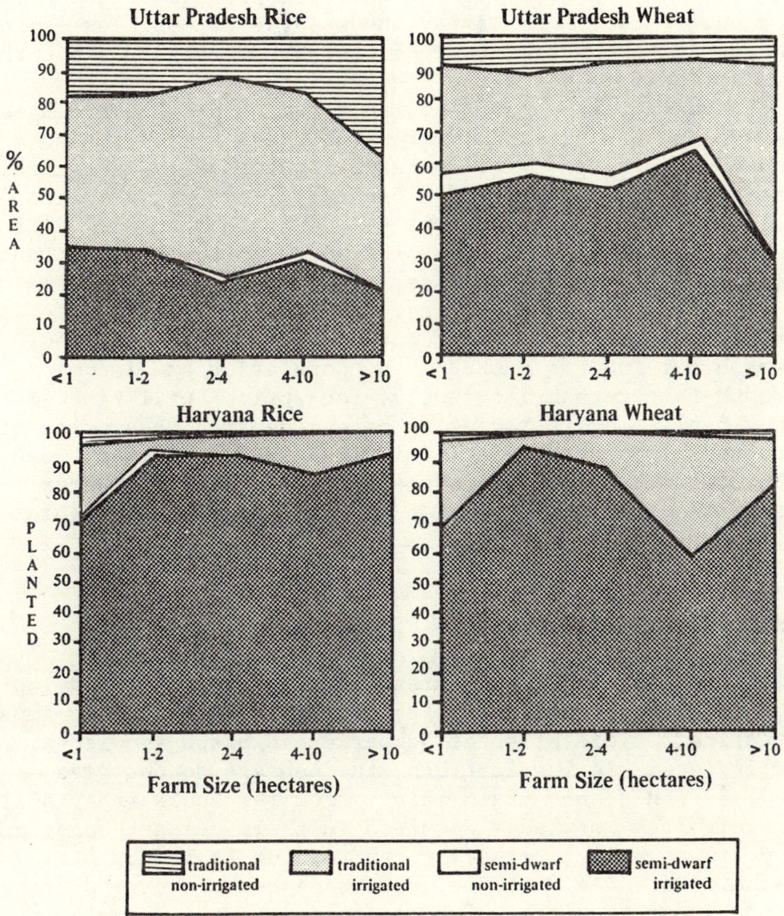

Figure 2.3. Percentage of area planted to semi-dwarf and traditional varieties on farms of five size groups, Uttar Pradesh and Haryana, 1976

varieties do succeed. The double cropping facilitated by low photoperiod sensitivity ensures a smoother flow of food throughout the year; thus the poor stand to gain since they can seldom save or borrow against lean seasons.

Modern varieties raise labor demand per hectare by a modest amount, especially around harvest time. But an ample, mobile, and growing labor force usually keeps real wage rates from rising significantly. Modern varieties raise the demand for land at a lower rate than they do for labor; however, land supply is fairly inflexible, so rents and land values usually rise.

In low-income countries, the poorest 20 percent of the population spend 60-75 percent of their income on food. If demand for food is pushed up by population and income growth faster than its supply, food prices will rise. Conversely, modern varieties have moderated this price rise and thus poor consumers (including farmers and farm workers who consume what they produce) gain relatively greater benefits through consumption than do the wealthy (table 2.7).

TRAINING AND OTHER IMPACTS

Each of the CGIAR centers has been training developing country researchers since it has had the capacity to impart practical, field-oriented, experienced-based knowledge. Unlike academic institutions, the centers do not give degrees and do not train mainly from textbooks--they train largely from knowledge acquired in their research programs and, in numerical terms, concentrate on training technicians. As a result, different centers have different kinds and intensities of training programs. Overall, the contribution of the training efforts are impressive, numerically and on the basis of the findings in the case studies.

Table 2.8 summarizes the numerical data on training at the centers. By far the largest number of people have participated in short-term, non-degree, group training courses that impart particularly technical, "hands-on" research skills--the skills that are essential to agronomic and plant breeding research. A significant number of people have also participated in individualized technical training programs aimed at imparting similar competencies. While not sophisticated, and not designed to make people into independent, scientific researchers, this training has been very useful in staffing the research programs of many

TABLE 2.7
Impact of a 10 Percent Decrease in the Price of Food on Real Income of Low and High Income Population Groups

Country	Percent Increase in Real Income		Source
	Lowest 10% per Capita Income	Highest 10% per Capita Income	
Sri Lanka	8.5	4.1	Sahn (1985)
Thailand	6.0	2.0	Trairatvarakul (1984)
Egypt	5.6	1.0	Alderman and von Braun (1984)
India	7.3	2.9	Murty (1983)
Funtua, Nigeria	7.7	6.5	Pinstrup-Andersen and Uy (1985)
Gusau, Nigeria	9.0	5.7	Pinstrup-Andersen and Uy (1985)
India[a]	5.5[b]	1.3[c]	Mellor (1978)

[a]Foodgrains only.
[b]For the lowest 20 percent.
[c]For the highest 5 percent.

TABLE 2.8 Number of Persons from Different Regions Participating in Various Training Programs at the CGIAR Centers, 1968-1983

Type of Training	Tropical Africa	Middle East North Africa	Asia	Latin America	Industrial Countries	Av. No. per Year, Recently
Group courses	6248	807	3453	2143	2155	2275
MS and PhD work	308	47	607	246	218	370
Individualized	344	71	518	427	155	355
Post-doctoral	106	7	282	77	210	105

developing countries. A significant number of these trainees have come from all parts of the developing world. While centers do not give academic degrees, a significant number of people completed advanced degrees in association with staff at a center--conducting research and writing a thesis at a center. There have also been a significant number of post-doctoral fellows who have done research at the centers.

Investigators in the impact study became aware of many, often subtle, institutional influences that the centers were either responsible for or with which they were closely involved. These ranged from a "hands on" work ethic induced in most trainees; the new professional respectability perceived from working on "humble" nonexport crops; research organizational formats modeled on some of the centers' structures; and many other small and large changes in procedures, priorities, and the commitment of resources.

CONCLUSION

If the present CG system did not exist, we believe that something like it would have to be invented. It provides institutions whereby the knowledge generated by advanced research can be translated into technologies or pre-technologies appropriate for developing countries--something that requires the capacity to be "on the ground" for a continuing period in the developing world. Thus they are part of a global research system that is needed to feed the world more effectively and to realize some of the opportunities for high returns to public technological investigation.

Technological advance, while critical to agricultural and economic development, is clearly a partial instrument, and certainly a poor one for solving perceived social ills such as maldistribution of resources. Still, the remarkable thing about the effects of the adoption of modern varieties associated with the centers is that beneficial impacts have been so widely distributed across societies, including many of those in greatest need.

Working alone is seldom easy. The collaborative arrangements between centers and research workers is demonstrably productive in many ways, but it is profoundly appreciated in developing countries for bringing workers into the global knowledge system and into a community of committed research scholars. Agricultural problems created by policies that are inimical to increased productivity are now receiving attention in the CG system largely through IFPRI, but only in trifling amounts given the transcending importance of the issues. These issues are intrinsically political, as well as economic, and the system must confront them more overtly if it is to contribute to accelerated progress.

Taking the simplistic view that the system is essentially a plant-breeding enterprise with decentralized management, it does a fair job of responding to the many (often inconsistent) forces. It may, however, be too conventional and myopic, paying insufficient attention both to fundamental germplasm issues, such as the wild relatives of crops, and to the risky business of biotechnology.

The human dimensions of technological advance were overlooked early on. The considerable recent investment in farming systems research has helped to correct initial problems, but there are still areas of neglect, e.g., the problems of female farmers in a male-dominated society and research infrastructure.

Even though returns to research are generally high, small countries, especially those with diverse agricultural zones, cannot mount productive research programs on all fronts. The centers play a critical role in assisting such countries with biological material and scientific resources--a role that will be needed well into the future on a wider range of crops and livestock than is presently serviced.

NOTES

1. The authors collaborated on the CGIAR Impact Study from which this paper is drawn and are grateful to many study colleagues on whose work this paper is built. A draft of the main report of the study was circulated to CGIAR members in late 1985, and was in the process of being published when this paper was written. Many of the country case studies and commissioned papers have been published for the CGIAR Secretariat by the World Bank, and are listed in the World Bank's publications catalogue.

2. This is supported by a review of the research on adoption that formed part of the Lipton and Longhurst monograph prepared as part of the impact study. Other reviews of this research include papers by Feder et al. (1982), Herdt and Capule (1983), and Byerlee and Harrington (1982).

REFERENCES

Alderman, Harold, and Joachim von Braun. The Impact of the Egyptian Food Ration and Subsidy System on Income Distribution and Welfare. Research Report 45. Washington, D.C.: International Food Policy Research Institute, 1984.

Byerlee, D., and L. Harrington. "New Wheat Varieties and the Small Farmer." IAAE Conference paper, Djakarta, August 24-September 4, 1982.

Dalrymple, Dana G. "Development and Spread of High-Yielding Wheat Varieties in Developing Countries." Washington, D.C.: Agency for International Development, 1986a.

________. "Development and Spread of High-Yielding Rice Varieties in Developing Countries." Washington, D.C.: Agency for International Development, 1986b.

Feder, G., R. E. Just, and D. Zilberman. Adoption of Agricultural Innovations in Developing Countries: A Survey. World Bank Staff Working Paper No. 542. Washington, D.C.: World Bank, 1982.

Herdt, R. W., and C. Capule. Adoption, Spread, and Production Impact of Modern Rice Varieties in Asia. Los Banos: IRRI, 1983.

Judd, M. A., J. K. Boyce, and R. E. Evenson. "Investing in Agricultural Supply." Economic Development and Cultural Change 35(1)(1986): 77-114.

Lipton, M., with R. Longhurst. Modern Varieties, International Agricultural Research, and the Poor. CGIAR Study Paper No. 2. Washington, D.C.: World Bank, 1985.

Mellor, John W. "Food Price Policy and Income Distribution in Low-Income Countries." Economic Development and Cultural Change 27(2)(1978): 1-26.

Murty, K. N. Consumption and Nutrition Patterns of ICRISAT Mandate Crops in India. Economic Program Progress Report 53. Patancheru: International Crops Research Institute for the Semi-Arid Tropics, 1983.

Organization for Economic Cooperation and Development (OECD). Development Cooperation. 1983 Review. Paris: OECD, 1983.

Pinstrup-Andersen, Per, and Thongjit Uy. "Nutritional Effects of Technical Change in Agriculture and Related Public Policies and Projects: Selected Case Studies." Washington, D.C.: International Food Policy Research Institute, 1985.

Sahn, David E. "The Effects of Price and Income Changes on Food-Energy Intake in Sri Lanka." Washington, D.C.: International Food Policy Research Institute, 1986.

Trairatvorakul, Prasarn. The Effects on Income Distribution and Nutrition of Alternative Rice Price Policies in Thailand. Research Report 46. Washington, D.C.: International Food Policy Research Institute, 1984.

3
Toward A Global Agricultural Research System*

Vernon W. Ruttan

In this chapter I address the task that remains of designing and implementing the global agricultural research system that will need to be in place by, at the very latest, the first decade of the twenty-first century. I will give particular attention to the special problems of the smaller countries in the emerging global system.

THE INTERNATIONAL AGRICULTURAL RESEARCH SYSTEM

It is useful to remind ourselves of what has been accomplished over the last several decades. The architects of the post-World War II set of global institutions included meeting world food needs and reducing rural poverty as essential to their vision of a world community in which all people could be assured of freedom from want and insecurity. They sought to achieve this vision by the creation of a set of global bureaucracies--the United Nations (UN) specialized agencies. The establishment of the Food and Agriculture Organization of the United Nations (FAO) was the initial institutional response to this concern (Hambridge 1955).

In spite of limited efforts by the FAO and several regional organizations, it was not until the late 1950s and early 1960s that a combination of (1) concern about meeting world food needs, (2) experience in advancing technology in food grain production in the tropics, and (3) a more adequate analysis of the role of advances in agricultural technology in the development process converged to provide the impetus for a major effort by several bilateral and multilateral assistance agencies and national governments to build the research capacity needed to sustain agricultural production in the poor countries of the tropics.

Organization and Impact

One of the most remarkable institutional innovations of the last two decades was the establishment of a new system of international agricultural research institutes (Chapter 2, table 2.1). The organization of these institutes drew on two historical traditions. One was the experience of the great colonial commodity research institutes that played such an important role in increasing the production of a number of tropical export commodities. There is a substantial English language literature on the development of British Colonial research institutes and botanic gardens (Masefield 1972; Brockway 1979). There is not a comparable history of French colonial research. But it is clear that research stations developed and maintained by France, during both the colonial and post colonial period made important contributions to oil palm, coconut and a number of other tropical export crops in West Africa (Eicher 1984). The Dutch made important contributions to the improvement of rice, sugarcane and a number of other tropical crops in Indonesia and Surinam. One of the greatest tropical research institutes during the colonial period was at Yangambi in the former Belgian Congo. In spite of its rather short colonial history Germany initiated important research programs in Cameroon, Togo and former German East Africa (now Tanzania). The second tradition was the experience of the Rockefeller Foundation in Mexico and the Ford and Rockefeller Foundations in the Philippines (Stakman, Bradfield, and Mangelsdorf 1967). The first four institutes in the system were the products of the joint efforts of the privately endowed Ford and Rockefeller Foundations. The system is now funded by a consortium of bilateral and multilateral assistance agencies and private foundations and operates under the oversight of the Consultative Group for International Agricultural Research (CGIAR).

An important innovation in the management of the CGIAR system is that each institute is governed by an independent board of directors and operates as an autonomous institution. This structure combines decentralized decision making (with respect to scientific program) with centralized oversight and judgments (with respect to funding, program direction, and system design and strategy).

Relations with Developed Countries' Research Institutions

The initial years of the new international institutes were characterized by a tendency to keep relationships at arm's length between the institutes and the developed countries' universities and research institutions. This relationship has changed over time. As the institutes have identified problems in which lack of knowledge in areas such as physiology, pathology, and other fundamental or supporting areas of science has constrained their ability to expand yield frontiers, they have taken steps to institutionalize their relationships with developed countries' research organizations.

Examples include the relationship between the Centre Internacional de Mejoramiento Maiz y Trigo (CIMMYT) and several Canadian institutions for work on triticale. The International Potato Center (CIP) has used contract linkages with institutions of developed countries for work on fundamental problems related to the CIP's mission more extensively than any of the other international centers. At the time of the 1977 quinquennial review mission, the CIP identified twelve such contracts with developed countries' institutions and seven with those of less-developed countries. In a number of cases, the CIP's contracts induced additional effort and expenditure on CIP-related problems by the developed country's contracting institution.

There are clear dangers in the growing relationships among the international centers and the centers of fundamental research in the developed countries. If the less-developed countries are to establish a viable base for self-sustained scientific effort leading to productivity growth in agriculture, it is important that they establish a capacity to work on the fundamental problems that are of particular significance in tropical environments.

System Impact

Evidence regarding the productivity of the international system is fragmentary and incomplete, yet there is little doubt that the rate of return to the investment in the system has been high--even in comparison with the more productive developed country (DC) national systems (Ruttan, 1982, pp 242-43). As early as the mid-1970s, evidence developed by Robert Evenson and colleagues at the University of the Philippines and the International Rice Research Institute (IRRI) indicated that the supply of rice in all developing countries was approximately 12 percent higher than it would have been had the same total of resources been devoted to the production of rice using only the varieties that were available before the mid-1960s (Evenson, Flores, and Hayami 1978). More recent studies by Nagy suggest that the gains to Pakistan alone from the wheat research conducted by CIMMYT would have been more than enough to cover the cost of the entire CIMMYT wheat program from its inception to 1980 (Nagy 1984, 1985). Stated another way, for the same amount of money, Pakistan could have profitably invested in a wheat research program of its own comparable in capacity and cost to the entire CIMMYT program.

In 1983, the CGIAR commissioned an independent study group to assess the productivity and distributional impacts of the technology developed at the CGIAR centers and at collaborating national centers. The study was directed by Jock R. Anderson, a distinguished Australian economist (Anderson 1985), and the study group's staff was drawn from the several social science disciplines from both developed and developing countries. The results of this study are summarized in Chapter 2.

The international system is particularly important for enhancing and sustaining the productivity of the smaller national agricultural research systems. Personal observation, evidence presented at the Wageningen symposium on research in small countries, and the evidence from the impact study (Anderson 1985, chaps. 4 and 5) indicate that the international system has provided a mechanism by which many smaller developing countries with only limited national research capacity obtain access to research results from the larger developing countries as well as the IARCs. The infrastructure for this function simply was not in place two decades ago in spite of efforts by organizations such as the FAO and the Inter-American Institute for Cooperation on Agriculture (IICA).

As the capacities of the less-developed countries' (LDCs) national research systems improve, the relative contributions of the international agricultural research institute system to the generation of knowledge and technology will decline. One possible outcome of this process is the loss of the institutes' distinct leadership roles. A viable model for the future of the institutes is an expanded role as centers for the conservation and diffusion of genetic resources and of scientific and technical information, relative to their role as producers of new knowledge and new technology. If they are careful to select staff members for their leadership capacities as well as for their scientific and professional competence, they will be able to continue to play a strategic role in establishing research priorities.

There is, however, what might be considered a natural history of research institutes (Ruttan 1982, 7). A new institute that is able to bring together a team of leading scientists tends to go through a period of high productivity that often lasts a decade or longer. After this initial period of creativity, there is a tendency for an institute to settle down to filling the gaps in the scientific literature and to fine-tuning incremental changes in technology. It is possible that the system of governance adopted by the system, program autonomy at the individual center level combined with centralized oversight, will enable the CGIAR centers to retain and enhance their vitality over a longer time. But the difficulties experienced by the Technical Advisory Committee (TAC) and the CGIAR in attempting to reform management and program at several "problem" centers do not lead to great optimism about the capacity of the individual centers to avoid cycles of creativity and stagnation.

A second factor that could contribute importantly to future vitality would be the incorporation of stronger LDC representation in both the governance of the system and, at the operational level, in research planning and collaboration. If the international institutes are able to strengthen their capacity to link the national systems to a carefully articulated international system, they will assure their own continued viability. If they become viewed as being competitive with national research systems, they could fade away into mediocrity.

A Continuing Need for International Support

When the system of international centers was being established by the Ford and Rockefeller Foundations in the early and mid-1960s, there was a general perception that over a period of several decades the foundations would withdraw and transfer the management and support of the institutes to the host countries. The two foundations have now withdrawn from anything more than token support of the system. But responsibility for oversight and support has been assumed, as noted earlier, by the CGIAR and its member institutions. Yet one still hears comments from both staff members of the DC donors and the LDC national research system that at some time in the future the responsibility for the system can be transferred to the LDCs or that the major units of the system (excepting the International Board for Plant Genetics) will eventually be phased out.

I find such discussion unrealistic! The system should be viewed as a permanent component of the global agricultural research system. This should not mean that every unit in the present system should be regarded as permanent. It is not difficult to visualize circumstances that would lead to the deemphasis of some programs and the initiation of new programs, but the international system itself should be regarded as permanent. The funding for the system should become part of the permanent commitment of the more developed countries to the agricultural development of the poorer and smaller countries in the system. In this respect there is a similarity between the national funding of a system of regional research centers, in larger countries such as China, Brazil, India, and the United States, even though the individual states or provinces also support state or provincial experiment stations.

An Incomplete System

The international system remains incomplete. There is a need to rationalize the management and oversight of a number of international agricultural research centers that have grown up outside the CGIAR system (table 3.1). I also see the need for greater capacity to conduct research on some of the difficult resource problems that continue to inhibit the development of agriculture in tropical environments. It also seems apparent that lack of basic scientific knowledge represents a serious constraint on the development of viable and sustainable technologies in many areas in the tropics.

The establishment of the International Fertilizer Development Center at Muscle Shoals, Alabama, in 1974 was an initial step in the development of an international capacity for research on resource development and management problems. The recent establishment by a group of CGIAR donors of an International Irrigation Management Institute (IIMI) in Sri Lanka and an International Board for Soils Research and Management (IBSRAM) in Thailand represents more recent initiatives. The establishment of an International Center for Research on Agro-Forestry (ICRAF) in Nairobi reflects a growing concern about the need for research capacity in the tropics on the development, management, and utilization of fast-growing trees to sustain the demand for biomass for fuel and other uses.

A beginning has been made in providing international support for the development of capacity to work on some of the problems where lack of basic knowledge acts as a constraint in technology development. Within the CGIAR system, the International Laboratory for Research on Animal Diseases (ILRAD) has been forced to direct much of its research to basic investigations. The International Centre for Insect Physiology and Ecology (ICIPE), initially established in 1970, has gradually evolved into an institution with very substantial research capacity.

The United Nations Industrial Development Organization (UNIDO) has sponsored exploratory studies leading to the establishment of an International Centre for Genetic Engineering and Bio-Technology. It is doubtful, however, that it will devote adequate attention to the work in molecular biology that will be most relevant for animal and plant protection in developing countries. In my judgment, there is also a very strong need for research to overcome the lack of knowledge about problems of fertility maintenance and enhancement of tropical soils. In many parts of Africa this lack puts a serious constraint on the ability to design viable short rotation systems to replace the more extensive slash-and-burn or other long rotation systems now in use.

Finally, there are serious deficiencies in the knowledge needed to develop economically viable technologies for the control of the parasitic diseases that inhibit the development of more intensive systems of agricultural production. In many cases, the relationship between disease and development appears to be symbiotic. Intensification of agricultural production enhances the environment for parasite disease, and parasite disease reduces the capacity of rural people to pursue more intensive systems of cultivation (Desowitz 1983; Walsh and Warren 1979).

It is not too difficult to generate agreement, at least in principle, on the need for greater international support for research on problems of resource development

TABLE 3.1
Some International Agricultural Research Activities Outside the CGIAR[a]

Center	Primary Focus	Location	Year of Initial Operation	Budget US$m	Budget (Year)	No. of Senior Staff	Programs
ICIPE	insect physiology and ecology	Nairobi, Kenya	1970	4.77	(1982)	46	crop borers livestock ticks tsetse fly plant resistance medical vectors insect pathology pest management
AVRDC	tropical vegetables	Shanhua, Taiwan, China	1972	3.60	(1983)	32	tomato Chinese cabbage sweet potato soybean mungbean
ICLARM	living aquatic resources	Manila, Philippines	1973	1.70	(1983)	14	aquaculture traditional fisheries resources development and management information services
INTSOY	soybeans	Urbana, Illinois, U.S.A.	1973	0.95	(1983)	8	soybeans

IFDC	fertilizer	Muscle Shoals, Alabama, U.S.A.	1974	6.70	(1982)	60	nitrogen research nutrient interaction phosphate research sulfur research potassium research economics research national programs technical assistance training
ICRAF	agro-forestry	Nairobi, Kenya	1978	2.20	(1983)	18	agro-forestry systems agro-forestry technology information training collaborative research
IIMI	irrigation management	Kandy, Sri Lanka	1984	5.00	(when operational)	10-12 in HQ, 3-4/ unit	collaborative research training information dissemination
IBSRAM	soils	not fixed	1985	4.54	(when operational)	5-10	headquarters soil management networks
INIBAP	banana and plaintains improvement	not fixed	1985	1.75	(initially)	small	headquarters regional networks

Source: Personal communication from Consultative Group on International Agricultural Research, World Bank, Washington, D.C., 1985.

[a] Activities currently using CGIAR meetings or in some other way related to CGIAR activities in 1984 (totaling about $30 million).

and management. But there is considerable skepticism among donor agencies about the need for international support for a series of basic research institutes in the tropics. The argument is frequently made that the basic research can be done in DC institutes, particularly in countries such as France, the United Kingdom, and the Netherlands, that have a tradition of tropical research and are now seeing that capacity erode as support adjusts to the disappearance of colonial responsibilities and to budget exigencies. Part of my answer is that the experience of the present IARCs indicates that intellectual commitment to the solution of even scientific problems is enhanced when the scientists working on a problem are located in the environment in which the problem exists. Basic research capacity in the tropics will also facilitate more effective dialogue with the basic research community in the developed countries.

Considerable thought will also have to be given to the appropriate governance for the emerging system of natural resource and basic science research centers. The present CGIAR system is already approaching severe strains on its financial and managerial capacity. There is a pervasive view among donors to the CGIAR system that it will be extremely difficult to push funding for core or base programs at the CGIAR centers much beyond $200 million (in 1985 U.S. dollars), yet subsets of the same donors have funded the new centers that have emerged outside the CGIAR system.

It would be a serious mistake if new natural resource and basic science institutes were to continue to emerge on an ad hoc basis. One of the great strengths of the CGIAR system is its planning and oversight role in welding the set of autonomous institutes into an international research system. The CGIAR and TAC secretariat infrastructure could perform the oversight functions for a much larger system than at present with only a modest expansion in staff. Nevertheless, donor funding considerations may make it desirable to consider the establishment of a new oversight body, perhaps a consultative group for natural resources research, to govern the new natural-resource-based institutes. It may also be desirable to establish a separate governance system for any new system of basic research institutes--a consultative group for biological sciences for tropical agriculture. As new internationally supported basic research units are established in the tropics, more attention should be given to the training role, particularly advanced training at the Ph.D. and post-doctoral levels, than was the case when the present international commodity institutes were established.

A Global System

Finally, I would argue that an effort should be made to assure that the international system becomes a truly global system. The new international system has been effective in building communication between LDC national research systems. The linkages of the international centers with DC research institutions are, however, generally filtered through the bilateral development assistance agencies. Direct linkages with the national research systems of the developed countries remain underdeveloped. The linkages between the national research systems of the developed countries are even less developed. It is my impression, for example, that there has not yet emerged any institutional capacity to rationalize or coordinate agricultural research between European Economic Community (EEC) or Organization for Economic Cooperation and Development (OECD) member countries. There is a modest program of information exchange between EEC and OECD countries, but these activities appear to be more ceremonial than substantive (FAO 1984). And we have barely begun to build effective linkages between either the national research systems of the developed countries or the international systems with agricultural research systems of the centrally planned countries.

NATIONAL RESEARCH SYSTEMS

By the late 1960s, many of the bilateral and multilateral aid agencies were recognizing serious shortcomings in the results of their efforts to support the development of national agricultural research systems. Most national systems in the less-developed countries were unprepared to absorb effectively large amounts of financial, material, and professional assistance. The capacity for scientific management and entrepreneurship of the newly trained scientific community was often underdeveloped. Many systems were plagued by cyclical sequences of development followed by erosion of capacity as budgetary priorities responded to changes in political regimes (Ardila, Trigo, and Piñeiro 1981).

Impatient staff members at aid agencies were often unaware of the history of their own national institutions. They had forgotten that the national agricultural research systems of the United Kingdom, Germany, the United States, and Japan had taken decades, not years, to acquire the research and training capacity needed to generate the new knowledge and technology needed to sustain agricultural development (Ruttan 1982, 66-115). Furthermore, the political support available to many national and international aid agencies was often so fragile that support for institution building was difficult to sustain unless a short-term payoff could be visualized. In addition to a sense of frustration with efforts to strengthen national research systems, there was a growing conviction of urgency about the problem of meeting food requirements in the poor countries. The initial success of IRRI's rice program and CIMMYT's wheat program combined to create a conviction that the international agricultural research institute, which could operate independently of the vagaries of the local political environment and could draw on the global agricultural science community for its staff, represented an effective instrument for the management of research resources and for the generation of new technology.

By the mid-1970s, it had become increasingly clear that the productivity of the international agricultural research system was severely constrained by the limited capacity of many national systems and that the adaptation and dissemination of the knowledge and technology generated at the international institutes were dependent on the development of effective national systems. It became widely accepted that the ability to screen, borrow, and adapt scientific knowledge and technology requires essentially the same capacity as is required to invent new technology (Evenson 1977a). Capacity in the basic and supporting biological sciences is at least as important as capacity in applied science. But the outreach programs of the international institutes, even when working through networks such as the international wheat research network, the inter-Asian corn program, and others, did not have the capacity to take on the role of strengthening national systems.

The bilateral and multilateral assistance agencies had no alternative, therefore, but to place the strengthening of national research systems high on their assistance agendas. Both the FAO and the Rockefeller Foundation played important entrepreneurial roles in this development.

After a series of consultations with the leaders of national research systems, the International Agricultural Development Service (IADS) was established, with initial funding from the Rockefeller Foundation, to provide contract research management and development services to national research systems. The FAO, through its Research Development Center, took steps to strengthen its capacity to support training in the field of research organization and management.

The initiatives of the Rockefeller Foundation and FAO influenced the CGIAR to intensify its own deliberations. In 1977, the CGIAR organized a task force to explore the possibility of establishing an international service for the strengthening of national agricultural research within the CGIAR's systems. These deliberations led to the establishment of the International Service for National Agricultural Research (ISNAR) in 1979. There had been some expectations that, in establishing ISNAR, the CGIAR might absorb IADS, much as it had incorporated IRRI, CIMMYT, the International Institute of Tropical Agriculture (IITA), and the Centre International de Agricultura Tropical (CIAT) under its umbrella in 1971. By 1979, however, the CGIAR had become somewhat sensitive about absorbing activities initiated before the CGIAR/TAC assessment and evaluation process. Some European donors were also sensitive about the fact that staffing patterns at the institutes had not drawn effectively on European professional capacity. The FAO, one of the CGIAR's sponsors, expressed strong concern that the new service was infringing on an area of traditional FAO responsibility.

IADS (recently merged into Winrock International) has now acquired substantial experience in managing projects funded by agencies such as the U.S. Agency for International Development (USAID) and the World Bank and designed to strengthen national agricultural research systems. ISNAR has acquired considerable expertise in diagnosing the problems that have inhibited the effectiveness of national research systems and in assisting national agencies in planning for research system reform and development. It is clear, however, that the strengthening of national research systems is only partially, and perhaps only marginally, amenable to the efforts of the assistance agencies. External funding agencies have often inhibited the development of national systems as a result of lack of sensitivity in their assistance efforts to the difficulties faced by a national research system in achieving political and economic, in addition to scientific and technical, viability.

As the efforts by the bilateral and multilateral assistance agencies to strengthen national agricultural research systems got under way, it became apparent that the 1970s were witnessing a remarkable expansion in agricultural research capacity in a number of important developing countries (Chapter 1, table 1.1).

When one examines the individual country detail, however, it is clear that most of this growth has occurred in a relatively few countries such as Brazil, Philippines, India, China, and Nigeria. In 1980, there were only slightly more agricultural research scientists in all of Latin America and Africa combined than in the U.S. federal-state system--and fewer than in the Japanese national-prefectural system. Even in those countries that have made substantial progress, the ratio of research expenditures to the value of production remains low--and it remains lowest for those commodities produced and/or consumed primarily by the poorest farmers and consumers.

During the last several years I have been involved in a series of studies of agricultural research systems in Asia (Ruttan 1981; Evenson, Pray, and Quizon 1986). The concerns about the development of national agricultural research systems that have emerged out of my own research and experience have been reinforced by the series of very useful reviews conducted by the World Bank (1983), the U.S. Agency for International Development (1983), and the UNDP-FAO (1984). Although the literature on the performance of national agricultural research systems is much more adequate for Latin America and Asia, the concerns expressed in this section impinge with particular force on many African agricultural research systems (Eicher 1984; Lipton 1985). Let me summarize some of these concerns.

1. <u>Excessive investment in research facility relative to development of scientific staff</u>. There are too many facilities without programs. Many of the premature facility developments are the direct result of the multilateral and bilateral assistance agency programs that find it easier to invest in facility development than in human capital development or program support. Premature facility investment represents a burden on the research system rather than a source of productivity.

2. <u>Excessive administrative burden that stifles both routine investigations and research entrepreneurship</u>. A major challenge to any national research system is how to achieve consistency between the personal and professional

objectives of individual researchers, research teams, and research managers and the social objectives of the research system. In many respects the individual scientist can appropriately be viewed as an independent contractor who makes his or her services available in return for professional and economic incentives. Bureaucratic efforts to achieve consistency between the objectives of the individual and the objectives of the system (or simply fiscal responsibility) are often carried to the point where they become an excessive burden on research productivity.

3. <u>The failure of location decisions for major research facilities, often made with the advice of assistance agency consultants, to give adequate weight to the factors that contribute to a productive research location</u>. These factors include (a) location in a community that includes related educational and professional infrastructure; (b) location in an agro-climatic environment that is representative of an important part of the area in which the particular commodity is grown or that is representative of a major resource (soil, water) problem area; (c) selection of a site with appropriate resources (soil, water) and infrastructure (electricity, transport, amenity).

4. <u>Lack of congruence between research budgets and the economic importance of major commodities or commodity groupings</u>. If new knowledge and new technology were equally easy (or difficult) to come by in each commodity area, a good rule of thumb would be to allocate research resources roughly in proportion to the value (or value added) of commodity output or resource input. It is easy to think of good reasons for departure from such a rule. In a small research system, critical mass (i.e. scale economies) implies the desirability of focusing resources on commodities that account for a large share of output (such as wheat in Northern India) or on a commodity where very large gains can be made in a short time (such as lowland irrigated rice in the 1960s). But extreme lack of congruence often suggests that little careful thought has been given to research resource allocation or that particular interest groups have biased research allocation to their own benefit.

5. <u>The apparent presumption in some national systems that it is possible to do research in agricultural science without scientists</u>. In too many national research systems, commodity program leaders often have neither the training nor the capacity to direct either scientific research or

technology development. Salary structures and noneconomic incentives are frequently so unattractive, relative to other national and international alternatives, that potential leadership is eroded, research programs become routine, and returns to research investment are low.

6. The cycles of development and erosion of capacity that have characterized a number of national agricultural research systems. Periods of rapid development have often been followed by the erosion or collapse of research capacity when external support has declined. Martin Piñeiro, Eduardo Trigo, and their colleagues have documented this pattern most thoroughly in a number of Latin American countries such as Argentina, Peru, and Colombia (Ardila, Trigo, and Piñeiro 1981; Piñeiro and Trigo 1983). But such cycles are also familiar to anyone who has followed the progress of agricultural research in developing countries in other areas of the world.

7. Lack of information and analysis that goes into establishment of research priorities. In the research planning staffs that have successfully struggled with the research resource allocation problem, it has become increasingly obvious that effective research planning requires close collaboration between natural and social scientists and between agronomists, engineers, and planners. This is because any research resource allocation system, regardless of how intuitive or formal the methodology employed, cannot avoid making judgments about two major questions. (a) What are the possibilities of advancing knowledge or technology if resources are allocated to a particular commodity problem or discipline? Such questions can only be answered with any degree of authority by scientists who are on the leading edge of the research discipline or problem being considered. The intuitive judgments of research administrators and planners are rarely adequate to answer such questions. (b) What will be the value to society of the new knowledge or the new technology if the research effort is successful? The intuitive insights of research scientists and administrators are no more reliable in answering questions of value than are the intuitive insights of research planners in evaluating scientific or technical potential. Many of the arguments about research resource allocation flounder on the failure of the participants to recognize clearly the distinction between these two questions and the differences in expertise and judgment needed to respond to them (Ruttan 1982, 262-64).

The perspectives and concerns that I have expressed about agricultural research in LDCs are not the exclusive problems of new and growing research systems. Don Hadwiger has provided evidence that in the United States the "pork barrel" approach to the location of agricultural research facilities resulted in 44 percent of all U.S. Department of Agriculture (USDA) research facility construction between 1958 and 1977 occurring in states represented by members of the Subcommittee on Agriculture of the Senate Appropriations Committee. He noted that this practice has forced "the federal Agricultural Research Service to operate a 'traveling circus' opening up new locations in current Senate constituencies, while closing some locations in states whose Senators are no longer a member of the subcommittee" (Hadwiger 1982).

SMALL COUNTRY AGRICULTURAL RESEARCH SYSTEMS

We are confronted with a remarkable paucity of data and analysis on the relationship between scale (or size) and productivity in agricultural research. And what evidence there is, even in the way of casual observation, often lacks precision as to whether the size-output relationship being referred to is with respect to the size of the individual research unit (team, laboratory, department), the individual research institution (center, institute, faculty), or the national or international research system. The views that "small is better" or that "big is better" have often been advanced with considerable heat but with relatively little precision in concept or definition and with even less empirical evidence. The issues discussed in this section represent an important opportunity for research to bring better theory, method, and data to bear in order to advance our understanding.

Size and Productivity in Research

What little knowledge we do have suggests that the optimum scale of the research is affected by factors both external and internal to the research process. The optimum level of resources devoted to a commodity research program is positively related to the area planted to a commodity in a particular agroclimatic region (Binswanger 1978). Therefore, determining the optimum scale of a research unit or program involves balancing the increasing returns associated with the area devoted to the commodity (or problem) on which the research is being conducted against the possible internal diseconomies of scale of the research process or system.

The data that we do have suggest that industrial research and development productivity, measured in terms of patents per engineering or scientific worker, is lower in the large laboratories of the largest firms than in the smaller firms in the same industry (Schmookler 1966; Kamien and Schwartz 1975). There is similar evidence for agricultural research (Pound and Waggoner 1972). There also are a number of case studies that suggest very high rates of return to individual public, philanthropic, and private research units, often with fewer than twenty scientific or technical staff members per unit (Evenson 1977b; Sehgal 1977). However, many of these small "free-standing" agricultural research units are engaged primarily in technology screening, adaptation, and transfer activities that depend only minimally on in-house capacity in such supporting areas as physiology, pathology, chemistry, and even modern genetics.

Evenson also noted that during the early stages in the development of national research systems, experiment stations tend to be widely diffused, to utilize primarily technical and engineering skills, and to be characterized by a strong commodity orientation. In the Chinese system, for example, decentralization includes not only a provincial research system but also autonomous prefectoral and county research institutions that are financed and governed at the local level. Evenson also pointed to a trend toward hierarchical organization and consolidation into a smaller number of larger units at later stages in the development of agricultural research systems. These centralizing trends are apparently motivated in part to take advantage of economies resulting from research activities in the basic and supporting sciences and to use economically the laboratory, field, communications, and logistical facilities.

The urge for consolidation can easily be overdone, however. In the United States, for example, there is now rather strong evidence supporting the value of decentralization even within individual states. For a given level of expenditure, a state system that includes a strong network of branch stations gets more for its research dollar than a state system that is more concentrated. What decentralization gives up in lower costs seems to be more than compensated for by the relevance of the research and the more rapid diffusion of results. There are, of course, limits to the gains from decentralization. Disagreement about the relative gains from centralization and decentralization, and about the relative emphasis that should be given to basic science, applied science, and technology development, has been the basis for much of the recent argument about the organization and funding of the U.S. federal-state agricultural research system (National Research Council 1972); Workshop on Critical Issues in American Agricultural Research 1979).

A Minimum National System

One of the most difficult issues related to size and productivity in agricultural research is the problem faced by the smaller countries in the development of their agricultural research systems. Most of the smaller countries--those in the four- to ten-million population range--do have the resources, or have access to donors' resources, that would permit them to develop, over a ten- or twenty-year period, an agricultural research and training capacity capable of staffing the nation's public- and private-sector agricultural research, education, planning, and service institutions. The fifty or so smallest low-income countries must, however, think of research systems that will often be little larger than a strong branch station in a country such as the Netherlands or Denmark or in a state such as Texas or Minnesota.

But how can the government of a small country decide on the appropriate size and organization of its national agricultural research system? For countries like Sierra Leone or Nepal, even the financial and professional agricultural research resources of a small American state or a Japanese prefecture are probably at least a generation in the future. The time required to achieve viable research systems for many of the smaller national systems must realistically be calculated in terms of more than a generation rather than the five- or ten-year project cycles used by most development assistance agencies.

One major focus of the research effort in these smaller research systems must be the direct support of agricultural production and rural development programs. This means a primary focus on applied research and technology development fields such as agronomy, plant breeding, animal production, crop production, farming systems, and agricultural planning and policy. Trigo and Piñeiro (1984) have estimated that a minimum research module for one product requires a team of four researchers trained at the M.S. and Ph.D. levels, complemented by eight specialists with graduate-level training, plus a complement of support personnel. They estimate that the total cost of such a program would run in the range of $250,000 current (1984) U.S. dollars (table 3.2). For a small country with six to ten major commodities and several important agro-climatic regions, this implies a research budget of $5 to $8 million U.S. dollars. When this effort is complemented by the non-commodity-oriented research in areas such as soil and water, pest management, cropping systems, and socio-economic aspects of agricultural production, marketing, and policy, the implications run into the $12 to $15 million range.

The viability of even a small nation's agricultural production also requires capacity for higher education, in agriculture, at least through the master's level, to support national programs of technology in transfer, rural development, and regulatory and service activities. When these activities are aggregated it is not difficult to arrive at a minimum level of professional capacity, with training at the M.S. and Ph.D. levels, of around 250 and with budget support somewhere in the $20-$30 million range for even the smaller (but not the smallest) countries. For the very smallest countries even this investment is not feasible in the foreseeable future. Any serious attempt to solve the problem of agricultural research and technology development in the smallest countries must face up to the difficult problem of designing a viable system of regional research collaboration (Wilson 1984).

Interdependent Systems

The idea of reducing or eliminating technological dependency generates strong emotional appeal. Yet even larger countries with advanced agricultural research systems--the United States, the Soviet Union, Japan, India, and Brazil, for example--are not self-sufficient in agricultural science and technology. An effective national agricultural research system must have the capacity to borrow both knowledge and materials from the entire world. The problem of how to link effectively with an increasingly integrated and interdependent global

TABLE 3.2
Estimated Cost of a Minimum Research Module for One Product (in thousands of current US dollars)[a]

I.	Direct Research Costs (60% of total budget)			306
	A.	Personnel		245
		1.	4 chief researchers, MS or PhD, 3 persons/year in plant breeding, agronomy and pest and disease control and 1 person/year equivalent in socioeconomics and other specializations, according to requirements (soils, physiology, etc). Total cost per person/year US $30,000[b]	120
		2.	8 specialists, university graduates. Total cost per person/year US $12,500	100
		3.	Training Calculated on the basis of 2 x 1 rate of retention; total rotation every 15 years; cost of US $100,000 per PhD (MS 60%). Total annual cost for a permanent team of 3 PhD and 1 MS (approximately)	25
	B.	Services and materials Calculated as 12.5% of direct costs.		
	C.	Equipment Calculated as 7.5% of direct costs.		
II.	General Costs and Administration (40% of total budget) Includes direction, support and services (administration, laboratories, library, communication, field, etc.)			204
	A.	Personnel Calculated as 60% of general and administrative costs		122
	B.	Services and materials 25% of general and administrative costs.		51
	C.	Investments and equipment 15% of general and administrative costs.		31
			Total Budget	510
	Percent summary by broad budgetary items (approximate)			
	A.	Personnel	72.5%	
	B.	Services and materials	17.5%	
	C.	Equipment	10.0%	

Source: Eduardo J. Trigo and Martin E. Pineiro, "Funding Agricultural Research" in Selected Issues in Agricultural Research in Latin America, eds. Barry Nestel and Eduardo J. Trigo. (International Service for National Agricultural Research, March 1984, The Hague, Netherlands, p. 85).

[a] The estimates were made using the budgetary structure of the international agricultural research centers as a guideline for determining the percent of each item of expenditure.

[b] US $30,000 was used as an average of the case for the different countries of the region. The sum includes salaries plus benefits. A variation of US $1,000 above or below this average figure implies an increase or decrease of US $4,250 in the total budget.

agricultural research system is difficult for the state and provincial research units in the larger national systems. It is even more difficult for the national agricultural research systems in the smaller countries.

One approach to this problem has been to attempt to establish cooperative regional research programs--for example, the West African Rice Development Association (WARDA) and the international crop research networks that are linked to the international agricultural research institutes. Other regional institutions not directly linked to the international (CGIAR) system include the Centro Agronomico Tropical de Investigacion y Ensenanza (CATIE), the Caribbean Agricultural Research and Development Institute (CARDI), and the Southeast Asian Fisheries Development Center (SEAFDEC). Networking has become the most recent theme in assistance agency jargon. But it is hard to find many outstanding success stories among these efforts. Program activities and cooperative efforts often appear stronger in the glossy pamphlets issued by the organizations than they do in practice (Venezian 1984). Experience suggests that such regional programs can succeed only with the commitment of long-term external support and with the participation of the external donors in the governance of such centers. Some of the most effective collaborative regional efforts have been organized around the research programs of the international research centers (Plucknett and Smith 1984).

The international crop research networks, centered around the international institutes, have not been without problems. When the institutes have had confident and effective leadership, they have often played an exceedingly useful role in creating opportunities for productive professional interaction and collaboration. But the institute research networks tend to be selective. At times they have found it hard to bend institute priorities to meet national priorities. Collaborative efforts tend to involve the strongest institutions and the leading scientists rather than those who have the greatest need.

A richer institutional infrastructure is needed to strengthen and sustain the capacity of the smaller national agricultural research systems. In spite of ideological considerations, many small countries have found it advantageous to encourage the transfer and adaptation of technology by the private sector genetic supply industry or by the multinational firms engaged in commodity production, processing, and trade (Pray 1983). Firms engaged in the production of crops grown under plantation systems and independent growers producing under contract arrangements with processors have at times provided their own research and development facilities. In other cases, associations of producers have been willing to tax themselves to support commodity research stations. Such arrangements have often been associated with discredited systems of colonial governance. A strong case can be made for reexamining and strengthening the legal institutions and financial incentives for private sector research, development, and technology dissemination in the developing countries.

The perspectives outlined in this section are highly tentative. Although they are drawn from considerable experience, they should be treated as hypotheses to be tested by further research rather than as conclusions. Institutions such as IADS, ISNAR, and IICA should devote a reasonable amount of analytical effort to attempts to understand the problem of developing and sustaining effective agricultural research in the smaller national research systems.

Some Generalizations

In spite of the limited available knowledge, there are a few generalizations about smaller agricultural research systems that can hardly be avoided. One is that the research investment per acre or per hectare will have to be higher in a small system than in a large system to achieve an equal level of effectiveness. This is because the cost of developing, for example, a new millet variety that will be grown on a million acres is not likely to be substantially greater than one that will be grown on half a million acres.

A second generalization is that the cost of developing productive farming systems for a small country with great agroclimatic variations will be greater than for a small country that is more homogeneous. For example, the cost per hectare of developing an effective agricultural research system for Sri Lanka is likely to be much larger than developing one for Uruguay. The issue of guns versus butter in national budgets is also likely to cut more sharply in a small country than in a large country.

Finally, there is no way that a small country can avoid being dependent on others--on the international agricultural research system, on the research systems of large countries in the same region, on multinational firms--for much of its agricultural technology.

Furthermore, a small nation with a strong research program but a limited agricultural or industrial base cannot capture as high a proportion of the benefits from its investment in basic research as can a large nation with a diversified economic base. Much of the benefit will spill over to other countries. If it has a weak agricultural research system, it will lack the knowledge needed to capture the benefits of research in other countries or to choose a technological path consistent with its own resource and cultural endowments. Even a strong agricultural research system cannot assure autonomy. But small countries do need to develop sufficient agricultural science capacity to enable them to draw selectively on an interdependent global agricultural research system. They need to be able to choose what is useful to borrow from other national systems and from the international system.

TOWARD A REFORM OF AGRICULTURAL RESEARCH SUPPORT

What can be done to replace the deficiencies that characterize assistance for the support of agricultural research, extension, and rural development programs in poor countries? A solution to the problems of "aid effectiveness" in support of research is particularly important at this time. I anticipate that the next decade will experience a decline in the real flow of aid resources and increasing competition among the several claimants on aid resources.

The basic thrust of the needed reforms is to move away from primary reliance on narrow project approaches. In supporting agricultural research the project system should be largely replaced by a "formula funding" or "revenue sharing" approach (Ruttan 1984). There have been many criticisms of the project approach followed by the major bilateral and multilateral development assistance agencies. The criticism most frequently heard is that the assistance agencies exert undue influence on the content of the national development programs (Faaland 1980; Salmon 1983). This criticism is partly justified. It is not too difficult to identify cases in which close patron-client bonds have been established between particular officers in the aid agencies and the leadership of favored national program agencies. Such relationships have often appeared to give particular national programs a degree of stability and continuity that would be difficult to achieve in the unstable political environments that characterize many developing countries.

In my judgment, cycles of development and erosion are inherent in the traditional project approach. The reason for this inherent contradiction is that external assistance provides an alternative to the development of internal political support. National research system directors have frequently found that the generation of external support requires less intensive entrepreneurial effort than the cultivation of domestic political support. Domestic budget support required by donors is often achieved by creative manipulation of budget categories rather than by increments in real program support--particularly when donor representatives are under pressure from assistance agency management to "move resources." Most existing project systems thus have built-in incentives for national research system leadership to direct entrepreneurial effort toward the donor community rather than toward the domestic political system.

Any effective alternative should attempt to reverse the perverse incentives that characterize existing development assistance instruments. The system should be reformed to provide incentives for national research system directors to redirect their entrepreneurial efforts toward building domestic political and economic support for agricultural development.

I am increasingly convinced that the long-term viability of agricultural research systems depends on the emergence of organized producer groups that are effective in bringing their interests to bear on legislative and executive budgetary processes. The support of finance and planning ministries for agricultural research is undependable. Their tenure in office is often short. And their support tends to fluctuate with the perceived severity of food crisis and foreign exchange demands.

A Formula-Funding Model

What alternatives to the existing system do I suggest? I do not want to be interpreted as completely negative with respect to traditional development assistance instruments. Project aid is often quite appropriate for physical infrastructure development projects. Program aid can be an effective way to provide macroeconomic assistance for structural adjustment or for sector development in a country with substantial capacity for macroeconomic policy analysis and program management. But neither the traditional program aid nor project aid instruments are fully effective in countries that have little financial or professional capacity for providing support for long-term institution-building efforts. New methods of combining the flexibility of program support, effective technical assistance, and sustained financial support for long-term development efforts must be sought.

One innovation that might be used effectively is for the donor community to move toward an approach in which the amount of external support is linked to growth in domestic support. An example of how such a system might work is presented in table 3.3. This implies the development of a "formula" approach in which the size of donor contribution would be tied to the growth of domestic support. The formula should include a factor that adjusts the ratio of external to domestic support to take into account differences in domestic fiscal capacity. Given the political considerations that impinge on the allocation of bilateral donor resources, implementation of the formula-funding model is probably unrealistic in the immediate future.

Country-Level Research Support Group

A second alternative might take its lead from the experience now accumulated with the CGIAR model and the various donor consortia that have been organized to coordinate assistance to some of the larger aid recipients. This could involve country-level research support groups, chaired by the chairman of the national agricultural research council or the director of agricultural research. The support group will need to have available to it relatively long-term program plans for the development and operation of the national agricultural research system. To produce and continuously update this program, the national research system may require external assistance, but in general the program should be the product of the national agricultural and general science policy system. Its focus, to help protect the program from vagaries of political change, would be on long-term agricultural research needs and goals and on the incremental steps required for implementation.

It is expected that the long-term program development and the priority setting would be done through an

TABLE 3.3
Illustration of a Funding Model for Agricultural Research Support

	Program Support and Assistance Level (in millions of U.S. $)					
	Low		Medium		High	
National Fiscal Capacity	National Support	Donor Assistance	National Support	Donor Assistance	National Support	Donor Assistance
Low (40% Assistance)	20	8	50	20	100	40
Medium (20% Assistance)	20	4	50	10	100	20
High (10% Assistance)	20	2	50	5	100	10

interactive process with the support group. Once the program has been accepted, donor members of the support group, it is hoped, would collectively agree with the host country to help provide the components essential to the execution of the program as a whole. The host country, in turn, would assume the responsibility for moving its national research program along the agreed-upon development path. Initial commitments might be for three to five years subject to annual review and course corrections suggested by the analysis and feedback from actual experience.

Use of an institution such as a support group has the potential of helping the country involved avoid many of the pitfalls of the project mode while retaining several of its desired attributes. Donor identity could be retained by relating grants to components of the agreed-upon overall program. These could even be called projects if, for administrative purposes, it were so desired. The support group, like the CGIAR, would likely involve bilateral grants developed in the framework provided by the forum of multiple donors and the host country. The impersonal process of contributing to a common fund is not envisioned. However, this would not preclude "incentive funding" of a formula type. At the same time, the danger that a single donor would dominate the priority-setting process or that essential program components would be ignored would be minimized.

The research support group has several other potential advantages. (1) It could contribute to building a national constituency by focusing from the onset on this essential ingredient for viability. The donors, for example, might agree to increase their contributions by some fraction of the rise that occurred in the real support provided by the nation involved. Other matching provisions might be agreed upon to provide incentives for nurturing and cultivating national constituencies. (2) It could provide reasonable continuity in support. Commitments should be fairly long term and subject to review and extension well in advance of termination dates to avoid risk of the excessive program fragmentation that is frequently associated with narrowly defined project funding. (3) It could reduce the administrative and management load on the host country through the planning and review process the RSG would follow. (4) It could place donors in a position of genuinely complementing and supplementing one another <u>and</u> the national program rather than competing for "good investment opportunities."

That such a support mode is often discussed but little used is evidence that implementation is not a simple, trouble-free task. The method has, however, been used successfully in Bangladesh and, somewhat more informally, in several other countries. An important element in its success in Bangladesh was that the support group meetings were chaired by the director of the Bangladesh Agricultural Research Council rather than by a donor representative.

A dialogue on donor assistance to national agricultural research programs was initiated by the World Bank in 1981. The dialogue has been continued by ISNAR in a series of meetings with directors of national agricultural research systems. It is imperative that these dialogues be continued. The issue of reform of agricultural assistance should be recognized as one of the most urgent items on the agenda.

NOTES

*A more complete version of this paper has been published in Research Policy (Ruttan 1986).

REFERENCES

Anderson, Jock R., ed. International Agricultural Research Centers: Achievements and Potentials, parts I-IV. Washington, D.C.: Consultative Group on International Agricultural Research, World Bank, August 31, 1985.

Ardila, Jorge, Eduardo Trigo, and Martin Piñeiro. "Human Resources in Agricultural Research: Three Cases in Latin America." Cooperative Research Project on Agricultural Technology in Latin America (PROTAAL), Document No. 50. Inter-American Institute for Cooperation on Agriculture (IICA), San Jose, Costa Rica, 1981.

Binswanger, Hans P. "The Microeconomics of Induced Technical Change." In Induced Innovation: Technology, Institutions and Development, edited by Hans P. Binswanger, Vernon W. Ruttan, and others, 91-127. Baltimore: Johns Hopkins University Press, 1978.

Brockway, Lucile H. Science and Colonial Expansion: The Role of the British Royal Botanic Gardens. New York: Academic Press, 1979.

Desowitz, Robert. New Guinea Tapeworms and Jewish Grandmothers: Tales of Parasites and People. New York: Avon Books, 1983.

Eicher, Carl K. "International Technology Transfer and the African Farmer." Department of Land Management, University of Zimbabwe, Harare, Zimbabwe, 1984. Mimeo.

Evenson, Robert E. "Cycles in Research Productivity in Sugarcane, Wheat and Rice." In Resource Allocation and Productivity in National and International Agricultural Research, edited by Thomas M. Arndt, Dana G. Dalrymple, and Vernon W. Ruttan, 209-36. Minneapolis: University of Minnesota Press, 1977a.

_________. "Comparative Evidence on Returns to Investment in National and International Research Institutes." In Resource Allocation and Productivity in National and International Agricultural Research, edited by Thomas M. Arndt, Dana G. Dalrymple, and Vernon W. Ruttan, 237-64. Minneapolis: University of Minnesota Press, 1977b.

Evenson, Robert E., Piedad M. Flores, and Yujiro Hayami. "Costs and Returns to Rice Research." Economic Consequences of New Rice Technology. Los Banos, Laguna, Philippines: International Rice Research Institute, 1978.

Evenson, Robert E., Carl Pray, and Jamie Quizon. Research, Extension, Productivity and Incomes in Asian Agriculture. Ithaca, N.Y.: Cornell University Press, 1986.

Faaland, Just. Aid and Influence: The Case of Bangladesh. London: Macmillan & Co., 1980.

Food and Agriculture Organization of the United Nations (FAO). Research in Support of Agricultural Policies in Europe, 31-32. Fourteenth FAO Regional Conference for Europe, Reykjavik, Iceland, September 17-21, 1984.

Hadwiger, Don. The Politics of Agricultural Research. Lincoln: University of Nebraska Press, 1982.

Hambridge, Gove. The Story of FAO. New York: Van Nostrand, 1955.

Kamien, Morton I., and Nancy L. Schwartz. "Market Structure and Innovation: A Survey." Journal of Economic Literature 13(March 1975): 1-37.

Lipton, Michael. "The Place of Agricultural Research in the Development of Sub-Saharan Africa." Washington, D.C.: Consultative Group on International Agricultural Research, World Bank, 1985. Mimeo.

Masefield, Geoffrey B. A History of the Colonial Agricultural Service. Oxford: Clarendon Press, 1972.

Nagy, Joseph G. "The Pakistan Agricultural Development Model: An Economic Evaluation of Agricultural Research and Extension Expenditures." Ph.D. diss., University of Minnesota, 1984.

________. "The Overall Rate of Return to Agricultural Research and Extension Investments in Pakistan." Pakistan Journal of Applied Economics 4(1985): 17-28.

National Research Council. Report of the Committee on Research Advisory to the U.S. Department of Agriculture. Springfield, Va.: National Technical Information Service, 1972 (The Pound Report).

Piñeiro, Martin, and Eduardo Trigo. Technical Change and Social Conflict in Agriculture: Latin American Perspectives. Boulder, Colo.: Westview Press, 1983.

Plucknett, Donald L., and Nigel J. H. Smith. "Networking in International Agricultural Research." Science 225(September 7, 1984): 989-93.

Pound, G. S., and P. E. Waggoner. "Comparative Efficiency, as Measured by Publication Performance of USDA and SAES Entomologists and Plant Pathologists." In Report of the Committee on Research Advisory to the U.S. Department of Agriculture, National Research Council, 145-70. Springfield, Va.: National Technical Information Service, 1972 (The Pound Report).

Pray, Carl E. "Private Agricultural Research in Asia." Food Policy 8(May 1983): 131-40.

Ruttan, Vernon W. The Asia Bureau Agricultural Research Review. Economic Development Center Bulletin 81-2, Department of Agricultural and Applied Economics, University of Minnesota, St. Paul, March 1981.

________. Agricultural Research Policy. Minneapolis: University of Minnesota Press, 1982.

________. "Toward a Global Agricultural Research System: A Personal View." Research Policy 15 (December 1986): 307-27.

Salmon, David C. Consequences of Agricultural Research in Indonesia 1974-1978. Economic Development Center Bulletin 83-1, Department of Economics and Department of Agricultural and Applied Economics, University of Minnesota, St. Paul, January 1983.

Schmookler, Jacob. Invention and Economic Growth. Cambridge, Mass.: Harvard University Press, 1966.

Sehgal, S. M. "Private Sector International Agricultural Research: The Genetic Supply Industry." In Resource Allocation and Productivity in National and International Agricultural Research, edited by Thomas M. Arndt, Dana G. Dalrymple, and Vernon W. Ruttan, 405-15. Minneapolis: University of Minnesota Press, 1977.

Stakeman, E. C., Richard Bradfield, and Paul C. Mangelsdorf. Campaigns against Hunger. Cambridge, Mass.: Harvard University Press, 1967.

Trigo, Eduardo, and Martin E. Piñeiro. "Funding Agricultural Research." In Selected Issues in Agricultural Research in Latin America, edited by Barry Nestel and Eduardo J. Trigo. The Hague, Netherlands: International Service for National Agricultural Research, March 1984.

United Nations Development Programme and the Food and Agriculture Organization of the United Nations. National Agricultural Research: Report of an Evaluation Study in Selected Countries. Rome: FAO, 1984.

U.S. Agency for International Development. Strengthening the Agricultural Research Capacity of the Less-Developed Countries: Lessons from AID Experience. AID Program Evaluation Report No. 10. Washington, D.C.: U.S. Agency for International Development, September 1983.

Venezian, Eduardo L. "International Cooperation in Agricultural Research." In Selected Issues in Agricultural Research in Latin America, edited by Barry Nestle and Eduardo J. Trigo, 99-124. The Hague, Netherlands: International Service for National Agricultural Research, March 1984.

Walsh, Julia M., and Kenneth S. Warren. "Selective Primary Health Care: An Interim Strategy for Disease Control in Developing Countries." New England Journal of Medicine 301(1979): 967-74.

Wilson, Lawrence A. Toward the Future: An Alternative Framework for Agricultural Research, Training and Development in the Caribbean. St. Augustine: University of the West Indies, February 1984.

Workshop on Critical Issues in American Agricultural Research. Science for Agriculture. New York: Rockefeller Foundation, October 1979 (The Winrock Report).

World Bank. Strengthening Agricultural Research and Extension: The World Bank Experience. Washington, D.C.: World Bank, September 1983.

PART 2

CRISIS AND REFORM IN THE U.S. AGRICULTURAL RESEARCH SYSTEM

INTRODUCTION

Vernon W. Ruttan

During the last several decades the U.S. agricultural research system has been criticized and defended from a diverse set of populist and scientific perspectives. At the risk of some oversimplification it may be useful to characterize the criticisms of the U.S. agricultural research system along the following lines.

The populist criticisms have viewed agricultural research and the technology generated by it as responsible for the displacement of small farms and farm workers, as a source of the decline of rural communities, as a cause of deterioration in the quality and safety of food, and as an assault on the quality of the environment. Thus, in the populist view, agricultural research is regarded as a powerful instrument of technical and social change that has been captured by organized agribusiness and has misdirected its energies against the people and the institutions that it was designed to serve. The influential books by Rachel Carson, Silent Spring (1962), and by Jim Hightower, Hard Tomatoes, Hard Times (1973), were important landmarks in the populist critique of agricultural research.

In contrast, a criticism that has often been directed toward agricultural research by the general science community is that agricultural research is not even good science. A central element in this negative perception of agricultural research seems to be that it has been funded primarily through institutional support rather than through competitive grants. A second element is that a relatively high share of agricultural research is directed toward technology development rather than to the generation of new knowledge. These criticisms by the general science community reflect an ambiguous attitude toward technology development. While it is generally conceded that the investment in agricultural research has paid high social dividends in the past, there is a concern that the system is losing its capacity to make comparable contributions in the future. The "Pound" report to the National Research Council (1972) and the "Winrock Report," sponsored by the Office of Science and Technology Policy and the Rockefeller Foundation (1982), were the landmarks in the criticism of agricultural research by the science community.[1]

The initial response to the Carson, Hightower, and Pound studies by the agricultural research community was largely defensive. By the time the Winrock Report was issued, however, the agricultural research community was engaged in an intensive effort of self-criticism and reform. Several of the papers in this section, particularly the papers by James T. Bonnen, Jean Lipman-Blumen, and Irwin Feller, are the products of these efforts.

Professor James T. Bonnen of Michigan State University has been one of the more acute observers and constructive critics of the U.S. agricultural research system for over two decades. In his paper Professor Bonnen draws from the history of U.S. agricultural research, a series of lessons from agricultural research experience for national science policy for agriculture. He is particularly concerned about the relationship between the agricultural and general science communities and with the implications of social and economic change for the future of agricultural research.

The paper by Jean Lipman-Blumen is based on a report to the assistant secretary for science and education of the USDA to assist the department in responding to both the scientific and populist criticisms. She draws from the general field of management to evaluate both the strengths and the limitations of the organization and management of the federal-state agricultural research system.

The paper in this section by Gary Heichel, a plant pathologist who holds appointments both in the USDA Agricultural Research Service and in the Department of Agronomy at the Minnesota Agricultural Experiment Station, illustrates a new methodology that can be used to track scientific developments and anticipate patented scientific advances within a specific discipline or field of technology. The methodology has very significant implications for research planning.

The paper by Richard Sauer and Carl Pray on the Minnesota Agricultural Experiment Station is included as a case study of the development of a state experiment station. Although the Minnesota station is certainly not "typical," its history is similar to many other state stations in the problem that had to be resolved and in its relation to a major research university. Particular attention is given to the process of mobilizing political and economic support for the station and the implications of this process for research program development.

The paper by Irwin Feller is based on a study commissioned by the Cooperative State Research Service to examine the effectiveness and the changing role of the relationship between federal and state agricultural research and the technology transfer and educational activities of the state agricultural extension services. It seems clear that the state agricultural extension programs are facing, in the 1980s, challenges stemming from changes in the technology and organization of the agricultural industry and the rural community that will be even more difficult to resolve than those that have faced the agricultural research community.

An important implication of the papers presented in this section is the responsiveness of the U.S. federal-state agricultural research system to the rapidly changing political, economic, and scientific environment in which it exists. In my judgment the system is in a substantially stronger position to meet the challenges of the rest of this century and beyond that it was a decade ago.

During the next decade I anticipate that the federal-state cooperative extension system will face challenges to both its legitimacy and performance comparable to the challenges faced by the research system during the 1970s and early 1980s. Comparable literature on the problems and responses of the extension system can be expected to emerge over the next several years.

NOTE

1. Appendix A is a reprint of an article by Nicholas Wade (1973) that reviews the findings of the Pound Report. Appendix B is a reprint of the Winrock Report (1982).

REFERENCES

Carson, Rachel. Silent Spring. New York: Houghton-Mifflin, 1962.

Hightower, Jim. Hard Tomatoes, Hard Times. Cambridge, Mass.: Schenkman, 1973.

National Research Council. Report of the Committee on Research Advisory to the U.S. Department of Agriculture. Springfield, Va.: National Technical Information Service, 1972. (The Pound Report)

Rockefeller Foundation and Office of Science and Technology Policy. Science for Agriculture: Report of a Workshop on Critical Issues in American Agricultural Research. New York: The Rockefeller Foundation, October 1982. (The Winrock Report)

4
A Century of Science in Agriculture: Lessons for Science Policy*

James T. Bonnen

". . . unapplied knowledge is knowledge shorn of its meaning."

Alfred North Whitehead

Today all of science seems to be in some political and policy difficulty. There is rising conflict over the funding for and the performance of science. In the agricultural sciences there has been a crescendo of external criticism in Congress and elsewhere ever since the National Academy of Sciences "Pound Report" in 1972 (NAS 1972). Repeated criticisms from the national science establishment suggest that agricultural science lacks a basic science foundation and is a third-rate enterprise. Their usual prescription for this problem is quite simplistic: eliminate all the "politically allocated" Hatch-type formula funding, substituting for it peer-reviewed, competitive grants--open to researchers anywhere, not just in colleges of agriculture.

Various advocacy groups, the media, and some politicians are also highly critical of the agricultural sciences. They focus on such dangers as uncontrolled new genetic technologies and the threats to health, safety, and the environment of other agricultural technologies. The public attitude toward science has shifted from unqualified support to a questioning ambivalence and even fear of its consequences.

At the same time, some state legislatures perceive their land grant college to have abandoned the land grant mission and agricultural problem solving for the glories of basic science. These land grant colleges of agriculture are in difficulty with their clientele and legislatures.

Still other colleges of agriculture have become so applied and isolated from many of the basic disciplines that they are losing scientific and intellectual vitality. After resisting the idea for over a decade, agricultural science now shows some sign of understanding that it must adjust its mission and adapt its institutions to a society and an agriculture greatly different from that of fifty or even twenty-five years ago.

It would appear that the national science establishment is also slowly beginning to understand that it too faces some fundamental questions. Since World War II, public sector national science policy, except in medicine and agriculture, has been focused only on the basic disciplines. Applied science and technology, when considered, is treated separately as primarily a private sector matter of industrial R&D plus a few federally funded R&D centers. Dissatisfaction with this policy posture is growing, especially in industry and politics (David 1986; Norman, 1986; Shapley and Roy 1985). The current administration and its recent science advisor, Dr. Keyworth, have argued, as others have before, that United States science policy should be directed to help achieve greater economic competitiveness (Keyworth 1983; Press 1982). This would require that science policy place more emphasis on mission-oriented, applied research, and on the coordination of disciplinary and applied research.

Science policy then involves not just disciplinary funding priorities--it must also face some fundamental questions: What kinds of science research (disciplinary and applied) should be funded? How should such a diverse scientific enterprise be institutionalized, funded, and managed? What role should the private sector play? And indeed, what philosophic values should inform the priority-setting process? The debate, however, is poorly informed and inflamed by parochial ideologies and self-interest--in science and out (Johnson 1984). Many scientists still believe there is only one problem--inadequate funding.

First, I will examine some of the lessons I believe we should have learned from a century of science in agriculture. I will then attempt to extract from those lessons some implications for science policy today.

My remarks reflect two theses. First, the future competitiveness of the U.S. economy will depend more on the performance of science than in the past. But national science policy is not currently well adapted to serve society in improving its competitiveness (Press 1982). Second, the experience of agricultural science holds important lessons not only for agricultural science, but also for national science policy and the institutional design and management of science. However, as Schuh has pointed out, the value structures and behavior pursued today in many land grant universities and their colleges of agriculture suggest that the land grant idea is being abandoned (Schuh 1986). We have little institutional understanding of our own historical experience, face different demands, and are not currently as capable of sustaining our traditional mission as once we were. If we do not understand our own past, we cannot learn from experience or explain its meaning to others.

I believe the problem in both agricultural and overall science policy in the United States arises out of their successes. The accomplishments of science have changed both science and society. The consequence is that both are very different and more complex today. This results in demands on science generally to expand its scope into a much more complex role--to something different from but more like the land grant research mission. The specific pressures on agricultural science suggest a role more like that of science generally, since a rapidly growing private sector is taking over some basic science but much more of the applied R&D and extension functions. It also appears that responsibility for coordination of agricultural science policy is shifting from a predominantly public function to one of more shared public and private responsibility, making both policy and its coordination more complex.

THE LESSONS TO BE LEARNED

Agriculture has been described as an exceedingly complex system of biological and institutional processes (Mayer and Mayer 1974). I have argued elsewhere that the primary driving force in U.S. agriculture is an articulated system of science-based developmental institutions (Bonnen 1983, 1986, 1987). There are lessons to be learned from a century of experience in these institutions that speak to the problems we face today in general science policy as well as agricultural science policy. We in agriculture need to understand these lessons, act on them, and explain them to our colleagues in science.

Lesson 1: The Four Prime Movers of Greater Societal Capacity

Science research is not the only source of increased productivity. The experience (and the literature) in international agricultural development leads one to the conclusion that the primary sources of increased societal capacity include not only technological change (only some of which is science based) but also institutional improvements, increases in human capability (human capital formation), and the growth of biological and physical capital.[1] While all four of these prime movers are considered in the development literature and in our research efforts, there is a frequent lack of balance in their treatment (Johnson 1986a).

Many biological and physical scientists and some economists focus on technological advance to the neglect of the other forces. Indeed, our society tends to exhibit a disordering "technological fixation." The development process is a search for the appropriate balance in complementary investments needed in all four. That balance is determined in good part by the nature and limitations of the specific social, biological, and physical environment within which development occurs.

There is as well the matter of the values we attach to the human purposes in which the four prime movers are used. Substantial changes in values (monetary and nonmonetary) or in a value's perceived importance transform the capacities of society and modify the mix of its activities as it changes their relative productivity and impacts on human welfare. An example is the clear change in valuation that has occurred with respect to environmental degradation and the many other negative external effects of agricultural production technologies.

Lesson 2: The Continuum of Knowledge and the Research Process

Thoughtful examination of the application of scientific knowledge to human problem solving exposes a continuum of institutions and processes involving several categories of knowledge they help to create and manage (fig. 4.1). We need to understand this continuum of knowledge.

The term basic research I take to refer to disciplinary knowledge, which is the theory, empirical measurements and measurement techniques, and methods used to explain the fundamental class of phenomena of concern in a discipline such as physics, botany, economics, or philosophy. This knowledge improves the capacity of a discipline. It includes research on values in the social science disciplines and humanities.

An applied, multidisciplinary mode of inquiry produces subject-matter and problem-solving knowledge. Subject-matter knowledge is multidisciplinary knowledge useful to a set of decision makers facing a common set of problems. This knowledge is organized under such labels as biotechnology, animal nutrition, agronomy, marketing, or farm management. Most departments in colleges of agriculture are more like broad multidisciplinary, subject-matter institutes (e.g. agronomy, animal husbandry, agricultural economics, horticulture, agricultural engineering) than the disciplines of traditional colleges and universities. Similarly, professional schools and institutes are also typically multidisciplinary, subject-matter organizations. These units organize the inquiry and knowledge from different disciplines needed to understand a subject. Their knowledge bases are necessary to support systematic and sustained problem solving.

Rarely can one go directly from subject-matter knowledge to a decision. Before even multidisciplinary, subject-matter knowledge has direct relevance to a specific problem, it must usually be fashioned into multi-disciplinary, problem-solving knowledge--that is, into a form that is relevant to a single decision maker with a specific problem (or set of decision makers, all with one specific problem). Problem-solving knowledge comes in prescriptions--that is, "should" or "ought" statements for which knowledge of values is essential (see Johnson and Rossmiller 1978, 29-33; and Johnson, 1986b).

Production of these three types of knowledge requires a complex research process. Disciplinary knowledge either does or does not have some known relevance when it is created. If that without known relevance is to have meaning or value, someone must devote himself to research developing the implications of the new disciplinary knowledge. After its potential becomes clear, one can begin to think about specific uses for it; that is, what

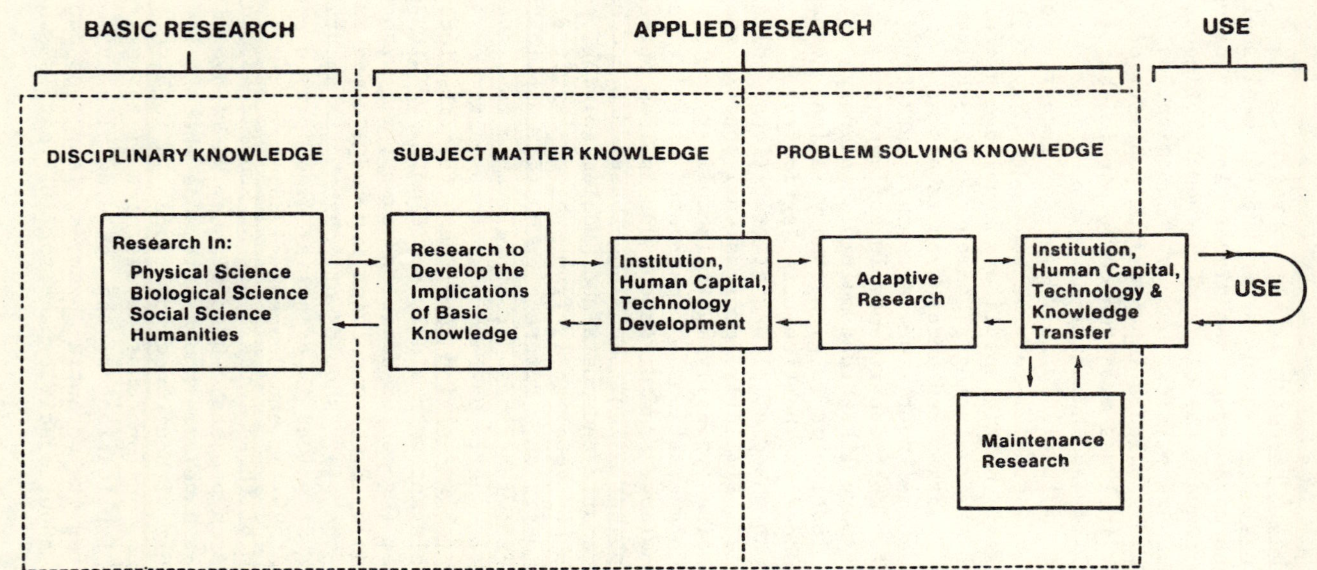

Figure 4.1. The creation-development-utilization of knowledge

kinds of technologies, institutions, and human capital are appropriate and useful to develop out of physical, biological, and social science and humanistic disciplinary knowledge?

After developing a new biological technology (or a new institution), one must face the problem of making that new technology (or institution) work in thousands of ecosystems (and social systems) across the United States. This requires research in every state to adapt its productivity potential to each state's unique, diverse environments. When first developed, hybrid corn production was limited to five cornbelt states. It took twenty years of adaptive research in all the states before it could be grown commercially across the entire United States and before its potential yields were fully realized, even in the corn belt. The location-specific character of agriculture makes adaptive research a central feature of agricultural research programs.

Creating an increase in productive capacity disturbs the ecosystems into which it is injected. New host populations and ecological niches are created, making it possible for pests and diseases to attack and destroy new productivity. Thus maintenance research across multiple ecosystems is necessary in perpetuity to defend the productivity that has been created. As scientific knowledge increases and with it agricultural productivity, the proportion of total R&D investment going to maintenance research must inevitably grow (Ruttan 1982, 60). Livestock is less location specific and thus requires less adaptive and maintenance research than crops. Given the location-specific nature of agriculture, a geographically dispersed institutional system with diverse ecosystem-specific capability is required to sustain adaptive and maintenance research as well as technology transfer activities. All of this seems fairly obvious, but I find many scientists, including some in agriculture, who do not understand that nature imposes these varied activities on agricultural science. This makes unique demands on agricultural science policy.

Another form of maintenance research is becoming increasingly important. Various external effects of agricultural technology are now creating a need for assessment of new technologies and institutional innovations. As we engage in R&D efforts we need to understand the potential effects of an incipient technology. Will it undermine rural communities, degrade the environment, or poison the food chain? What positive and negative values do we place on the impacts of an institutional adaption to the new technology? Who gains, who loses? Assessment research is needed to guide the research process and to legitimize and protect its integrity. Like problem solving, assessment involves research on values.

The agenda of issues in agriculture today strongly suggests that more social science and humanities knowledge is needed. This is due to growing needs for the modification of old institutions or the development of new institutions, for the adaptation and transfer of technologies, for the resolution of ethical problems, and in the creation of new human capital. The mix of relevant disciplines varies with the problems addressed. When the agenda of problems changes as drastically as it has in the 1980s, so does the appropriate mix of disciplines. Implementing the right mix is imperative to future public support of agricultural science. The growing complexity of the industry, and thus the chance of error, as well as the increasing expectation that science should minimize deleterious impacts on society, means that we may not ignore current criticisms and expect continued and adequate public funding. The need for social science and humanistic research on agriculture and rural life is growing.

Lesson 3: A System of Interactive, Coordinated Linkages

It has been found necessary in agriculture for the continuum of knowledge, from creation to use, to be not only institutionalized but also coordinated and focused on problems. The literature on returns to investment in U.S. agricultural R&D, education, and extension/technology transfer demonstrates that a large part of the productivity achieved in agriculture arises out of the interactive linkage coordinating these various investments and the institutions that manage them, rather than flowing directly from separate investments in the four prime movers themselves (Evenson et al. 1979; Johnson 1986a; Ruttan 1982). This is because the four prime movers are complements in production. Each is necessary but not sufficient to achieve an optimum output.

The same conclusions can be drawn about investments in the different kinds of research. Investment in disciplinary research, although absolutely necessary, is not sufficient to achieve high levels of capacity and productivity. The full potential of productivity from disciplinary research is not realized until it is complemented with investments in multidisciplinary, subject-matter and problem-solving research. The reverse is also true; without a continuing basic science investment the productivity and economic return to applied science will decline.

The same principle applies to many elements that affect productive capacity: they are often individually necessary but not sufficient. It is the investment in all relevant complementary factors plus systematic coordination of decisions about the combinations and timing of the various factors influencing productivity that is most important. Failure to link together in the same goal-driven system the public and private decisions about investments in disciplinary research, various types of applied research and technology development, the development or modification of institutions, extension, education, and other human capital and conventional capital slows the achievement and reduces the level of productive capacity that can be extracted from a given investment in agriculture. When an investment is potentially profitable, productivity deferred is in some part lost forever (Knutson and Tweeten 1979). The near exclusive focus on disciplinary research in U.S. science policy has led to a disconnection with subject-matter and problem-solving research that is a drag on productivity.

For a set of institutions to be a developmental system, the individual institutions and functions must be interlinked or articulated, so that they communicate and cooperate in action to achieve some common goals. Interlinkage and coordination speed interactions and the setting and achievement of goals. This linkage is iterative and interactive and is a major source of the system's adaptive capability. Successful systems of science-based development are not conceived or planned as a whole and then put into place. No one knows enough to do that successfully. Scientific inquiry, policy, and institution building decisions are all made under great uncertainty, with imperfect knowledge. Many failures and mistakes occur. Thus, institutional behavior and science must be iterative and interactive in their mode of both inquiry and action to sustain the learning necessary to maintain adaptive capability. Substantial adaptive capacity is necessary to deal not only with uncertainty and mistakes but also with the tension and conflict between institutions and multiple goals within the system.

Lesson 4: Decentralized Decision Capacity

Another characteristic of the U.S. agricultural research system closely related to its interactive nature is decentralization. While the U.S. system is a national system, authority is not concentrated solely at the national level. The conditions of agricultural production are highly varied and location specific. Decentralization is necessary for successful adaptation of science knowledge and technology to the many varied, local ecospheres that characterize agricultural production. There are, in addition, all sorts of local political, cultural, and social traditions that make it necessary to accommodate institutional structures to local politics and resources to ensure a legitimized and coordinated system.

Lesson 5: Consensual System Decisions

It follows from the decentralized nature of the U.S. system that policy decisions affecting all or large parts of it must be developed by consensus, if they are to be accepted as legitimate and implemented effectively. The goals and major initiatives of the system historically have evolved out of a debate on issues ending in a bargained consensus. Unilateral power plays to achieve something that substantially affects the whole system generally create excessive conflict, reduce cooperation and interlinkage, and end in system failure.

Lesson 6: The Integration of the Agendas of Science and Agriculture

The institutions of agriculture combine and manage in a single system societal problem solving and the pursuit of the agenda of science. The pragmatism and political expediency necessary to sustain effective societal problem solving involve organization, values, and expectations that are in some degree inconsistent with and, in the same system, in perpetual tension with the organization, values, and expectations of science, especially the goal of knowledge for its own sake. Much of the societal support as well as the productivity generated by agricultural science has arisen out of the sustained interlinkage of these functions and the management of the resulting tensions to maintain a working balance between the agenda and capacities of science and the agenda of problems in agriculture. Effective science-based problem solving requires coordination and integration of science with any economic sector in which science is expected to drive major increases in productivity.

Lesson 7: Chronic Underfunding and Spillover

What have we learned about funding agricultural science? Measures of the annual rate of return on public investments in agricultural research run three to five times the rates on most alternative public investments (Ruttan 1982, 248). While the causes are not fully understood, this suggests that agricultural R&D is substantially underfunded by normal investment criteria (Oehmke 1986). Today one must deal with the counter argument that the world's markets are awash with the products of excess agricultural capacity because of science research. This is pure nonsense. The creation of this excess capacity is the consequence of mistaken investment decisions by farmers based on excessively optimistic expectations, induced in part by subsidies and foolish national policies (Johnson 1985). These same expectations were subsequently destroyed by the growth in world supplies, increased industrial nation subsidies of their agricultures, and a U.S. macroeconomic policy that has destroyed U.S. agriculture's export potential while escalating its costs and leaving it in the deepest financial crisis since the Great Depression.

We are in a contest for international markets that depends in part on a healthy agricultural research enterprise, including social science research on policy, markets, other institutions, human capital formation, and our capital and natural resource bases, to maintain comparative advantage in production costs and market access. The argument that R&D causes surpluses arises out of the equally fallacious reverse argument of scientists who urge increased biophysical R&D to solve world hunger, when hunger is with few exceptions due to inadequate income or its maldistribution (Johnson 1985).

Today, state appropriations greatly exceed federal funding for the state experiment stations (OTA 1981). In real terms, federal funding of the state system all but ceased to grow after 1967. This raises serious questions about the federal commitment to its historic partnership with the states in agricultural research. The issue is fundamental to the long-term performance of the system since a large part of the benefits of research funded by one state spill over into other states. The empirical evidence on spillover of the benefits from research financed by one state accruing to farmers and consumers in other states is strong (Havlicek and White 1983). While spillover varies greatly from state to state, typical spillover losses range from one-half to two-thirds on basic science and one-third to one-half of technology-oriented investments (Evenson et al. 1979). Losses of state level benefits deter state investment in agricultural science. Without compensating federal funding, the states, acting alone and rationally, will never achieve an optimum level of national investment in agricultural research. This is the classic public finance problem faced by federal systems with two or more levels of government.

Ruttan argues that the primary rationale today for federal support of state agricultural research is to compensate the states for spillover--in order to achieve a socially optimum national rate of research investment (Ruttan 1982, 251-59). Thus, each state's share of federal support for agricultural research should at minimum approximate its spillover losses. This suggests that the federal government should be matching state funding on an open-ended basis rather than the reverse. The only other way to achieve the optimum level of investment is to fund all agricultural R&D at the federal level.

Despite some secular decline in spillover rates, the continued existence of spillover means that the private sector, although large and growing, cannot be expected to perform all of the applied or basic research necessary to reach an optimum level of investment in agricultural science (Evenson et al. 1979).

Spillover also occurs internationally. The benefits of European and U.S. agricultural research have flowed to both developed and developing countries over many decades. The rising perception of this today in the midst of a ferocious international battle for export markets gives rise to attempts, such as that of the U.S. soybean producers, to eliminate U.S. international aid in agriculture in a futile effort to monopolize U.S. agricultural research results. The only real long-run hope for increased demand for U.S. farm exports lies in higher incomes in developing countries due to development. In addition, agricultural science is an international enterprise today. We are about as likely to benefit from research elsewhere, as the reverse. Withdrawing from international cooperation in agriculture would be suicidal for our own productivity and export markets.

Lesson 8: Stability of Funding

The research on the optimum rate of investment in agricultural science demonstrates substantial losses of productivity when the funding of R&D is highly variable from year to year and decade to decade--that is, when we are unable to sustain the pursuit of the inherent long-term goals of R&D and development (Knutson and Tweeten 1979; White and Havlicek 1982). Disciplinary research, much of technology development, and major adaptive research would appear to be especially vulnerable. This principle applies to R&D generally, although the characteristics of aggregate demand and supply response accentuate the problem in agriculture. Thus, this is a race that goes to the tortoise, not the hare--that is, to sustained, long-term institutional support of agricultural research, not the jerking around that has been imposed on the agricultural system nationally for almost two decades through inflation, stagnant and uncertain appropriations, inconsistent goals, political abuse, and inattention.

IMPLICATIONS OF THESE LESSONS FOR POLICY IN A CHANGING WORLD

What, then, are the implications of these lessons for science policy today? In making this assessment we must recognize that both science and society have changed since the system of agricultural science institutions matured.

The Search for a National Science Policy

The Funding Debate. The conflict over research funding is a debate of the deaf. The national science establishment argues that only competitive grants can be used effectively to allocate resources for science purposes; any other approach produces poor quality science. The agricultural science establishment has responded by defending Hatch or formula funding as essential to science in agriculture. Some agricultural leaders clearly fear the effects of exclusive use of competitive grants on the stability and long-term vitality of their institutions.

Competitive grants, of course, are now a small but regular part of the federal funding of agricultural science. The funding argument tends to be put in either/or terms. If one is good the other has to be bad for science. This is wrong and obscures the real funding problem.

Formula or institutional funding was originally established to induce development of a decentralized state system of agricultural science. It has been responsive to the need for sustaining the large fixed or overhead costs of science and the mission-oriented, largely applied nature of experiment station research. As Ruttan points out, the great productivity achieved by this system of institutional support places the burden of proof on those who would change it (Ruttan 1982, 231). The cost of entrepreneuring and managing formula funding falls on the administrators in the agricultural research system, not on the researchers (Bredahl et al. 1980). The quality of science produced depends not on the funding process but on the quality of individual scientists hired; how supportive the research institutions and their incentive structures and academic freedom are of creativity; and the quality and amount of administrative support, especially for the more administratively demanding multidisciplinary, subject-matter, and problem-solving research (Berry 1980; Johnson 1986b, chaps. 13-16).

Peer-reviewed, competitive grants are a centralized system that is reasonably well adapted to allocating disciplinary research resources. The cost of entrepreneuring and managing competitive grants falls mostly on the researcher (Bredahl et al. 1980). This is why one finds senior scientists who no longer have time for anything but developing grant proposals and managing a laboratory. This cannot be the best use of a creative scientist's time. Short-run, project-by-project grant proposals do not add up to coherent long-term research programs necessary in much applied research and technology development. Competitive grants often do not cover the total costs of research (Ruttan 1982, chap. 9). Consequently, the current battle over the efforts of the Office of Management and Budget (OMB) to reduce the overhead rates universities receive on research grants is extremely heated (Werner 1986a, 1986b). OMB believes that some universities are abusing the system.

Experiments have been done on the performance of peer review panels. Panels composed of different scientists with comparable background and ability produce quite different rank orderings of the same set of project proposals (Cole and Cole 1981; Lyon 1985). While there is an unavoidable subjective element in the awarding of peer-reviewed, competitive grants, the experiments do not suggest any systematic bias (Cole et al. 1978; Lloyd 1985). The recent flurry of attempts to use political power to go around the peer review system in obtaining federal research support nevertheless appears to be based on the belief that an "old-boy system" does dominate peer review and rewards the long established and discriminates against smaller and less well-established institutions (Greenberg 1986; Norman 1986; Norman and Marshall 1986). I do not believe this is true but the perception has become a growing problem in legitimizing science budgets.

Neither allocative device is perfect. Each works reasonably well for some purposes but not for all. The real problem, which tends to be obscured, is that science and its purposes have become so complex that research funding requires some integrated mixture of funding devices including, but not limited to, formula or institutional funding and competitive grants. The type of research and purpose should control the mix of allocation devices. The purpose and criteria for funding, evaluation, administration, and conduct of disciplinary research differ from those for subject matter research, which in turn differ from those for problem-solving research (Johnson 1986b, chaps. 13-16). The appropriate peers for evaluating problem-solving research, for example, are quite different from those for disciplinary research.

New ways of funding science are needed to deal with the increasing complexity of science and the growing demands of society. We need to stop our senseless arguments and look at science and its multiple purposes (public and private) and examine pragmatically the ways in which we might best fund different types of research. First, however, we must agree on the role of science in society. Without clarity of purpose, very little else can be decided.

The Scope of U.S. Science Policy. The overall science establishment needs to take a more comprehensive view of science and its role in society. It functions today like a special interest pleader, since its only effective goal is support of basic (i.e., disciplinary and academic) science. This leaves academic science isolated from and failing to make its greatest contribution to society. Society's support of science is consequently not as strong as it should be. Somehow the private sector is supposed to cover all the applied research and development and coordinate the continuum of knowledge. With little or no public policy direction and substantial public good elements in applied R&D, this has never been realistic. The change in science policy after World War II, in my judgment, has led not only to stronger basic science research but also to a weaker linkage between basic science and technology development and transfer, and thus to lagging productivity. A comprehensive national science policy would include not only disciplinary research but also subject-matter and problem-solving research in which there is a substantial national interest. A comprehensive national science policy would attempt to prioritize all federal science expenditure, not leaving priorities segmented as we do now by setting priorities within but not between disparate categories such as federally funded R&D centers and labs; large, unique "big science" facilities; National Science Foundation (NSF)-National Institutes of Health (NIH) grants to individual researchers; federal funding of state agricultural research; and funding of internal federal agency research programs (Koshland 1986). Such a policy would recognize appropriate public and private roles in R&D and contribute to guiding its coordination. This may be the counsel of perfection, but it is the only path I can see to lower levels of conflict within science and to an improved performance of science.

The necessity for addressing the complete continuum of knowledge, even within the university, was put in perspective fifty years ago by the philosopher Alfred North Whitehead. At Harvard's tercentenary celebration of its founding he said:

> In the process of learning there should be present, in some sense or other, a subordinate activity of application. In fact, the applications are part of the knowledge. For the very meaning of things known is wrapped up in their relationships beyond themselves. Thus unapplied knowledge is knowledge shorn of its meaning. The careful shielding of a university from the activities of the world around is the best way to chill interest and to defeat progress (Whitehead 1936).

Much of the creativity in any discipline comes from the intellectual stimulation of confronting disciplinary knowledge with the test of application, knowledge from other disciplines, and the challenge of societal problem solving. We badly need to recapture this catholic view of science and again make it central to the ethic of science. Failure to do so will leave science less creative and productive, whether viewed from science or society's needs.

Many colleges of agriculture appear to be abandoning their classic responsibility for the full continuum of knowledge in agricultural research. They should reflect on the historical lessons from agricultural science and adapt their behavior and institutions to recapture their original vision, which included the same catholic view of science.

The pressures society is putting on science are pushing us toward a modern science and technology policy that would provide a conscious, coordinated balance of public and private investment across the entire continuum of knowledge from creation to use in areas critical to society. Today, only in medicine and agriculture can full science-based systems of developmental institutions be said to exist, and both of these systems need some institutional rethinking and reform (Bonnen 1986, 1987). Implied are changes on campus, in industry, and in government to interlink basic science and its policies with a limited number of long-term, science-based missions of significance to society. Consensus-based coordination of effort and policy across disciplines, government, and industry is involved. We need to face the fact that subject matter and problem-solving research involve more complex and costly administrative support. The lessons learned from science in agriculture are clearly relevant to any U.S. science policy that focuses on raising U.S. productivity. Many of the same lessons can be seen in the history of government involvement in industrial R&D (Nelson 1982; Nelson and Langlois 1983; Pavitt and Walker 1976; Shapley and Roy 1985).

The Expanding Continuum of Knowledge. Coordinating the continuum of knowledge from creation to use is made even more difficult today by the growing stock of more and more complex knowledge and by the progressive specialization of science. In terms of organizational distance, the extreme ends of the continuum of knowledge are moving away from each other as knowledge increases and science grows more specialized. Any policy that hopes to extract greater productivity from science through a coordinated attack on the practical problems of some economic sector must recognize that the problems of coordination, administration, and management of research have become far more complex. This, combined with the growing complexity of most economic sectors, makes the coordination and linkage of public and private sector problem-solving research and technology and knowledge transfer (extension) much more difficult to conceptualize and manage, especially in what must for the most part be a decentralized system (Feller 1986). Only a decentralized system is likely to have the adaptive capability to work well in such complexity. Clearly, policy for it must be arrived at through bargained consensus.

The Changing Paradigm of Science. Another reason for rethinking the scope of science policy and its funding system is that the modal paradigm of science appears to be changing. For example, as the frontiers of knowledge have expanded, once reasonably separate domains of disciplinary inquiry have begun to overlap extensively and interpenetrate each other. Vast areas of physics and chemistry are now common to both disciplines. Indeed, the identity of chemistry is all but lost in the rest of science (Browne 1985). The last two Nobel Laureates in chemistry were not chemists but mathematicians. This overlap results in new disciplines, or at least separately organized units. We have departments today of biophysics and biochemistry. Advanced math and statistics are essential to the cutting edge in most disciplines. The boundaries of biology and its subdisciplines are transcended not only by physics, chemistry, and other disciplines but by technologies and techniques as well--e.g. microbiology and molecular biology. Technological capability drives the biophysical sciences as much as science drives technological capability (Price 1983; Shapley and Roy 1985). In the social science and humanities this two-way causation includes not only technology but also institutions. The social science disciplines have long had large overlapping domains.

The point is that to practice at the cutting edge in almost any discipline today, even in the biological and physical sciences, requires not only command of a discipline but also of major components of knowledge from related disciplines well beyond mathematics and statistics. As science grows more complex and interactive, a growing proportion of disciplinary inquiries pursued to completion takes one through multiple disciplines and techniques. This overlap has become so extensive that creating new disciplines or departments has ceased to be the best or only response. Thus, paradoxically, to practice a discipline today one must increasingly collaborate with other disciplines or become in some degree multidisciplinary. This is not really news. The change is only one of degree, but it is so fundamental that it suggests a change is under way in the modal paradigm of science.

All of this is without considering applied science, which is inherently multidisciplinary. The scale, scope, and complexity of applied research in the private sector and nonprofit institutes as well as government have grown immensely since World War II, as the importance of applied research and technology development has become more critical to national productivity.

An ever larger part of the economy is more and more dependent on science for profitability. The pace of innovation in some sectors and the integration of world capital and commodity markets place a premium on getting from basic science ideas and technological innovation to a successful product or service with the smallest lag in time. Firms and nations who become laggards tend to get squeezed out. Success requires selective command over the full continuum of knowledge and excellent management and coordination of the R&D, market intelligence, and marketing activities involved. National research and technology policy has become a critical element in maintaining a nation's comparative advantage in world markets.

The idea of how science is practiced that evolved out of the nineteenth century and around which the policies and funding of science, especially basic science, have been organized is predominantly that of the individual scientist surrounded by a few graduate students or laboratory assistants. The importance of this idea of science has been magnified by the mistaken but common belief that all technology arises out of basic science. The rapid development of the scope and importance of mission-oriented, private sector R&D, even in some basic science areas, now combines with the growing scale, scope, and complexity of applied science and the interpenetration and overlap of one discipline with others to erode the relevance of the old paradigm. What one sees increasingly are R&D consortia and cooperative research endeavors of various sorts. In sessions on science policy or university strategic planning one sees an increased emphasis on the need for more multidisciplinary research--although the difference between problem-solving, subject-matter, and disciplinary research is usually not made clear. The pressure for collaboration among disciplinary researchers in universities is growing. The numbers of multiuniversity consortia have grown. So have those that combine university and industry R&D efforts to span the full continuum of knowledge addressing some agreed purpose.

Industrial R&D consortia have existed for decades. Specific cases of cooperative research involve quite different purposes and portions of the continuum.

The motives for these arrangements include cost sharing where there are economies of scale (often involving large, specialized research facilities or tools) as well as the need to assemble diverse disciplines for fundamental research or to bridge some part or all of the continuum of knowledge from disciplinary through applied, subject-matter and problem-solving inquiry.

This shift toward more collective or cooperative research consortia makes the funding question very much more complex for everyone from NSF and NIH to the foundations and industry, as well as the Congress and state legislatures. It opens up the question of what are appropriate funding mechanisms and puts the emphasis on the very different purpose or goals of disciplinary, subject-matter and problem-solving research and thus the differing mixes of these types of research presided over by different kinds of collaborative research efforts. I believe both the competitive grant, basic science experience as well as the industrial and agricultural science experience are relevant--if all parties to the debate will listen to the others and think objectively about the problems we face in common.

The Growing R&D Role in the Private Sector. The private sector presence in R&D has been growing rapidly. This presence is especially significant in economic sectors where vertical integration produces large oligopolistic firms with substantial influence and control over the industry's demand and supply functions. Such firms know they will be around ten and twenty years from now. They can and increasingly do invest in both basic and applied science to guide and control the conditions of that future. Even so, few of these firms can afford the scale of basic and applied science investments they might like or that society needs if it is to be effective in international competition. This creates a growing interdependence between public and private R&D that generates pressures for collaboration and joint ventures. University basic science is being pulled into coordinated efforts that cover large parts or all of the continuum of knowledge and into new arrangements for funding both problem-solving and subject-matter as well as disciplinary research.

In agriculture, private sector R&D has developed more slowly than in industry because of the more atomistic nature of agricultural production. However, concentration is proceeding rapidly in agriculture and especially in agribusiness today. With the ability to patent plant material and biotech processes and with concentration has come a more rapid growth of private agricultural R&D. The private sector is now taking over many areas of applied research that have been a public responsibility. The consequence is that agricultural science and the land grant system are becoming somewhat less problem- and product-specific in some areas and more a general science wholesaler than a retailer. Agricultural science is being pressed toward a mix that is relatively heavier on basic science combined with a somewhat different set of applied science and extension activities. Agriculture is moving rapidly toward a considerably more complex, vertically integrated, high technology sector.

While still rather different, agricultural science and academic disciplinary science are being propelled toward a more common set of responsibilities, problems, and activities. In common they face the need to redefine the changing boundary between public and private R&D responsibility. The growing role of private R&D means that the private research institutions will have to play a far more significant role in the coordination of science policy. With greater intermixing of public and private motives, the public sector (Congress, universities, science professions) must find new ways to assure the integrity of science and its decision processes. Private purposes can easily dominate joint ventures, thus forfeiting much of the larger social benefit that might otherwise be achieved from collaboration between the public and private sectors (Ulrich et al. 1986). The public institutions are responsible for assuring that the public interest in science is served. Much applied R&D remains a public good that will be ignored, if applied science is ceded to the private sector without thought.

A mature industrial nation's comparative advantage in international markets rests on high technology and high human capital industries such as electronics, computers, communications systems, education, finance and, in many cases, agriculture. The U.S. economy is increasingly dependent on a coordinated scientific effort to remain competitive in markets for high technology products and services.

Our ability to sustain the kind of R&D policies that will support a high technology economy is being undermined by the drift of the academic community toward the view that the only research of importance and the only research worth financing and doing is basic science (disciplinary) research (Shapley and Roy 1985). At the same time that the academic community's capacities have shifted to the disciplinary end of the research spectrum, the problems of society have become more specialized, interactive, and complex, requiring (besides disciplinary research) greater coordination with and investment in applied, multidisciplinary research of a subject-matter and problem-solving nature. If all the applied research could be done by the private sector this would only be a problem of coordination. But most early technology, human capital, and institutional development and much of the adaptive and maintenance research in biology are clearly a public good and beyond the private sector's capability. Thus, the training and values of much of academic science are undermining society's capacity for problem solving, while the need for such capacity grows more intense.

Agricultural Science Policy

Until 1916, agricultural research activities accounted for one quarter or more of the USDA budget. Today, a far larger research enterprise accounts for less than 2 percent of the USDA's budget and about the same percentage of total federal R&D expenditures (OTA 1981).

Today the private sector accounts for about two-thirds of all agricultural R&D expenditure (Ruttan 1982, 181-86). Two-thirds of this is concentrated in physical science and engineering and only a small but growing part can be described as basic science. At the state experiment stations about three-quarters of the research is in the biological sciences and technology. According to Ruttan, social science research accounts for less than 3 percent of private sector R&D and less than 10 percent of public sector R&D in agriculture (p. 186).

The funding of science began to change after World War II with the creation of the National Science Foundation (NSF) and the great expansion of the National Institutes of Health (NIH). These institutions today support a large public and private academic science structure, mostly disciplinary in nature and, with a few exceptions, largely outside of and unconnected with the land grant-USDA system of agricultural science institutions. This means that the bulk of basic biological, physical, and social science and humanities research, some portion of which is undoubtedly relevant to agriculture, today lies outside the system of agricultural linkages.

The impact is more pronounced because of the fragmentation of university disciplinary science into a progressively greater number of separate academic units as scientific knowledge has expanded and become more specialized. The consequence in the land grant colleges is a steadily increasing organizational distance between applied research in agriculture and some of its disciplinary roots. It has increased the difficulty involved in interlinkage and coordination of the continuum of knowledge from its creation to use. In part, this is why we are having difficulty maintaining a balance across this continuum in agriculture. Some land grant colleges have worked to maintain effective linkages, others have not.

Complicating this is the dominant value belief of the academic science establishment that only disciplinary research in the biological and physical sciences is academically respectable and justified. The support given the social sciences tends to be limited to the behavioral sciences and to positivistic inquiry within them. Questions about what has value are not considered to be in the domain of science and are treated as subjective and nonscientific. The response in some colleges of agriculture to the changing values and distribution of power in academic science has been a parochial, nearly exclusive focus on applied problem-solving research for agriculture.

Other colleges of agriculture, many land grant universities, and some agricultural professional associations have absorbed as their ideal the academic science establishment's focus on disciplinary research. Their "search for academic excellence" is denaturing the land grant tradition of problem solving and service to all people, irrespective of wealth or position. A near exclusive focus on basic discipline depreciates applied, multidisciplinary research, denies admission of problem solvers and prescriptive analysis to the academic pantheon, and turns good land grant universities into second-rate, private academies. Such an environment destroys the basis for effective extension education and problem solving, and lowers the potential productivity of any agricultural science investment. Today these two parochialisms of "pure" and "applied" science constitute an obstacle in the search for an appropriate balance of investment across the continuum of knowledge necessary to achieve greater national capability (Johnson 1984).

Agricultural research is nationally of minor political concern today. Over recent decades the congressional interest in USDA research budgets has focused primarily on applied commodity research and the proliferation and location of regional research laboratories in selected congressional districts. The narrowing of farmer interest to immediate farm program benefits combined with the lack of scientific vision in either congressional or USDA political leadership has over several decades contributed to a confusion of purpose and to an erosion, isolation, and fragmentation of the USDA's national research capability in the biophysical and social sciences.

A once-effective priority-setting process has been undermined by abuse of the R&D function by Congress and USDA political leadership, the erosion of USDA research dominance, a decline in the dependence of the colleges on the USDA, as well as the rise of new public and private R&D actors of varying importance to agriculture but outside the agricultural science system (Bonnen 1987). Relevant research activities are not as well interlinked and coordinated. Agricultural research is in large part a public good. In a policy process dominated by highly organized economic interests with destructively narrow, short-term views of self-interest, public goods are of little concern. Why worry? The agricultural cornucopia will always flow. But will it? Or will the public interest be served?

Institutional changes since 1977 attempt to deal with part of this problem. The establishment of the Joint Council on Food and Agricultural Sciences, the National Agricultural Research and Extension Users Advisory Board, and more recently an Assistant Secretary for Science and Education creates a potential for greater clarity of purpose and coordination of priorities.

As an industrializing nation's agriculture develops, its production and marketing processes inevitably become highly specialized and its welfare and performance increasingly vulnerable to disruptive forces from outside the sector. The result is growing government policy intervention in agriculture, an expanding private-sector interest in public policy outcomes, and ultimately, severe fragmentation of economic interests as development proceeds. This fragmentation leads to rising levels of political conflict and disorder among the institutions of agriculture, along with the domination of the policy process by progressively narrower economic interests that make it far more difficult to pursue long-term goals, especially research that promises to provide only diffused or problematically distributed benefits. As a consequence, as long-term, steady support for agricultural science research has become absolutely essential to the future of a high technology agriculture, the increased fragmentation and narrowing of the economic interests in agriculture make it increasingly difficult to mobilize broad support for long-term goals. This can be seen in both Europe and the United States.

Thus, most industrial nations with highly productive agricultural sectors face an eventual political-organizational crisis in deciding whether or not, and in what form, they will sustain the science-based developmental system in agriculture that with varying degrees of success they have created. Failure to maintain that system will substantially disadvantage an industrial country both internally and in international affairs. Food will always be a strategic necessity, whatever a country's resource base.

Since agricultural sector political power would appear inadequate to sustain a modern, balanced science base for agriculture, agricultural science needs to become more nearly an integrated part of the science establishment. But the integration and cooperation needed between the two science establishments will not come unless there is greater mutual appreciation of the strengths each would bring to a common, more coordinated endeavor. The old land grant model exhibits many of the desired characteristics of such a system, but even it is in need of institutional rethinking to adapt it to the modern political environment of science and agriculture.

The developmental system of institutions of agriculture, although in some disarray, now seems to be adapting to these changes. There is enough disorder to raise questions about the system's continuing viability as a system. Most of the institutions will survive, but will the system? I believe it will.

Outside of agriculture most scientists are both ignorant and critical of the agricultural research system. Indeed, it is complex and not easily understood. National commitment to these institutions is in question. Yet, if past experience means anything, we must provide some kind of system for coordination and management of the complex of relevant national and local, public and private institutions, if we are to continue to have an agriculture that is competitive in world markets. Agricultural science differs in its needs from medicine and other parts of science. Such differences must be recognized and accommodated in science policy and its funding. Agricultural scientists must be able to explain those differences.

The Social Sciences in Agriculture

The current agenda of issues in agriculture and the rising criticism of the impacts of science suggest strongly that inadequate investment has been devoted to social science and humanistic inquiry in agriculture. Successful biological and physical science-based technological growth has created externalities and imbalanced investments in the four prime movers. We now face needs for technology assessment, new institutional innovations, and research on ethics and values. The latter two especially are the domains of the social sciences and humanities for which we have failed to provide adequately in science policy, public or private. There is no market incentive for industry to assume the costs of dealing with externalities. Many influential agricultural scientists either do not want to believe that their scientific inquiries involve such problems or simply see their social responsibility as limited. The public sees it otherwise.

Look at the larger agenda of issues now facing agriculture. It includes the largest financial crisis in agriculture since the Great Depression; complex and poorly understood national and international macroeconomic impacts on agriculture; major international trade issues ranging from protectionism and an immense trade deficit to the impact of obsolescent international monetary institutions on exchange rates and market stability; and the impacts of national deregulation (in finance and banking, petroleum-based energy sources, and transportation) on agricultural and rural welfare and property rights. These problems all fall in the domain of the social sciences.

Look at the need for new or modified institutions. The institutional structure supporting science in agriculture and nationally is in transition to some new configuration with almost no research on the issues involved--such as research funding systems, or the means for interlinkage and coordination of R&D actors, or on science policy itself. The experiment stations, USDA, NSF, and NIH all should be targeting this area of R&D. The nation's farm policy is in shambles but continues for lack of convincing alternatives to constrain instability and periodic excess capacity. The new genetic technologies are changing the way agricultural science is funded and managed and are raising new issues in property rights to genetic material as well as to processes for genetic manipulation. Information age technologies are changing the way we receive, process, store, and use information in farm and agribusiness decision making. Adapting agriculture and its institutions and policies to these and other new technologies presents a substantial research and education challenge to the social sciences. A major issue that must be faced is the future social performance of a farming and agribusiness sector that will be much more highly concentrated and vertically integrated. One could go on.

These are all issues that science policy must address seriously if it is to meet society's expectations and needs. In agriculture this requires more than economic expertise. Major sociology, political science, and legal research is necessary. In addition, some social psychology and cultural anthropology research is also relevant.

As for the humanities, the lack of any systematic historical perspective on agriculture disorders agricultural leaders' views. The history of the development of agriculture and its institutions needs to be taught and researched. The growing set of ethical issues and value-related problems in agricultural policy and in science requires philosophic attention.

Agricultural economics is often the only established social science department in colleges of agriculture. As a consequence, I believe agricultural economists have a responsibility to make the case for the missing social sciences and humanities in the colleges. We have neither the personnel nor the expertise to meet this flood of issues by ourselves. We also have to improve our own sense of purpose and performance before we can provide much leadership.

Since World War II agricultural economics has been drifting toward an antiempirical and a disciplinary outlook--away from the great empirical tradition around which the profession was built and upon which its reputation still rests. Today we celebrate theory and statistical methods while ignoring the data collection and problem solving necessary to validate our theory and models. Any profession becomes what it celebrates and rewards (Bonnen, forthcoming).

Why is this happening? First, we are emulating academic economics, which, with some distinguished exceptions, now exhibits little commitment to the empirical (Leontief 1971). Another source of the problem, I believe, is the search for "academic excellence" in agricultural economics that places excessive or sole emphasis on rewarding the development of disciplinary knowledge, almost to the exclusion of the development of subject-matter and problem-solving knowledge, both of which are essential outputs of an effective agricultural economics department.

The model of an agricultural college department as a collection of "pure disciplinarians" producing disciplinary and some applied disciplinary knowledge is a pathological distortion of the land grant mission. Yet that is the model some colleges and many agricultural college departments are now following. Just as pathological is the purely applied model of a subject-matter organization unconnected to the appropriate range of supporting disciplinary capacity in teaching and research.

The sciences in agriculture face great challenges and opportunities. But we must be able to address them with a balanced command over the full continuum of knowledge if we are to be successful. We need to see our role as one of producing disciplinary, subject-matter and problem-solving knowledge. We need to do this in an enterprise that brings into an integrated focus both the agenda of science and the agenda of problems in agriculture.

Many colleges are now striving to regain a better balance in their science capability. We only need to keep a clear focus on the problems of agriculture and rural society and on the capability and potential of science to drive our enterprise back toward a better balanced command of the continuum of knowledge. Agriculture needs leadership today with a broad vision for the future of a science-based agriculture. It needs strong leadership from the social sciences.

NOTES

*Revision of Fellows address, American Agricultural Economics Association meeting, Reno, Nevada, July 29, 1986. Published in American Journal of Agricultural Economics 68(5) (December 1986).

The author is indebted to colleagues at Michigan State University for critical reviews, most especially Glenn L. Johnson, Eileen van Ravenswaay, Stanley Thompson, Lester V. Manderscheid, Larry J. Connor, James F. Oehmke, Lindon Robison, James D. Shaffer, Allan Schmid, David Schweikhardt, Jack McEowen, and Larry Hamm.

1. Institutions I define to include both the rules of behavior that govern patterns of relationships and action as well as public agencies, private firms, families, and other decision-making units (Commons 1950; Knight 1952).

REFERENCES

Berry, R. L. "Academic Freedom and Peer Reviews of Research Proposals and Papers." American Journal of Agricultural Economics 62(1980): 639-44.

Bonnen, James T. "Agriculture's System of Developmental Institutions: Reflections on the U.S. Experience." L'Agro-Alimentaire Quebecois Et Son Development Dans L'Environment Economique Des Annees 1980. Department D'Economie Rurale, University of Laval, Quebec, Canada, 1983.

________. "The Institutional Structures Associated with Agricultural Science: What Have We Learned?" Proceedings of the Agricultural Science Workshop, Agricultural Research Institute, Wayzata, Minnesota, May 1-2, 1986.

________. "Improving the Data Base." In Agriculture and Rural Areas Approaching the 21st Century: Challenges for Agricultural Economics. Edited by James Hildreth, Katherine Lipton, Ken Clayton, and Carl O'Connor. Ames: Iowa State University Press, forthcoming.

________. "The Role of Science-Based Technology, Human Capital, and Institutions in United States Agrarian Development." In United States-Mexico Relations: Agriculture and Rural Development. Stanford, Calif.: Stanford University Press, 1987.

Bredahl, Maury E., W. Keith Bryant, and Vernon W. Ruttan. "Behavior and Productivity Implications of Institutional and Project Funding of Research." American Journal of Agricultural Economics 62(1980): 371-83.

Browne, Malcolm W. "Chemistry Is Losing Its Identity." New York Times, December 3, 1985, p. 18.

Cole, Jonathan R., and Stephen Cole. Peer Review in the National Science Foundation: Phase Two of a Study. Washington, D.C.: National Academy Press, 1981.

Cole, Stephen, Jonathan R. Cole, and Leonard Rubin. Peer Review in the National Science Foundation: Phase One of a Study. Washington, D.C.: National Academy Press, 1978.

Commons, John R. The Economics of Collective Action. New York: Macmillan, 1950.

David, Edward E., Jr. "An Industry Picture of U.S. Science Policy." Science 232(1986): 968-71.

Evenson, Robert E., Paul E. Waggoner, and Vernon W. Ruttan. "Economic Benefits from Research: An Example from Agriculture." Science 205(1979): 1101-7.

Feller, Irwin. "Research and Technology Transfer Linkages in American Agriculture." In The Agricultural Scientific Enterprise: A System in Transition, edited by L. Busch and W.B. Lacy. Boulder, Colo.: Westview Press, 1986.

Greenberg, Daniel S. "The Race for Science Money." Detroit News, July 24, 1986, p. 19A.

Havlicek, Joseph, Jr., and Fred C. White. "Interregional Transfer of Agricultural Research Results: The Case of the Northeast." Journal of the Northeastern Agricultural Economics Council 12(Fall 1983): 19-30.

Johnson, Glenn L. Academia Needs a New Covenant for Serving Agriculture. Mississippi State: Mississippi Agricultural Experiment Station, July 1984.

________. "Agricultural Surpluses--Research on Agricultural Technologies, Institutions, People, and Capital Growth." In Crop Productivity: Research Imperatives Revisited, edited by M. Gibbs and C. Carlson. East Lansing: Michigan Agricultural Experiment Station, Michigan State University, 1985.

________. "Institutional Frameworks for Agricultural Policy Monitoring and Analyses." Paper presented at the Economic Development Institute, World Bank, Washington, D.C., July 10, 1986a, pp. 34-39.

________. Research Methodology for Economists. New York: Macmillan, 1986b, pp. 11-29.

Johnson, Glenn L., and George E. Rossmiller. "Improving Agricultural Decision Making: A Conceptual Framework." In Agricultural Sector Planning, edited by George E. Rossmiller. East Lansing: Department of Agricultural Economics, Michigan State University, 1978, pp. 23-51.

Keyworth, G. A. "Federal R&D and Industrial Policy." Science 220(1983): 1122-25.

Knight, Frank H. "Institutionalism and Empiricism in Economics." American Economic Review 42(1952): 45-55.

Knutson, Marlys, and Luther G. Tweeten. "Toward an Optimum Rate of Growth in Agricultural Production Research and Extension." American Journal of Agricultural Economics 61(1979): 70-76.

Koshland, Daniel E. "To Lift the Lamp Beside the Research Door." Science 233(1986): 609.

Leontief, Wassily W. "Theoretical Assumptions and Non-Observed Facts." American Economic Review 75(1971): 293-307.

Lloyd, James E. "Selling Scholarship Down the River: The Pernicious Aspects of Peer Review." Chronicle of Higher Education, June 26, 1985, p. 64.

Lyon, Jeff. "Focus on a Distorted Image: Do We Expect Too Much from the Experts?" Chicago Tribune, March 25, 1985, sec. 5, pp. 1, 4.

Mayer, Andre, and Jean Mayer. "Agriculture the Island Empire." Daedalus 103(Summer 1974): 83-95.

National Academy of Sciences. Report of the Committee on Research Advisory to the U.S. Department of Agriculture. Washington, D.C.: Division of Biology and Agriculture, National Research Council, 1972.

Nelson, Richard R., ed. Government and Technical Change: A Cross-Industry Analysis. New York: Pergamon Press, 1982.

Nelson, Richard R., and Richard N. Langlois. "Industrial Innovation Policy: Lessons from American History." Science 219(1983): 814-18.

Norman, Colin. "House Endorses Pork Barrel Funding." Science 233(1986): 616-17.

Norman, Colin, and Eliot Marshall. "Over a (Pork) Barrel: The Senate Rejects Peer Review." Science 233(1986): 145-46.

Oehmke, James F. "Persistent Underinvestment in Public Agricultural Research." American Journal of Agricultural Economics 1(1986): 53-56.

Office of Technology Assessment (OTA), U.S. Congress. An Assessment of the United States Food and Agricultural Research System. Washington, D.C., 1981, pp. 201-6.

Pavitt, K. and W. Walker. "Government Policies toward Industrial Innovation: A Review." Research Policy 5(1976): 11-97.

Press, Frank. "Rethinking Science Policy." Science 218(1982): 28-30.

Price, Derek deSolla. "Sealing Wax and String: A Philosophy of the Experimenter's Craft and Its Role in the Genesis of High Technology." Sarton Lecture, American Association for the Advancement of Science, May 1983.

Ruttan, Vernon W. Agricultural Research Policy. Minneapolis: University of Minnesota Press, 1982.

Schuh, G. Edward. "Revitalizing the Land Grant University." Choices, 2nd Quarter, 1986.

Shapley, Deborah, and Rustum Roy. Lost at the Frontier: U.S. Science and Technology Policy Adrift. Philadelphia: ISI Press, 1985.

Ulrich, Alvin, Hartley Furtan, and Andrew Schmitz. "Public and Private Returns from Joint Venture Research: An Example from Agriculture." Quarterly Journal of Economics 101(1986): 103-29.

Werner, Leslie M. "Administration Plans Cutbacks in Research." New York Times, March 20, 1986a, 1, 16.

________. "Changes in Research Funds Challenged." New York Times, July 27, 1986b, sec. 4, p. 24.

White, Fred C., and Joseph Havlicek, Jr. "Optimal Expenditures for Agricultural Research and Extension: Implications for Underfunding." American Journal of Agricultural Economics 64(1982): 47-55.

Whitehead, Alfred N. "Harvard: The Future." Atlantic Monthly 158(3)(1936): 267.

5 Priority Setting in Agricultural Research*

Jean Lipman-Blumen

The American agricultural research system[1] is a highly complex and successful enterprise, which paradoxically has encountered mounting criticism in recent years.[2] Much of this criticism focuses on the system's capacity to establish and maintain priorities for publicly funded agricultural research. As public coffers have dwindled, Congress and key executive branch offices have insisted that public monies spent on agricultural research must be securely linked to national goals.

THE HISTORICAL CONTEXT

The Fragmented Farm Block

In part, the difficulty arises from historical precedent and an earlier definition of priority setting. For close to a century, the USDA budget sailed on favorable congressional waters. Its good fortune flowed largely from the support of a strong farm bloc representing a substantial rural population. By mid-century, the rapid demographic shift from rural to urban centers had left the congressional farm bloc in serious disarray. To bolter its fragmented base by attracting critical budget support from their urban colleagues, the farm bloc introduced new urban programs. This strategy, however, alienated some traditional supporters, who questioned USDA's purpose and loyalty to its original rural constituency. With only flagging help from influential rural groups, the USDA budget faced even rougher sailing.

Unexpected Criticism

In the ebullience ushered in by the Green Revolution, agricultural research approached a zenith in public acclaim and support. It handily garnered a substantial share of a robust federal budget. That excitement gradually dissipated as environmentalists drew public attention to serious negative health and environmental consequences unanticipated by agricultural scientists. The consumer movement also fed the growing disenchantment with agricultural research. Public support of agricultural research began to give way to public suspicion of unrestrained science. The agricultural research community, reeling from this unexpected reversal of fortune, reacted with dismay and defensiveness.

The Pound Report (NRC 1972) compounded this unprecedented external criticism by a stinging critique of agricultural research. Now, the agricultural science community was beset from all sides. Congressional demands grew for more stringent oversight of agricultural research (OTA 1981; GAO 1977, 1981, 1982).

New Definition of Priority Setting

Congressional insistence on agricultural research programs linked to long-term planning and national goals occurred within a larger context of federal belt tightening. This set the stage for redefining priority setting in agricultural research. Previously, "priority setting" had meant adding new programs to the annual "wish list." The new definition required evaluating and rank ordering research programs, eliminating some and adding others. Still, the agricultural research system only dimly sensed the new meaning.

Nor were the calls for serious quality improvement and priority setting limited to the congressional branch. The Office of Science and Technology Policy (OSTP) in the White House and the Office of Management and Budget (OMB) expressed similar concerns. Joining forces with various congressional offices and concerned private foundations, OSTP and OMB staff encouraged upgrading the quality of agricultural research by more stringent review processes and scientific competition. The drawn-out controversy over competitive grants vs. formula funding for agricultural research often served as the battleground in this "quality of science" debate.

The timing of these events contained several ironies. First, the federal demand for control over priority setting increased just as the federal contribution to agricultural research budgets was diminishing. Second, critics wanted agricultural research linked to long-term national goals, even though these goals were themselves ill-defined.

DECENTRALIZATION AND PARTNERSHIPS

The decentralized character of the agricultural research system constituted a great strength, enabling it to respond rapidly to local crises. Decentralization, however, also complicated the task of setting system-wide parameters. Rigorous priority setting within a loosely coupled system poses a special challenge. The centralized authority required for such a comprehensive plan grates against the structured autonomy jealously guarded by each component.

Historically, the agricultural research system enjoyed a multilayered partnership between federal and state agricultural agencies, as well as between the land-grant institutions and government agencies. The partnership encompassed not simply the agricultural research components, but the entire extension system as well. Various mediating bodies, such as ARPAC, the Joint Council, the Users Advisory Board (UAB), the Committee of Nine, the Regional Associations, and the Committees on Organization and Policy (COP) within the National Association of State Universities and Land-Grant Colleges (NASULGC), were created at different times to maintain the harmony and coordination of this vast, complicated system. This complex decentralized partnership, running from the state to the federal level, had enabled the agricultural research community to respond rapidly and effectively to divergent local demands, as well as to agricultural emergencies.

Strained Partnership

The "partnership" within the agricultural community rests on a web of relationships spun over a century in rural communities and land-grant institutions. Common backgrounds and experiences within a tightly knit agricultural world generated shared understandings, facilitating traditional priority-setting strategies.

Even the most congenial partnerships, however, are heir to tensions that intensify when mutual help and trust are rubbed raw by change. Such tensions magnify when change is imposed, rather than sought. Moreover, when new conditions evoked unfamiliar, uncomfortable, internal competition for needed research support, the partnership faced imminent jeopardy. Still, internal competition alone did not constitute the total problem.

In this case, partnership static was exacerbated by the political jolt to the traditional independence and professionalism of science. Previously, scientific decisions had been the sacred domain of agricultural researchers. Now, the call for stringent political oversight underscored an unpalatable reality: the political process was impinging on the decision-making territory of agricultural scientists.

Not only were the large policy issues of agricultural research priorities and national goals at stake, but the narrower political ones also came into play. While scientists were being urged to make hard scientific choices, many of these decisions, encoded in budget items, subsequently were reversed by legislators (and their staffs working behind the scenes) to meet the special interests of their constituents. The political signals were complex, contradictory, and confusing even to the most astute science administrators. Suddenly, the dynamic tensions between the scientific enterprise and the political process took a quantum leap.

THE PLETHORA OF PRIORITY-SETTING ACTIVITIES

Perhaps the greatest irony of all was nested in the complex priority-setting activities that actually existed throughout the sprawling agricultural research system. Both formal and informal structures and processes crisscross the system from the local to the federal level.

These activities occur within the USDA, in the cooperating institutions in the regions, states, and counties, in groups with oversight responsibility, in various mediating and coordinating structures, as well as in ad hoc task forces. Both long- and short-term priorities are formulated within this network of organizations, which, at first glance, appear uncoordinated (see Lipman-Blumen and Schram 1984 for a detailed description).

Priority Setting in the U.S. Department of Agriculture: An Overview

At the upper administrative echelons of USDA, priority setting occurs within the Office of Budget and Program Analysis, which reports to the deputy secretary of agriculture. The separate agencies within the USDA--for our purposes, the Agricultural Research Service (ARS) and the Cooperative State Research Service (CSRS)--have their own planning/budget staffs, which develop overall program plans from documents sent forward from lower bureaucratic levels. Selection and rank ordering of priorities are carried out within general substantive and budget parameters set at the higher administrative levels. To some degree, these parameters reflect OMB and congressional signals regarding priorities and funding levels.

Thus, beginning with every project officer, annual program plans move through successive bureaucratic echelons of ARS and CSRS at the federal level. Explicit rankings are often matched to budget figures set by top administrators. The CSRS administration of Hatch formula funds requires a slightly different procedure, involving the approval of plans submitted by the State Agricultural Experiment Stations (SAES).

The Office of Budget and Program Analysis coordinates the budgets of the various components into an overall USDA budget. This comprehensive budget is then presented to the deputy secretary and secretary of agriculture and, through them, to OMB.

The complex annual budget process proceeds relentlessly, shaped and reshaped by endless discussions. Negotiations occur at all levels, before each USDA agency makes its annual presentation to the various assistant secretaries of agriculture, the deputy secretary, his staff, and the secretary of agriculture. Within USDA, the major determination of each agency's budget level is the responsibility of the deputy secretary, whose final choice commonly reflects the top overriding USDA priority: productivity. More recently, production efficiency and profitability have emerged to form a troika of key priorities.

These federal agency budgets that express the current priorities for agricultural research represent the distillation of a complex set of processes within CSRS and ARS. Within CSRS, their roots reach down to regional and state levels. The budget development process below the federal level represents an elaborate series of negotiations within and between the COP[3] as well as between the COP and the respective federal agency partners. The COP legislative and budget committees, as well as the Division of Agriculture Budget Committee (NASULGC), play an important role in this stage of budget formulation.

Priority Setting in the Agricultural Research Service (ARS)

The Agricultural Research Service (ARS) is USDA's intramural research arm. It is generally acknowledged to be the world's largest and most diverse publicly funded research organization. ARS has demonstrated long-term stability as the federal home of agricultural research. Its stability has been both shaken and reestablished by recurrent internal reorganizations, sometimes several within a decade. In fact, reorganization has been ARS's favorite mechanism for signaling critics that it is heeding their demands for new or streamlined priorities by reallocating resources. The most recent ARS reorganization, linked to a "continuum of actions," specifically redirected resources to programs raised to national priority status.

ARS's Mission

ARS's mission is to provide an enduring scientific organization to ensure continuity for long-range, high-risk research of national, even international, scope. The long-term existence of the structure (1) promotes responsiveness to national and regional emergencies; (2) supports a permanent repository of scientific taxonomic collections; and (3) provides a base for developing scientific knowledge. In changing political climates, ARS's continued presence offers structural protection for long-term research projects.

By the early 1980s, adherence to ARS's national mandate had become somewhat problematical. Some researchers questioned the usefulness of differentiating national from local research priorities, particularly when certain so-called "local" research problems have national implications. To compound the problem, ARS routinely confronts contradictory congressional demands: strict attention to long-term, high-risk basic research of national scope <u>and</u> continued support for local agricultural problems in individual congressional districts.

The distinction between federal and national research is another resistant issue troubling the ARS mission. The question remains whether more precise organizational filters should be designed to address distinctions between national, federal, and local, as well as basic and applied, research.

In the face of recent budgetary cutbacks, OMB has urged ARS and other federal research agencies to conduct only that research that is national in scope, long-term, high-risk, and unlikely to be performed by private industry. There is little evidence to suggest that private industry, without consultation, coordination, or meaningful incentives from the public sector, will undertake certain short-term projects that fall outside federal research parameters.[4] If the history of "orphan drugs" is any example, this seems rather unlikely. The danger remains that some important short-term, low-profit projects will simply fall between the cracks.

Reorganizing to Address Priorities

ARS's history is replete with efforts to shape its structure to address clearly articulated research priorities. Consonant with the general structure of the agricultural system, ARS's research activities are decentralized into four regions, yet overall management is directed from Washington. At the federal level, ARS is managed by an administrator who reports to the assistant secretary for science and education. The administrator is charged with speaking on behalf of the department's national agricultural research enterprise.

During the Reagan administration, ARS has undergone a series of organizational changes designed to improve its capacity to address national research priorities. Within this new organizational plan, ARS has sought to (1) redirect resources to high priority research; (2) clarify and streamline reporting relationships; (3) augment top management's capacity to focus on important ARS and major program issues; and (4) increase both centralization of a national program implementation and decentralization of daily program management.

As part of this reorganization, ARS reshaped the National Program Staff (NPS), the unit whose primary mandate has been the technical excellence of ARS-funded research programs. Administered by a deputy administrator, NPS is responsible for planning, coordinating, monitoring, reviewing, and evaluating all ARS programs. Formal priority-setting responsibilities rest with the deputy administrator and the National Program Staff. Considerable impetus for reshaping the NPS stemmed from widespread concern within the agricultural research community over the staff's eroding quality.

The need for an organizational structure better suited to address national agricultural priorities led ARS to redesign its nation-wide area and center system, which previously had encompassed more than 150 work sites across the nation and abroad. These sites, ranging from small laboratories and stations to large, internationally recognized research centers, are generally devoted to interdisciplinary efforts. Many ARS facilities are located on university campuses and State Agricultural Experiment Stations, presumably ensuring mutually beneficial utilization of equipment, expertise, and resources. This co-mingling of personnel and other resources is consistent with the partnership framework described earlier.

The latest reorganization plan consolidated thirteen areas and twelve centers into eleven areas with 135 work sites confined within four regions (i.e., northeastern, north central, southern, and western). Four previous positions of regional administrator were elevated to deputy administrator, with added national responsibilities. Eleven area directors were charged with managing the agency's research activities.

An Administrator's Council was established as an executive committee to consider major "policies and decisions," such as (1) the direction of agricultural research programs; (2) "coordination and joint planning with other federal, state, and private institutions"; and (3) central management activities, including, "planning, budgeting, information, and human resource management." Administrative management functions also were streamlined, with personnel and financial resources reallocated to "high priority" research needs.

Formal and Informal Mechanisms and Processes

ARS encompasses a broad spectrum of formal and informal priority-setting procedures. Close working relationships between ARS and State Agricultural Experiment Station (SAES) scientists[5] influence the priority-setting process. Direct and indirect "signals" from congressional staff significantly affect the process. Occasional informal discussions between congressional staff and ARS personnel transmit "signals" that substantially influence priority setting. Informal communication between ARS personnel working within the same discipline may influence the perceived importance of specific research projects. Research interests of individual scientists, as well as overall administrative shifts in research emphasis, play important roles in setting and sustaining research priorities in ARS.

FORMAL PRIORITY SETTING IN ARS

The ARS Program Plan

In the early 1980s, ARS came under severe criticism for failing to engage in long-term planning. Moreover, a widely held perception outside ARS suggests that top-down planning has traditionally characterized the agency, despite disclaimers by ARS insiders.

Attempting to address both issues simultaneously, ARS undertook a bottom-up strategic planning process in the early 1980s, which queried over 500 ARS scientists in the field, as well as university and private industry scientists. Structurally, the planning process eliminated the regional filter and reached out to individual scientists in the field. Coordination was assigned to the National Program Staff, which reportedly confronted considerable difficulty in interpreting the results. Ultimately, however, six major mission-oriented objectives were identified. Although close to 100 top regional, area, and center managers set first targets, the final determination of priorities rested with the deputy administrator.

The ARS program plan is based on a six-year strategy. It includes a yearly review and adjustment for unforeseen research needs and changes in priorities, appropriations, new research findings, personnel shifts, and other research and political outcomes.

Many CSRS and SAES staff insist that they were not consulted during the design stage of ARS's Strategic Plan. This unilateral activity increased tensions between ARS and CSRS/SAES, triggering post facto, conciliatory briefings by ARS to assuage the problem. Prior consultation with only a limited set of Hill contacts sparked strong reaction from unconsulted congressional members and staff. ARS defenders countered Hill complaints with three arguments: (1) the plan represented the scientific, rather than political, judgments about national priorities; (2) prior consultation with the Hill would have jeopardized ARS's capacity to close or relocate ARS installations in individual congressional districts; and (3) orders from the highest USDA echelon had prohibited consultation. At the time, observers predicted congressional reaction to this "end run" maneuver would have a long-term impact on ARS's autonomy, outweighing the short-term gain. This prediction has not been substantiated. In fact, ARS's more recent consultative efforts toward the Hill have created an unusually congenial ARS-congressional rapport, marked by substantial budget gains.

Long-Term Priority-Setting Practices: Appointing Scientists

For ARS, appointing scientists is a long-term priority-setting activity. Most ARS scientists are civil servants, engaged in mission-oriented research. Appointing a scientist with specific expertise and interests signals a long-term commitment to that area. Although scientists ordinarily are not expected to engage in research extraneous to their own disciplinary training, substantive shifts in individual scientists' research foci occasionally occur. These shifts may represent an intellectual change or a pragmatic adaptation of research interests associated with perceived changes in funding or political demands.

Peer Reviews

Peer review practices, or lack thereof, have been a target of long-term concern to ARS critics. Despite ARS insistence that ARS laboratory research undergoes peer review, non-ARS scientists tend to discount this claim. Critics contend that peer review procedures are either nonexistent, or, at best, less rigorous than university-based reviews. Moreover, despite ARS's proclaimed interest in interdisciplinary research, peer review panels often fail to represent interdisciplinary expertise. Social and behavioral scientific expertise is largely nonexistent within ARS panels, not to mention staff.

ARS scientists defend their peer review process as rigorous, insisting each new project's methodology is reviewed by two scientists selected by the research leader.

As in all non-blind reviews, however, bias may be introduced inadvertently by reviewer selection, as well as by other procedures.

Subsequently, the proposal is reviewed by the center director, NPS staff, and the regional administrator. Research leaders annually examine the projects within their programs, including every experiment within each project. Individual ARS scientists undergo annual performance reviews, which include evaluations of their scientific publications.

INFORMAL PRIORITY-SETTING FACTORS

Political Factors

Because of its emphasis on long-term basic research, ARS occasionally has been caught in the cross fire between research priorities and political agenda. For example, some research facilities have been funded to accommodate a mix of political and scientific considerations.

At times, the lack of fit between scientific and political agenda has resulted in the establishment of facilities lacking adequate staffing funds. At other times, research centers have been located in areas too remote from other scientific resources simply to satisfy a legislator's constituent.

Professional Associations

High status scientific societies and associations have an indirect, but significant, informal input to priority setting. The American Chemical Society, the American Association for the Advancement of Science, and other professional scientific societies facilitate the exchange of professional and research information. Professional societies also offer formal legitimation of scientists' work by selection for symposia presentations, honorific awards, and committee appointments, as well as by election to society offices.

The National Academy of Sciences plays a significant role, primarily when it focuses specifically on agricultural research issues. Special reports, ironically often undertaken because of available funding or special interests of the Academy staff, bear considerable weight. Unfortunately for purposes of independent priority setting, the NAS budget is a line item in the federal budget. Moreover, NAS staff must often seek outside funding (including USDA support) for their research. These circumstances are fraught with the potential for subtly subverting the integrity of the priority-setting process in agricultural research.

Priority Setting in the Cooperative State Research Service and Cooperating Institutions

The Cooperative State Research Service's (CSRS) primary mission is to administer federal appropriations (Hatch, Special Grants, and Competitive Grants) for agricultural research. CSRS performs a brokerage function vis-à-vis the land-grant universities and experiment stations. With the establishment of the Competitive Grants Program, CSRS extended this role to encompass non-land-grant institutions as well.

Because CSRS essentially represents a "holding company" for SAES, or a pass-through structure for formula funds (Hatch), its own leadership role in establishing substantive priorities is considerably limited. In fact, the CSRS budget, with its associated priorities, reflects primarily the efforts of the cooperating institutions, the State Agricultural Experiment Stations (SAES), and their allied departments in colleges of agriculture and life sciences and home economics. These cooperating institutions, in turn, have consulted with local commodity groups and other farm organizations in developing their budgets.

The priority-setting processes in the State Agricultural Experiment Stations and land-grant institutions are ultimately linked to the state and federal budget processes. The Experiment Station Committee on Organization and Policy (ESCOP),[6] through its budget/legislative subcommittee activities, plays a critical role in formulating priority-expressing budgets that become the core of CSRS priorities/budgets. The Division of Agriculture Budget Committee (NASULGC) integrates the individual COP budgets before they are submitted to their respective federal agencies.

Regional Research Priorities and the Committee of Nine

Regional and national coordination of priority-setting activities within SAES was strengthened by the Research and Marketing Act of 1946, which created a Regional Research Fund to support collaborative research undertaken by SAES on "problems that concern two or more states." Twenty-five percent of the annual Hatch appropriation is allocated as Regional Research Funds.

Regional research projects submitted by four Regional Associations must receive the approval of the Committee of Nine,[7] which recommends projects for the CSRS administrator's approval. The Regional Associations have more than 200 committees variously engaged in the priority-setting process.

The Regional Experiment Station Associations and ESCOP

Within the four geographic regions, complex arrangements exist for establishing and aggregating priorities that reflect each region's agricultural research concerns. Huston's (1982) detailed description traces the numerous planning committees that prepare reports based on research program groups and research programs identified in the CRIS (Current Research Information Service) database. Regional summaries approved by Regional Associations are reviewed by numerous groups, including the Regional Planning Committees. These regional reports are submitted to the National Research Planning Committee, which then synthesizes them into a national summary.

The Regional Associations are linked to the Experiment Station Committee on Organization and Policy (ESCOP), which, as noted above, participates extensively in developing priorities and formulating the federal budget for the SAES. ESCOP is composed of three representatives and an alternate, elected by and from each of the four Regional Associations of SAES directors. The chairpersons of four other groups--Home Economics, Forestry, 1890 Institutions, and Veterinary Medicine--act as ESCOP liaison members.

After consulting relevant groups and individuals, ESCOP's Legislative/Budget Committee develops the SAES budget, which it then submits to CSRS. The ESCOP-developed budget serves as the core of the CSRS budget, although it is subsequently reviewed by several USDA and planning groups before receiving approval by the assistant secretary for science and education. From there, the budget is presented to the deputy secretary of agriculture and then signed off by the secretary of agriculture. Ultimately, this budget is negotiated with OMB before submission to Congress as part of the president's budget.

From the time that ESCOP submits the budget to CSRS (i.e., when the budget "goes behind the curtain") until its reemergence in the president's budget, ESCOP officially remains uninvolved. After its submission to Congress, however, ESCOP engages in active lobbying. ESCOP, the Extension Committee on Organization and Policy (ECOP), the Resident Instruction Committee on Organization and Policy (RICOP), and the Council of Administrative Heads of Agriculture (CAHA) work collegially as the NASULGC Division of Agriculture to elicit congressional support for the entire Agricultural Science and Education (S&E) budget. In promoting the CSRS budget and its embedded priorities, ESCOP reaches out to industry and public sector groups, as well as to research units within USDA.

The Land-Grant System and the State Agricultural Experiment Stations (SAES)

Both formal and informal priority-setting activities abound in each state. Still, the complex processes underlying the expression of priorities in the final budget are so ingrained in the very rhythm of land-grant and experiment station interaction that they often escape recognition as legitimate priority-setting activities. Variation by state contributes to the problem.

In its formal priority-setting activities, each State Agricultural Experiment Station attempts to establish and sustain a balance between (1) the size and mix of scientific and administrative staff, (2) resources provided for different purposes by state and federal legislatures, as well as by commodity groups and private foundations, and (3) an organizational structure sufficiently flexible to meet both predictable, ongoing needs and the inevitable short-term crises that characterize agriculture (Huston 1982). To a large extent, the SAES rely on bottom-up priority-setting procedures, beginning with the bench scientist.

Most researchers and science administrators insist that specialization and the esoteric nature of scientific fields make the bench scientist the best judge of specific research initiatives. Nonetheless, individual scientists often claim that their role in shaping national research priorities is minimal. Certain highly esteemed senior scientists receive carte blanche to pursue whatever research they desire; however, they may or may not be involved in a broader spectrum of priority-setting activities.

Experiment Station scientists and their departmental colleagues[8] learn of new research directions and emerging scientific concerns from formal and informal interaction with their peers, commodity groups, and other users within the state. Traditionally, extension personnel provided the conduit for transmitting users' needs to researchers. This route suffers diminished importance in light of: (1) contemporary users' increased educational levels; (2) their personal links to former professors; (3) more sophisticated technology available to users; and (4) young researchers' disciplinary (vs. user) orientation.

Within the Experiment Stations and their allied faculty departments, several formal processes occur--some direct and others indirect--that bear on research priority setting. The most significant formal processes are (1) faculty appointments; (2) annual departmental reviews of individual scientists' current accomplishments and future research plans; and (3) annual reviews of commodity groups' needs and priorities. These three processes commonly are intertwined. To complicate matters further, the Experiment Station budget must be negotiated into an overall School of Agriculture and Life Sciences budget.

Faculty Appointments. As in ARS, the appointment of scientists to the CSRS and SAES staff and/or faculty of a related department is a fundamental long-term priority-setting mechanism. The decision to appoint a scientist in a given field represents what department heads perceive as a "30-year commitment." Delayed retirement policies could extend this considerably longer. Over time, the individual scientist's interests may (or may not) take new directions, sparked by individual curiosity, or by priorities or "trends" emanating from science and/or agricultural needs.

When a faculty position becomes vacant, extensive consultations occur between the department chair, members of the department, the college, and the Experiment Station.

Through these discussions, a collegial decision is made either to continue the commitment (by appointing a candidate in the previous incumbent's field) or to initiate a new commitment to a different research area--to a new priority. Thus, faculty appointments, as well as promotions and tenure decisions, serve as crucial long-term priority-setting mechanisms.

Annual Departmental Reviews. Annual departmental reviews take place within the academic departments of the colleges of agriculture and life sciences. These reviews are based upon annual reports of current completed research activities and descriptions of work in progress prepared by each faculty member. Following commodity board meetings and annual faculty reviews, department heads rank order current and desired research projects within their own departments. The dean of the School of Agriculture distributes these compiled lists to the directors of the Experiment Station, Extension, and Resident Instruction. Each director reviews the lists and sorts his/her own recommendations into program areas. These compiled lists serve as the basis for the dean's budget.

Commodity Group Reviews. In most states, Experiment Station directors and department heads conduct annual commodity reviews. In states with many large and significant commodity groups, these reviews may be held on a rotating basis. Faculty members engaged in commodity-related work make presentations that describe five-year research plans. Commodity board members, the Experiment Station director, and relevant department heads evaluate these presentations and identify research gaps. Commodity groups may write marketing orders to support research that is unlikely to be funded by other sources because of its applied, commodity-specific nature.

Commodity groups engage in other formal and informal aspects of the subsequent budget process. Commodity representatives, through close ties to researchers, science administrators, and legislators, wield a significant influence on both state and federal budgets.

College of Agriculture (Academic Departments) Annual Priority Setting. Academic department heads, directors of the Experiment Station, Extension, and Resident Instruction, as well as commodity group leaders, meet to discuss project cuts. After several additional iterations, the three directors coordinate their individual programs into an overall program, pared to the dean's budget mark. At a final meeting of commodity group leaders, department heads, and university administrators, the "prioritized" budget is discussed and approved.

By the time the budget receives the dean's rather pro forma approval, it has become "community property." The Advisory Council of the School of Agriculture and Life Sciences (composed of state agricultural leaders representing commodity groups, the Farm Bureau, the Grange, and other relevant components of the agricultural community) then works with the state legislature to negotiate the final budget.

To the stateside partners, national priorities are less salient than local interests. Support for clearly articulated regional and state programs seems reasonably easy to attain. At the state level, differentiating between state, regional, and federal/national priorities is not a major issue. This lack of focused concern for national priorities at the state level may emanate from the absence of any highly visible body able to articulate national agricultural research priorities.

This thumbnail description does not adequately capture the collegial nature of the process, particularly the interaction between university and Experiment Station administrators and faculty. Unlike the more hierarchical decision-making practices of the federal bureaucracy, the land-grant system provides the context for participatory democracy.

Shaping the Individual Scientist's Priorities. The young researcher's interests are subtly articulated with the institutional/system priorities by formal and informal rewards. These include promotions, tenure, salary increases, institutional and extramural funding, as well as praise, advice, and encouragement. Discipline-oriented scientists seek out mentors who can guide them in high-status basic research projects. Researchers oriented less to their disciplines and more to agricultural users may select topics reflecting applied, rather than basic, research priorities.[9]

Beyond the individual agricultural scientist's immediate work environment of colleagues and users, other important priority-setting signals are communicated by major funding sources: NSF, NIH, USDA, other public agencies, and private foundations. Related guidelines describe the funder's domains of interest, clearly identifying "fundable," "hot," and "cutting edge" topics. Grants from prestigious funding sources are a key component in the complex process that shapes scientists' interests to conform to priorities articulated by the agricultural research system.

In addition to research grants, other rewards highly prized by discipline-oriented scientists include accepted journal articles and published books, as well as committee appointments and honorific membership in professional societies. Consultations to industry and marketing orders from commodity groups provide parallel incentives for user-oriented researchers. Key legislators' research interests reinforce the complicated communication and reward systems that socialize the young and urge the mature scientist to address "cutting edge" problems.

Priority-Setting Activities of the Joint Council and the Users Advisory Board

The Joint Council. The Joint Council, composed primarily of research "providers,"[10] has been assigned specific priority-setting activities by Congress. Originally designed as an advisory microcosm of the agricultural research and extension system, the Joint Council has been reconstituted even in the few short years since its establishment. Its original membership included representatives of (1) land-grant institutions; (2) private institutions engaged in agricultural research; (3) commodity groups; (4) the USDA; and (5) other federal agencies with agricultural programs. A reorganization in 1981, initiated to strengthen the Joint Council, reinstated land-grant preeminence. In the 1985 Farm Bill, the only change in Joint Council membership was the inclusion of a food technologist.

The Joint Council is charged with "improving planning and coordination of publicly and privately supported food and agricultural science activities" and "relating federal budget development and program management to these processes."[11] The Farm Bill instructs the Joint Council to prepare (1) an annual priorities report, including required levels of financial and other support to sustain these programs; and (2) an annual accomplishments report, covering research, extension, and teaching programs. The 1981 Farm Bill called for a report, due June 1983, outlining a five-year plan for food and agricultural sciences, which would be updated every two years. In addition, the secretary of agriculture delegated to the Joint Council responsibility for conducting a long-term needs assessment of U.S. capabilities to meet needs for food, fiber, and forest products through and beyond the year 2020.

In its first half dozen years, the Joint Council encountered serious difficulty in gaining the agricultural community's acceptance. Several sources of strain were evident: (1) structurally induced turf problems between university administrators protecting their home institutions' interests, (2) tense interactions between land-grant and non-land-grant members based on land-grant representatives' desire for membership strength commensurate with their agricultural budget levels; (3) the concerns of the COP that a strong Joint Council would undercut its own priority-setting role; (4) the agricultural community's discomfort with non-USDA federal agency representatives on the Council; (5) the Joint Council's complex committee substructure; (6) certain members' limited independence deriving from their government employee or government grantee status; and (7) congressional mandates potentially unattainable owing to inadequate funding and inherent complexity, as well as the Joint Council's inexperience.

In the early 1980s, under the personal tutelage of the newly appointed assistant secretary for science and education, the Joint Council received a vital boost in stature. End runs around the Joint Council were explicitly discouraged. The chair of the Joint Council, working closely with the assistant secretary, contributed valuable political acumen to the council's workings. Increased representation by Experiment Station directors, powerful members of the land-grant system, spurred the agricultural community's acceptance of the Joint Council. This move, however, encouraged the informal withdrawal of high status non-land-grant members.

In later years, the Joint Council eventually gained gradual acceptance within the agricultural research and extension system. This recognition came only after the successful submission of the congressionally mandated reports, reports that the Joint Council delegated largely to external sources. More recent assessments concede the Joint Council has developed some measure of genuine legitimacy among land-grant scientists and administrators.

Acceptance has come more reluctantly from USDA administrators and the federal science establishment. Potentially supportive constituencies outside of this rather closely knit circle have yet to be drawn into the Joint Council's orbit.

More serious, however, is the fact that top administration officials, as well as many Hill staffers on key committees, are surprisingly unfamiliar with the work of both the Joint Council and the Users Advisory Board (UAB). As one congressional observer recently noted, "When things get to the Hill, where final decisions are made, there is not a lot of impact from the Joint Council or the UAB."

The Users Advisory Board. The UAB's membership includes a broad spectrum of groups regarded as "users" of agricultural research and extension products. Producers of agricultural, forestry, and aquacultural products, from four U.S. geographic regions, as well as food marketing and environmental representatives, hold UAB membership. Rural development and human nutrition representatives, as well as members from foreign and domestic agricultural transportation industries, a labor organization, and a private sector organization, complete the official roster.

The UAB has broad review and advisory responsibilities, including "reviewing and providing consultation to the Secretary on national policies, priorities, and strategies for agricultural research and extension for both the short and long term" (Farm Bill, 1981, Subtitle B, Sec. 1408, f, 2 (F)). The UAB fulfills its mandate by producing two specific annual reports: (1) a report evaluating the president's proposed food and agricultural sciences budget; and (2) an assessment of the responsibilities and allocations of research and extension funds made for the preceding fiscal year by the organizations represented on the Joint Council.[12]

The UAB members, all political appointees, generally have less experience in federal priority setting and budget/legislative processes than their Joint Council counterparts (who include deans of colleges of agriculture, used to public budget processes). Despite their greater collective inexperience, the UAB initially managed to project an image of independence, forthrightness, and relative competence within the agricultural system. Some critics, however, complain that the UAB reports have been rather heavy-handedly directed by sophisticated in-house USDA staff, beset by congressional deadlines.

The UAB's credibility edge stemmed largely from the fact that, unlike their Joint Council counterparts, its members were neither government employees nor direct recipients of government research funds. This perception beclouded an important reality: the priorities the UAB selects could ultimately benefit the commodity groups or other organizations they represented. This potential conflict of interest, however, did not seem to undermine the UAB's credibility.

Much of the UAB's independent image emanated from its original bipartisan character, doggedly protected by its first executive director and chair. More recently, the UAB's bipartisan composition has been jeopardized, leading some observers to express doubts about the UAB's political autonomy and credibility.

Agricultural Emergencies

Agricultural emergencies create preemptive, usually short-term, priorities that easily subvert the most carefully drawn plans. Fraught with drama and disaster, agricultural emergencies demand immediate attention.

The system's decentralized character enables the appropriate component to respond rapidly to such unexpected crises. Depending upon the pervasiveness, intensity, and solvability of the crisis,[13] scientists are drawn away from their regular work for periods of unpredictable length until the crisis is resolved. Those agricultural emergencies sufficiently threatening to mobilize public support may even become part of the longer-term national or state priorities agenda.

CONGRESSIONAL PRIORITY SETTING: FORMAL AND INFORMAL

Formal Mechanisms. The Congress wields an impressive influence on priority setting in agricultural research. The most important formal mechanisms for congressional priority setting are: (1) congressional hearings; (2) authorization and budget votes; and (3) congressionally mandated reports.

Hearings before Senate and House committees--agriculture, budget, and appropriations committees, particularly--serve several purposes. First, they highlight congressional members' research priorities. Second, they identify priority differences between the legislative and executive branches. Third, hearings provide a forum for considering the priorities of various constituent groups.

Congressional hearings held every four years by the House and Senate Agriculture Committees in preparation for the Farm Bill represent a crucial priority-setting process.

The Farm Bill establishes broad priorities that generally set the research parameters for the next four years. Annual hearings on budget and appropriations serve to maintain and "fine tune" the broad priority areas set forth in the previous Farm Bill.

Congressional authorization and appropriations votes establish priorities and potential funding levels for various programs and projects. On the Hill, regional and local interests of the chair, the ranking minority member, and politically adroit key committee members substantially shape research priorities. Occasionally, congressional members' broader scientific concerns are scuttled by their other constituent-based concerns. The natural "give and take" between members representing constituent interests often creates "deals" that effectively preclude subsequent pursuit of priorities "linked to national goals."

Congress also commissions various reports, oftentimes by appropriating funds for specific investigations. Congressional research and reporting bodies, such as the General Accounting Office, the Congressional Research Service, and the Office of Technology Assessment, issue formal reports on various aspects of agricultural research. These assessments are taken very seriously by the agricultural research community.

Informal Mechanisms. Informal congressional priority-setting activities are manifold. Contacts with constituents, including commodity groups, farm organizations, and administrators from land-grant institutions, as well as science administrators from USDA and other federal science agencies, sustain a complex network influencing research priorities. Many Hill staffers insist, however, that they depend largely on informal, word-of-mouth information from a rather limited set of constituents. Individual land-grant administrators and researchers with strong "Hill connections" are easily identified, admired, and even envied by their less politically active counterparts.

Grass roots concern, expressed through farmers' and farm-related organizations, is frequently cited as the most effective route for moving certain issues to national priority prominence. Without grass roots support, congressional aides contend, little change occurs. In fact, congressional insiders claim that, while USDA input is important, it may be counteracted by constituents' opposing judgments. USDA positions have greater "clout" when the congressional member shares party affiliation with the current administration. Although constituents' special interests drive many Hill staffers' efforts, as noted earlier, senior congressional members with key committee assignments and long-term scientific interests tend to take a broader view of national agricultural issues.

This brief review necessarily sketches only the most obvious priority-setting activities within the agricultural research community.[14] Still, it is apparent that a virtual overabundance of such activities exists. Some community members perceive this plethora of priority setting as time-consuming window dressing. Given the enormity of time and resources devoted to priority setting, it is hardly surprising that the agricultural research community expresses frustration when science-driven priorities are reshuffled in the political process.

MAJOR ORGANIZATIONAL ISSUES AND SELECTED RECOMMENDATIONS

The U.S. agricultural system is large, complex, and essentially strong, with a substantial history of unprecedented triumphs. As large, complex systems mature, they inevitably generate organizational issues whose resolution is imperative for effective functioning.

The agricultural research system is no exception. It currently confronts three major types of organizational issues: (1) structural; (2) human; and (3) communication issues. Although elsewhere we have examined these organizational issues and the forty-six related recommendations in considerable detail (Lipman-Blumen and Schram 1984), here we limit our discussion to a relatively small group of key structural issues and recommendations with long-term systemic implications.

Structural Issues

#1: No structure or process currently exists that could forcefully and systematically sustain crucial national priorities across political administrations and other political perturbations, as well as in the face of congressional and executive branch disagreement and public/private sector criticism.

Recommendation: A stable mechanism should be designed to articulate and sustain those long-term priorities that are linked to national goals. A priority-setting body with national scientific stature and policy experience should be created to deal with this problem. No existing structure meets these criteria. Not even the National Academy of Sciences Board on Agriculture does, since its staff must seek external research support from funders, who, in turn, set priorities. At first glance, this recommendation seems to threaten the current power balance of the agricultural research establishment. Properly understood, the recommendation offers a Tocquevillian solution[15] in which each component relinquishes a minor degree of power to achieve a significant increase in the group/system's power, a situation that would ultimately be more beneficial for all.

One possibility is the creation of a National Agricultural Priorities Commission, composed of science and policy elder statespersons, who would identify emerging agricultural issues and assess the scientific possibilities for addressing them (figure 5.1). This Commission should be housed within the National Academy of Sciences or the American Association for the Advancement of Science, but not within the federal government. It should have liaison to those parts of the agricultural research system charged with priority setting (i.e., Assistant Secretary's Executive Committee, the Joint Council, the UAB, etc.). The Commission would not represent an additional bureaucratic layer, however, since no system component would report to it.

The Commission members should be nominated through a rigorous procedure within the scientific community. By the scientific and policy stature of its members, the Commission would insulate the system from scientifically indefensible disruptions. The apolitical complexion of the Commission would increase the probability that its scientific judgments would create stability for agricultural research priorities across the cyclical perturbations created by changing administrations. Its respected scientific recommendations would sustain established research priorities despite political election or defeat of congressional gatekeepers.

Protected from external and internal political pressures, the Commission would identify long-term, national, even international, agricultural issues and knowledge needs. It would issue a major report through the president of the NAC or the AAAS every four years, in the year prior to the national elections. This timing is designed to maximize implementation and impact, by ensuring that a lame duck administration would not oppose it and an aspiring presidential candidate or savvy incumbent would

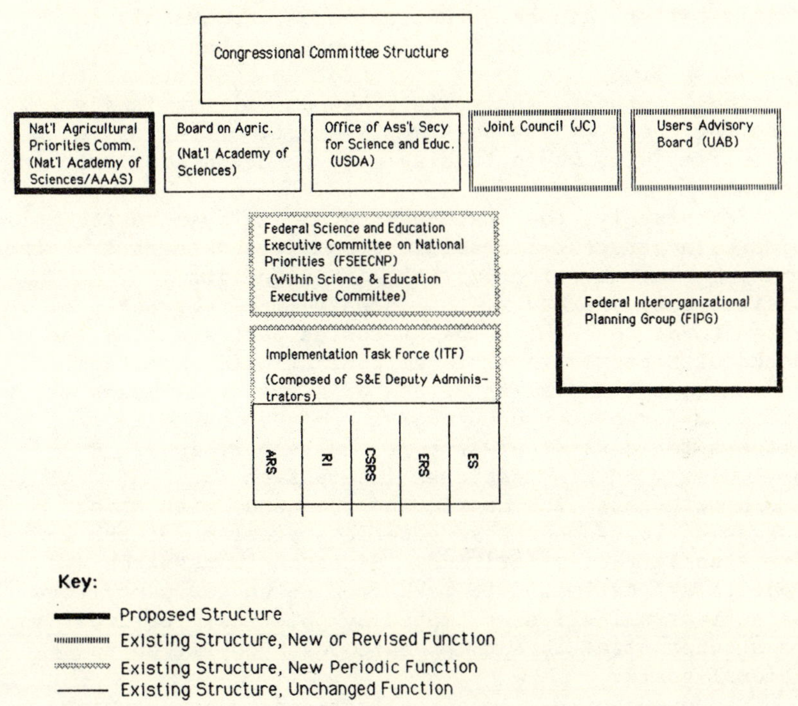

Figure 5.1. Existing and proposed priority-setting structure

endorse it. During each of the three subsequent years, the Commission would issue a minor report to reaffirm previous priorities and/or to make relevant course corrections.

The Commission's mandate would be to identify the overall priorities. Then, other priority-setting bodies that advise federal departments and agencies engaged in or funding agricultural and food research would carve out the components their department or agency should undertake. The Joint Council should assume responsibility for determining the appropriate USDA assignment from the Commission report. (See #4 below regarding the articulation of objectives, strategies, and implementation.)

#2: Structural reorganizations have been used repeatedly to accommodate changing priorities. Some reorganizations are designed to demonstrate the USDA's determination to appease congressional criticism, even when there is substantial disagreement with the critique. Other reorganizations occur routinely with the advent of each new administration. These structural changes commonly are driven more by a desire to "clean house" than by any systematic effort to shape organizational structure to functions, purpose, and goals. New administrations are noted for their desire to stamp their tenure by dramatic new initiatives, whose funding must be drawn from existing programs.

Fortunately, the state partners have been better able to sustain structural stability, buttressed in part by the norms of academic freedom within the land-grant institutions. Moreover, the decentralized character of the system tends to cushion the stateside partners from the shocks of repeated federal reorganizations. Nonetheless, the constant reshuffling of the USDA structure makes it difficult for stateside groups to rely upon the federal partners for consistent support and leadership, particularly in congressional interaction.

Nonetheless, the degree and frequency with which reorganizations occur pose a serious problem for the long-term stability of priorities. Since basic research projects are estimated to take from seven to twenty-five years, reorganizations prompted by political perturbations are a substantial threat to basic research linked to national goals.

Recommendation: Any organizational restructuring should be planned to involve the least amount of disruption to long-term research priorities. When reorganizations are unavoidable, all efforts should be bent to protect effectively functioning, priority-linked units. Less effective units or those that have accomplished their goals should be dismantled without undue organizational angst.

Structure, alone, does not define an organization. Thus, major reorganization is not necessarily the most productive response to external criticism. Alternatives to major structural change include:

- Creating temporary internal task forces with defined and time-limited objectives;
- Reallocating budgetary resources to existing organizational components that can assume additional responsibilities to respond to the criticism;
- Improving management information systems to encourage organizational responsiveness to problems;
- Disseminating information on the organization's performance;

- Designing forums for ongoing dialogues between the organization and its critics.

#3: The peer review system is an important mechanism for maintaining high standards of excellence, as well as priority continuity. By charging the scientific community with responsibility for judging the value and relevance of research projects, priorities can achieve another measure of insulation from political perturbations. Given the impetus for interdisciplinary research, peer review panels should also reflect interdisciplinary expertise.

In the early 1980s, external peer review panels established for federal projects were seriously jeopardized by the introduction of non-scientists on the grounds of "diversity." Moreover, scientists' political leanings were scrutinized before appointment to peer review groups. Serious public outcry placed the new administration on notice that neither science, nor the body politic, was well served by subverting the peer review process.

Recommendation: Internal and external peer review processes within ARS and the CSRS land-grant system should be strengthened and vigorously insulated from the political process. Given the importance of interdisciplinary research, internal peer review panels should reflect interdisciplinary expertise. In addition to including the broadest spectrum of the natural sciences, peer review panels should remedy the serious lack of behavioral and social scientific expertise. Internal review panels should include panelists from external institutions.

External peer review panels should be rigorously protected from political pressures. Only researchers with relevant scientific expertise should be considered as panelists. Scientists' political backgrounds should not constitute a criterion for peer review panel membership.

#4: The Joint Council and the UAB perform priority-setting functions. Recently, the Joint Council has begun to emerge as a credible priority-focusing structure within the agricultural research community in the land-grant system and USDA. With the publication of its congressionally mandated needs assessment report, the Joint Council finally began to assume a meaningful role within the agricultural research community. It has begun to provide a sense of cohesion and coordination between the USDA, its stateside partners, and constituents. Still, the council and its work are virtually invisible except to a few knowledgeable members in the Congress, where final priority-setting decisions are made.

Currently, the Joint Council is required to produce two annual reports: one on priorities, the other on research accomplishments. The annual priorities report is an oxymoron, since no major research priority can be adequately addressed within one year. The annual accomplishments report is used to convince Congress that the agricultural research system has "done much for it lately." Both reports consume inordinate staff time and resources. Despite this massive effort, the Joint Council's membership has neither the stature within the larger scientific community nor the political clout to protect its selected priorities from political disruption.

Recommendations:

- If the Joint Council is to continue in its priority-setting capacity, it should be given more visibility vis-à-vis key congressional members and staff.
- In the event that the National Agricultural Priorities Commission is created, the Joint Council's activities should be reshaped to mesh with those of the Commission and the Science and Education (S&E) Division.
    - Based upon the Commission's priorities report, the Joint Council should develop a Four-Year Strategic Plan for S&E, delineating objectives for addressing those Commission priorities that can be dealt with most appropriately by USDA and the land-grant system. The budget justification function would be better served by a Four-Year Strategic Plan, supplemented by a streamlined annual assessment of S&E's previous year's strategies.
    - The assistant secretary and the S&E agency administrators, working as an S&E Executive Committee on National Priorities, should then design and coordinate strategies for meeting the Joint Council's articulated objectives.
    - An implementation task force, composed of deputies to S&E agency administrators, would bear responsibility for implementing these strategies.
    - The Joint Council's annual assessment should then (1) evaluate the effectiveness of the previous year's strategies in meeting the articulated objectives; and (2) identify strategies requiring redesign, different implementation, or elimination.
- The Joint Council should address agricultural emergencies that impinge on long-term priorities.

The Council should consider developing innovative approaches to cope with enduring and emerging scientific issues that S&E and its stateside partners confront.

#5: The UAB is similarly invisible. Although UAB members claim they learn much from their federal responsibilities, it is not entirely clear that their efforts reap adequate benefits for the agricultural research system at large, or even the USDA alone. The resources required to increase each UAB group's policy sophistication sufficiently to meet its evaluation responsibilities, without massive inputs from the USDA staff, make this operation questionable. Moreover, the UAB's credibility is currently seriously questioned because of the widespread perception that the member selection process has been highly politicized.

Recommendation:

- The major benefit of the UAB seems to be an indirect one: the UAB members' increased understanding of the federal process. It would be more efficient to address this goal directly. A more effective mechanism would be a USDA-Users dialogue process that would bring the communities represented on the UAB into an ongoing discussion/briefing with their federal partners. This makes far more organizational sense than giving UAB members a "crash" federal budget course and requiring them to write a congressionally mandated report that ultimately becomes the USDA staff's responsibility.
- If the UAB is to continue in its current form, then its credibility must be restored by providing a more balanced bipartisan membership.
- Again, assuming UAB's continued existence, two alternative courses should be considered: (1) given the UAB's lack of federal budget experience, it should be relieved of its budget-evaluation responsibilities; or (2) the UAB should receive timely and adequate training in the federal budget process, and the members' terms should be both sufficiently long and staggered to ensure a core of reasonably experienced members at all times. If the first alternative is acted upon, a useful alternative UAB responsibility would be to react to the Joint Council's reports. This would provide top USDA officials with several perspectives, a useful strategy when key decisions are being considered.

#6: Many congressional and executive branch actors involved in priority setting and planning still work in compartmentalized ways. They meet primarily "after the fact" and in structured adversarial forums, such as budget and appropriations hearings.

While the budget coordination between the stateside and federal partners is clear and strong, at the federal level there are serious interorganizational discontinuities. The tensions generated between the congressional and executive branches often result in priority and budget "checks and imbalances." This particularly characterizes the situation between relevant congressional and executive branch units. From mid-July until the following February, USDA policy makers may not discuss budget details outside the department. Thus, any capacity for interorganizational continuity must be based on understandings that are developed through previous and ongoing informal dialogue and interaction.

OMB's role in priority setting is particularly sensitive in relation to the USDA, the White House, and the Congress. Oftentimes, OMB is caught in the struggle between various parts of the system; other times, USDA is pinned between OMB, acting as the administration's proxy, and congressional committees.

At the present time, there is no regularly convened forum where key federal planners concerned with agricultural research[16] meet to develop joint plans, as well as to address the inevitable tensions generated between the scientific enterprise and the political process. After one initial effort sponsored by S&E in the early 1980s, no regular systematic executive/congressional branch forum has been established.

Recommendation: An ongoing Federal Interorganizational Planning Group, composed of planners from ARS, CSRS, Extension, OMB, and relevant congressional committees, GAO, OSTP, and OTA, should be established to develop strategies for ensuring that the priorities articulated by the National Agricultural Priorities Commission Report and the Joint Council's Four-Year Strategic Plan are implemented through budget and other processes. An ongoing forum should be created so that in the intervening years between Farm Bills, key issues bearing on agricultural research priorities can be identified and negotiated for the good of the entire system.

#7: Although cooperation has been the leitmotif of the agricultural system, tightening federal budgets and demands for rigorous priority-setting activities have tended to undermine this ethic. ARS, CSRS, and other S&E units have found themselves pitted against one another in efforts to prove their worthiness in the eyes of congressional authorization and budget committees. More lip service than reality is involved in the commitment to coordinate priority-setting activities of ARS and CSRS.

Recommendation: The translation of longer-term priorities into USDA policies requires front-end and continuous collaboration between ARS, CSRS, and other S&E units. The assistant secretary's Executive Committee, composed of the S&E agency administrators and the assistant secretary, should meet regularly as an S&E Executive Committee on National Priorities. In this capacity, the Executive Committee should focus exclusively on coordinating S&E agencies' efforts supporting national priorities. For structural cohesion, this group should have liaison members from the Joint Council and the Implementation Task Force (composed of the agency administrators' deputies). Priority-setting activities of ARS and CSRS should be coordinated wherever possible.

#8: For the public sector to set agricultural research priorities without any systematic knowledge of comparable priorities in the private sector research system is a formula for waste and inefficiency. Until recently, public sector planners worked virtually in the dark regarding private sector priorities. Patent rights, competitive commercial advantage, and other serious, but not insuperable, issues do exist. Only in the mid-1980s did tentative and limited communication regarding agricultural research priorities begin between the public and private sectors. Some critics perceive these efforts as band-aids.

Recommendation: An ongoing Public/Private Sector Agricultural Research Committee should be created to address the thorny issues that inhibit public-private sector coordination in priority setting. The committee should (1) consider how to develop a reliable data base on private research efforts, and (2) negotiate the research and technology components to be undertaken by each sector.

SUMMARY

By the early 1970s, the agricultural research community had become the object of unrelenting criticism for its priority-setting performance. A history of unprecedented scientific successes, a strong farm bloc, and substantial budget support had ill-prepared the agricultural research system to cope with the onslaught of castigation. To complicate matters, agricultural research priorities were routinely derailed by political perturbations, leaving agricultural science administrators frustrated and still besieged.

During the 1980s, the agricultural research system has engaged in a plethora of priority-setting exercises. Despite some encouraging recent headway, much remains to be done. Eight resistant areas of organizational weakness are delineated, and alternative solutions are recommended.

NOTES

*This paper is based on chapters 7 and 8 of Jean Lipman-Blumen and Susan Schram, The Paradox of Success: The Impact on Priority Setting in Agricultural Research and Extension. The author is indebted primarily to Susan Schram, as well as Michael Hoback, A. Barry Carr, Mike Phillips, and Lowell Lewis for help in updating this paper.

1. For purposes of this paper, the agricultural research system is defined as including the USDA Agricultural Research Service (ARS), the USDA Cooperative State Research Service (CSRS), and the network of land-grant institutions at which agricultural research is conducted. The more general agricultural research community also includes various related organizations and structures, such as the Division of Agriculture of the National Association of State Universities and Land-Grant Colleges (NASULGC), that participate in setting priorities for publicly funded agricultural research.

2. For a discussion of the broader framework of paradoxes within which agricultural research has been conducted, see Lipman-Blumen and Schram (1984).

3. The Division on Agriculture Committees on Organization and Policy, within the National Association of State Universities and Land-Grant Colleges.

4. A 1983 study produced collaboratively by OMB and USDA indicated that the private sector is unlikely to assume research and development (R&D) responsibilities traditionally shouldered by USDA. The validity of the report's findings has been the subject of some controversy.

5. The working relationships are established through formal contracts, such as joint research projects, cooperative agreements, and extramural projects, as well as through informal collaboration.

6. ESCOP is a standing committee of the Experiment Station section of the Division of Agriculture, National Association of State Universities and Land-Grant Colleges (NASULGC).

7. The Committee of Nine is composed of eight experiment station directors and one home economics research administrator, the latter elected by the Experiment Station Committee on Organization and Policy (ESCOP).

8. Experiment Station scientists may have departmental faculty appointments.

9. Busch and Lacy (1982) suggest that user-oriented researchers are more likely to come from farm backgrounds, in contrast to their discipline-oriented peers with non-farm upbringing.

10. The Joint Council was originally mandated by the 1977 Farm Bill. Its membership and priority-setting responsibilities have been altered somewhat by subsequent Farm Bills.

11. 1981 Farm Bill, Sec. 1407, C.

12. Additional recommendations and assessments are included in these reports.

13. For a discussion of the relationship between the nature of a crisis and possible system responses, see Jean Lipman-Blumen (1973).

14. For a more detailed description of priority-setting activities in the agricultural research and extension system, see Lipman-Blumen and Schram (1984).

15. In 1835/1840, Alexis de Tocqueville (1956) described "self-interest rightly understood" solutions in which each action relinquishes a minimal individual benefit to achieve a greater group benefit.

16. Extension and Resident Instruction planners should be included.

REFERENCES

Busch, Lawrence, and William B. Lacy. Science, Agriculture, and the Politics of Research. Boulder, Colo.: Westview Press, 1982.

Carr, A. Barry, Genevieve J. Knezo, Edith Fairman Cooper, and Christine Matthews Rose. Critical Issues in Agricultural Research, Extension, and Teaching Programs, 84-774 ENR. Washington, D.C.: Congressional Research Service, Library of Congress, Oct. 31, 1984.

de Tocqueville, Alexis. Democracy in America (originally 1835/1840). Edited by Richard D. Heffner. New York: New American Library, 1956.

Huston, Keith. "Priority Setting Processes in the State Agricultural Experiment Stations." In An Assessment of the United States Food and Agricultural Research System. Vol. 2, Part B, Commissioned Papers, Congress of the United States, Office of Technology Assessment, Washington, D.C., April 1982.

Lipman-Blumen, Jean. "Role De-Differentiation as a System Response to Crisis." Sociological Inquiry 43(2)(1973): 105-29.

Lipman-Blumen, Jean, and Susan Schram. The Paradox of Success: The Impact on Priority Setting in Agricultural Research and Extension. Washington, D.C.: USDA, Science and Education, Office of the Assistant Secretary, 1984.

National Research Council (NRC). Report of the Committee on Research Advisory to the U.S. Department of Agriculture (The Pound Report). Springfield, Va: National Technical Information Services, 1972.

U.S. Congress. Office of Technology Assessment (OTA). An Assessment of the United States Food and Agricultural Research System. Washington, D.C.: U.S. Government Printing Office, December 1981.

U.S. General Accounting Office (GAO). Management of Agricultural Research: Need and Opportunities for Improvement, CED-77-121. Washington, D.C.: GAO, August 1977.

________. Long Range Planning Can Improve the Efficiency of Agricultural Research and Development, CED-81-141. Washington, D.C.: GAO, July 1981.

_______. Food, Agriculture, and Nutrition Issues for Planning, CED-82-27. Washington, D.C.: GAO, February 1982.

Womach, Jasper, and Geoffrey Becker. "Agriculture and the Budget." Major Issues System Issue Brief. Washington, D.C: Congressional Research Service, Library of Congress, January 1987.

6
Technology Transfer, Public Policy, and the Cooperative Extension Service*

Irwin Feller

Since its formal establishment as a federal, state, and local partnership under the Smith-Lever Act of 1914, the U.S. Cooperative Extension Service (CES) has come to represent the best of both an articulated but decentralized political arrangement and a technology transfer system. Extension has fared well in several forms of evaluation. Econometric studies point to high social rates of return on public expenditures for agricultural research and extension and, indeed, to a possible public underinvestment in these activities (Evenson, Waggoner, and Ruttan 1979; Ruttan 1982, 254-55). These studies also describe multiple ways in which extension aids farmers in learning about and adopting new technologies as well as in making improved resource allocation decisions (Huffman 1984). Warner and Christenson, in their national survey of extension users, report that "more than nine out of ten persons who had used the Cooperative Extension Service were satisfied with the services they received (Warner and Christenson 1984, 76), a level above that reported for most other public services. Finally, the "extension system" in its integration of user needs, research findings, and the dissemination of new practices has become a model for emulation, both domestically and internationally.[1]

---

*This chapter is a revised version of Irwin Feller, "Technology Transfer, Public Policy and the Cooperative Extension Service-OMB Imbroglio," Journal of Policy Analysis and Management 6(3) (Spring 1987):1-21.

Despite this approbation, cooperative extension is a system under challenge. As noted by Daryl Hobbs in his review of the Warner-Christenson study,

> Evaluating the Cooperative Extension Service is like assessing the strength of a tradition or myth.... The Service is beset by insecurity regarding its role and its direction. It is also hostage to its own structure of local county-based programs. In a sense the service is experiencing doubts as to its purpose and mission at the national level while it does the same old things locally to remain alive (Hobbs 1985, 275).

A number of the challenges relate to the coherence and relevance of extension's missions. The General Accounting Office (GAO 1981) and the U.S. Office of Technology Assessment (OTA 1981, 1986) have each criticized extension's ability to set priorities among constituencies and activities.[2] Extension's most direct threats, however, have come at the federal purse. President Reagan's budget proposals for FY 1986 and FY 1987 recommended sizable reductions in federal support for extension and a radical reduction in stated federal responsibilities for extension's programmatic coverage.

By themselves, proposed cuts in any single program in a period of Gramm-Rudman are not analytically interesting. Moreover, the budgetary disfavor of cooperative extension does not appear striking in face of the outright antipathy expressed by the executive branch toward other programs, such as Legal Services Corporation, the Economic Development Administration, and Urban Development Action Grants. As demonstrated in FY 1986 and as projected for FY 1987, CES still retains sufficient support within Congress to fend off significant cuts; at best (or worst) it confronts static absolute levels and declining real levels of federal support.

What is of interest is the stated rationale for the cuts and the responses by extension to these proposals. Analytically, an understanding of the debate initiated by the proposed cuts includes the changing locus of political influence at national and state levels, the intellectual history of the Office of Management and Budget's (OMB) delineation of extension's missions, and the intraorganizational dynamics within CES, whereby its portrayal of its mission simultaneously affects its ability to secure resources from the external environment and to produce an internal marshalling of resources in order to provide services.

Schematically, the analysis considers two interconnecting relationships (figure 6.1): (1) a "vertical" system, focusing on the effectiveness of CES as a technology transfer organization in agricultural production, and (2) a "horizontal," or comprehensive, CES system. Although commencing with a treatment of the political and organizational stresses generated by attempting to maintain the interconnections between these two systems, the paper's more focused interest is on how various stresses, now aggravated by the OMB rationale, affect extension's performance as a technology transfer organization in its agriculture and natural resources program area. There are issues concerning the effectiveness of extension in each of its vertical systems (e.g., the adequacy or content of the research base of extension programming in home economics; the organizational cohesion within land-grant universities among home economics, human nutrition, and extension programs), and in its horizontal layering (e.g., the stresses engendered by having extension specialists satisfy departmental promotion and tenure research criteria while performing a diverse array of extension activities; high turnover rates of extension personnel induced by "quantitative" and "qualitative overload," which form part of the milieu of "insecurity" surrounding extension). These issues, however, are too numerous to be covered in this paper (see USDA 1980; Lipman-Blumen and Schram 1984; Feller et al. 1984a; Marshall and Summers 1985; Sims et al. 1986; Patterson and McCubbin 1985; St. Pierre 1984; and Schuh 1986).

The paper is in five parts. Part 1 describes the nature of the debate precipitated by the FY86 and FY87 budget messages. Part 2 juxtaposes changes in the locus of decision-making within the federal government and within state governments with changes in extension's constituencies. Part 3 traces the evolution of technology transfer as a discrete programmatic activity of federal agencies, and its emergence as the primary lens through which extension has come to be viewed by OMB. Part 4 examines the various ways in which CES has described its mission, both at the national and at the state level. Part 5 addresses the policy implications of these debates.

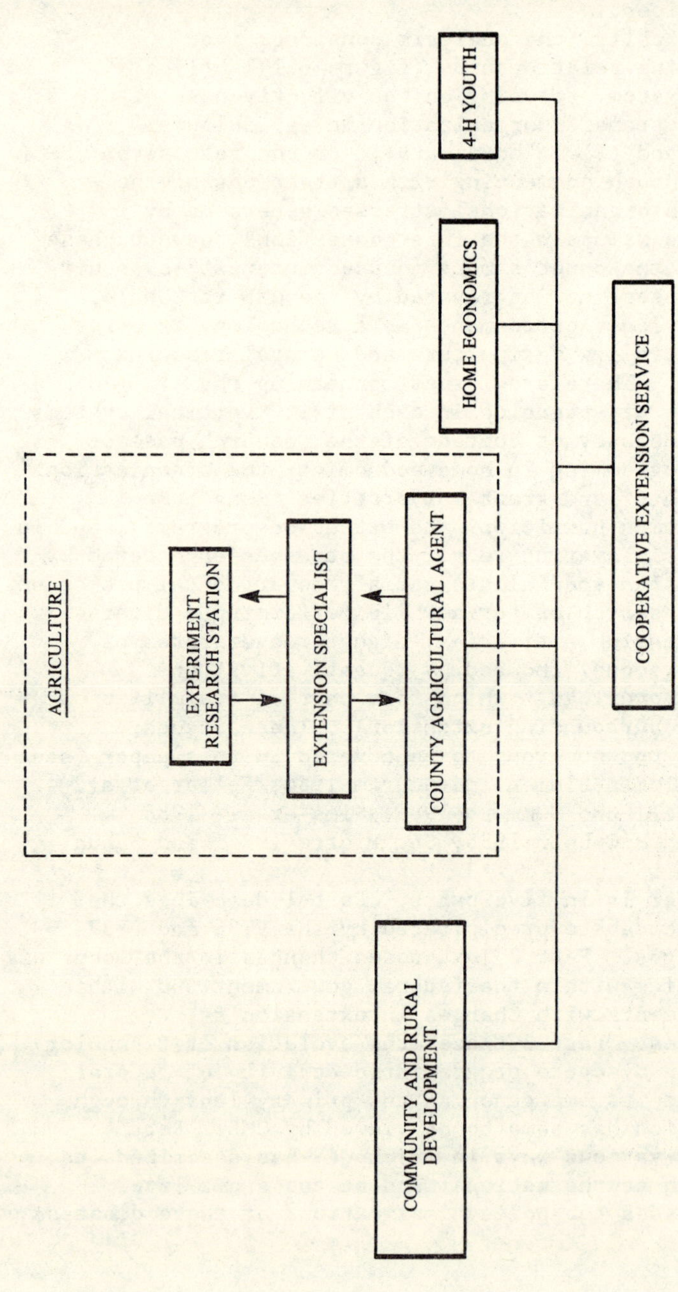

Figure 6.1. The Cooperative Extension Service system

THE MISSIONS OF EXTENSION: OMB AND COOPERATIVE EXTENSION

Cooperative extension is a federal/state/local system. Federal appropriations under the Smith-Lever Act, as amended, are made on a formula basis to each state for support of a cooperative extension service that is administered by memoranda of understanding between the U.S. Secretary of Agriculture and a state's 1862 and 1890 land-grant universities.

Nationally, in FY 1986, federal funds provided $329.5 million, or 31 percent, of the $1.04 billion of extension's funds; state governments contributed $486.2 million, or 47 percent; county governments contributed $193 million, or 15 percent; and other sources contributed $32 million, or 3 percent, of this total. Relative to state and local governments, federal support of extension has been declining since the 1960s (table 6.1).

Against this backdrop of an eroding federal contribution, President Reagan's budget message for FY 1987 proposed a $127.5 million appropriation for cooperative extension. This amount, representing a 47 percent decrease from the $240 million appropriated in FY 1986, was itself the survivor of an initially proposed 23 percent reduction. Beyond the dollar levels, the president's budget messages proposed new limits to the federal stake in cooperative extension. The FY 1986 budget message described extension's missions as disseminating agricultural research and other information to farmers and consumers, as supporting programs in home economics and agriculture-related youth programs, and as providing technical assistance to governments of local communities. Eschewing the traditional view of the "historic" partnership between the federal government (represented by the U.S. Department of Agriculture), state governments (as represented by state appropriations to land-grant universities), and local governments (through cash and in-kind support of county extension agents) in maintaining the spectrum of extension programming, the recommended level of support was primarily intended to permit extension to continue its activities related to production agriculture, and little else. As stated in the FY86 budget message:

> Program priorities should focus on the transfer of basic agricultural information to farmers. To the extent that other extension service programs are deemed needed at 1985 levels, the additional costs can be met by the other cooperating parties in this

TABLE 6.1
Extension Funding by Source, Selected Years

Year	Federal	State and county appropriations	Total
Millions of current dollars:			
1966	75	126	201
1975	179	269	448
1982	315	539	854
1986 (estimated)[a]	326	679	1038[b]
Millions of constant dollars:			
1966	76	128	204
1975	110	164	274
1982	112	192	304

Sources: 1966-1982: U.S. Office of Technology Assessment, *Technology, Public Policy and the Changing Structure of American Agriculture*, p. 267. 1986: U.S. Extension Service.

[a]Estimate for 1986 appropriations does not reflect sequestering as of March 1986.

[b]Includes $32 million in other, non-tax sources of revenue.

program--states, counties, and private donors (OMB 1985, 5-56).

A similar theme appeared in FY87 budget documents. The proposed reduction in budget authority was intended to reduce federal funds distributed through Smith-Lever formula grants "to the amount normally allocated to farm services" (OMB 1986, 66). Support for extension programs for education in home economics, nutrition, 4-H clubs, and community development programs was considered "no longer necessary or appropriate" for meeting national priorities. Extension support of these activities, if desired, was to be met by enhanced state or local contributions, increased use of volunteers, or from federal funds only after farmers' needs had been provided for. This rationale applied both to the general level of support for extension under Sections 3b and 3c of the Smith-Lever Act and to the proposed zeroing out of designated programs, such as food and nutrition education, pest management, farm safety, and farm financial management.

The view depicted by OMB (and OTA)[3] of extension as a technology transfer organization is indeed a cardinal element in every self-description and self-assessment produced by cooperative extension. Reports authored by CES, such as Extension in the '80s, note with pride that extension has contributed significantly to long-term increases in agricultural productivity, which are largely "the result of management skills, production inputs, and new technology embodied in modern farming practices" (Extension in the '80s, 1983, 8). As extension is essentially a state-organized system, specific states (e.g., California) could note that their programs had the essential characteristics called for by OMB.[4] Changes in a number of states that are designed to keep or strengthen ties between agricultural research and extension activities are also evident (e.g., organizational placement of extension specialists on tenure tracks within academic departments; establishment of regional centers housing researchers and specialists). Moreover, the National Agricultural Research, Extension, and Teaching Policy Act Amendments of 1985 explicitly added "development of practical applications of research knowledge" to extension activities, thereby formally legitimizing an activity engaged in by county agents in some states that had run afoul of extension service review of state plans.

Finally, program priorities within extension have shifted toward greater emphasis on agriculture and natural resources. Between 1974 and 1985, the relative share of extension personnel engaged in this program area increased from 37.4% to 45.7%. Over this same period, the personnel commitment to home economics remained stable, increasing from 21.0% to 22.0%, while that allocated to 4-H youth decreased from 32.7% to 25.9% and that in community and rural development fell from 8.9% to 5.9% (table 6.2).[5]

Cooperative extension has complained forcibly about the magnitude of the proposed cuts (NASULGC 1986; McDonald 1986). It also has complained that OMB's portrayal of its mission is both descriptively and prescriptively too narrow. It notes repeatedly, as in Extension in the 80's, that the technology transfer mission delineated for it in the president's budget mission is narrower than the "far reaching and extremely broad" mandates provided for it in the Smith-Lever Act of 1914 and Bankhead-Jones Act of 1935, its principal enabling legislation.[6]

Extension representatives contend as well that the technology transfer vocabulary is inadequate even to describe important extension activities within agriculture, such as informing farmers about the provisions of new governmental support programs. They note with pride extension's ability to bring together subject matter expertise in a variety of disciplines, such as recent program initiatives (e.g. Minnesota, South Carolina) to aid farmers and their families in responding to the economic and family stresses associated with recent financial stringencies. They note, too, that the "grassroots," state-and county-oriented characteristics of extension have produced a justifiable diversity of organizational arrangements and program priorities that cannot be compressed into any single "national" model.[7] Finally, they believe that OMB has caricatured some of its program elements, particularly home economics.[8]

Ironically, however, even extension's supporters speak in the constrained language of the OMB characterization. They speak repeatedly of agricultural extension rather than cooperative extension and of extension's historic contribution to the productivity of U.S. agriculture. For example, Representative Jamie Whitten, chairman of the House Appropriations Subcommittee on Agriculture, Rural Development and Related Agencies, expressed opposition to the proposed cuts as follows: "It just doesn't make any sense to devastate the Extension Service. This group has been doing a good job for years. At a time when farmers really need help from extension agents, this would eliminate their jobs."[9]

TABLE 6.2
Extension Professional Resources Expended by State Extension Services for Major Program Areas, 1974-1985

Program Area	1974	1975	1976	1977	1978	1979	1980	1981	1982	1983	1984	1985
	%	%	%	%	%	%	%	%	%	%	%	%
Agriculture and Natural Resources	37.4	38.7	38.5	40.7	41.3	41.7	42.5	43.2	43.7	43.6	45.4	45.7
Home Economics	21.0	21.5	23.1	22.6	22.8	22.5	22.2	22.0	21.9	21.9	22.6	22.5
4-H Youth	32.7	32.2	30.5	28.4	27.9	27.8	27.7	27.3	26.8	27.0	25.9	25.9
Community and Rural Development	8.9	7.6	7.9	8.3	8.0	8.0	7.6	7.5	7.6	7.5	6.1	5.9
FTE's-Total Percentages	100.0	100.0	100.0	100.0	100.0	100.0	100.0	100.0	100.0	100.0	100.0	100.0

Source: U.S. Department of Agriculture, Extension Service

The OMB's proposal, however, is quite explicit in its protection of extension's agricultural programs, at least, that is, as related to the transfer of technology. Thus the current situation is one in which critic and friend alike use cooperative extension and agricultural extension as synonymous terms and presumably homogeneous organizations. Cooperative extension simultaneously seeks to garner the political goodwill it enjoys from its being seen historically as an effective contributor to agricultural productivity and yet attempts to present and protect the full range of its program activities (and staff).

THE CHANGED POLICY ENVIRONMENT

Questions of the missions of cooperative extension and of federal responsibilities for maintaining traditional partnerships have arisen because long-standing congruence between the concepts and language that extension has employed to articulate its multidimensional activities has been moved askew by these factors: changes in the locus of federal budgetary decisions; extension's adaptations at the state and county levels to changes in the number and mix of their clientele; by the emergence of technology transfer as a distinctive policy field within the federal government; and, by specific changes in the character of extension activities in agriculture.

There is general agreement that the power of the "iron triangle" that supported agricultural programs has been shaken (Bonnen 1980; Paarlberg 1978). This has occurred as a result of the attainment, as Johnson has written, of a "new position of power and importance" by budget units (Johnson 1984, 510-11), most notably OMB's gain at the expense of more traditional congressional committees; the decrease in the number of elected officials from farm districts;[10] and, for a period at least in the 1970s, the increased influence of what Hadwiger has termed the "externalities/alternatives" coalition (Hadwiger 1982).

New loci of power also mean new people. Interviews with research and extension administrators in land-grant universities and with federal and state officials conducted in the Feller et al. study indicate that the number of elected officials with personnel experience in extension programs has decreased and that the role of budget units at the federal level has increased. As implied in the tenor of the president's budget message, these officials are less familiar (or imbued) with the concept of the "extension family," that is, a nuclear unit in which each individually (as producer, as homemaker, as parent or child, and as member of a community) and collectively benefits from the spectrum of extension's principal program areas. Extension, too, has received consistently poor reviews for its refusal to articulate a clear set of priorities, and its perceived unwillingness to make internal changes.[11] It also has received criticism from traditional agricultural supporters who feel that their needs have been neglected as the system has broadened its constitutencies.[12]

This overall climate has produced a situation in which extension is having to work hard to protect (1) its direct federal funding, (2) that portion of its state funding that is induced by federal grants-in-aid, and (3) its state funding. As noted above, the prospects for increased real levels of federal funding are doubtful in light of Gramm-Rudman, the president's budget message, and OMB's view of extension.

There also are linkages, as found by Rose-Ackerman and Evenson, between federal budgets and state-level support of extension. Their work considers the effects of changes in the political structure of state governments, particularly the reapportionment of state legislatures following the 1962 Baker v. Carr decision, the secular decline in the relative importance of the farm population and farm incomes in states, and the role of federal grants. It is the latter relationship that is of interest here. States must match federal contributions for "core" extension activities, but most states spend above this amount. This leads to the expectation that a "marginal increase in federal dollars would have an income effect but no price effect if governments responded as if they were rational individual consumers" (Rose-Ackerman and Evenson 1985, 11). In fact they find that federal money seems to stimulate state expenditures for extension (and research). They note that "why this happens is not clear," and account for this occurrence in terms of its consistency with the fiscal "flypaper" effect found in other intergovernmental grants-in-aid arrangements (Courant, Gramlich, and Rubinfield 1979). Their conclusion is that

> Since farm population continues to fall as a share of the total, this structural shift in political influence implies an overall decline in public spending as time passes. Unless federal funding for agricultural research and extension increases

> markedly, the proportion of state budgets invested in pubic sector research and extension will continue to decline. This may mean an absolute decline as well, unless total state budgets increase at a rapid rate (Rose-Ackerman and Evenson 1985, 11).

Similar if less evident pressures on extension's budget exist within state governments. As indicated in table 6.1, extension increasingly has become a state- and county-funded system. As such, the level and direction of extension's programmatic support is the product of influences on state budgetary decisions; the importance assigned to extension-related activities by land-grant university officials in the formulation of their requests for state funds; a state extension service's own program priorities; and the program preferences, organizational responsiveness, and influence of client groups at the state and county levels.

These influences are evident in each of the recent national studies of extension, but they have not been systematically analyzed. Where addressed, the determinants of state-level support for extension activities have been presented primarily in terms of accounting for interstate differences in levels of support (Guttman 1979; Huffman and Miranowski 1981). The analysis has been cast largely in terms of per capita expenditures for research and extension, using various measures of the importance of farming in a state as the principal explanatory variables. This approach implicitly treats extension as a technology transfer system in agricultural production. It provides little data on the allocation of extension efforts (dollars, personnel) by program areas, and does not describe the interrelationships between program allocations among program areas and the absolute levels of extension appropriations within a state. Clearly, though, what has been shaping extension's program efforts are its endeavors at the state and county levels to broaden its constituent base in order to be able to maintain public support. As stated by J. Orville Young, director of Cooperative Extension, Washington University:

> Over time our audiences have changed. In recent times we have found that urban citizens seek agricultural information for urban gardening, home economics information for food preservation, and numerous other living skills--information traditionally sought by rural citizens. To deny any citizen information which

> we possess is political suicide. No government agency can refuse to provide information that they possess when asked for that information by any citizen regardless of the location of their residence. If we decide that we are going to serve only production agriculture, it is unlikely that 97 percent of the population will be willing to provide tax support dollars to serve the other 3 percent (Young 1982, 6).

More generally, recent studies serve to make explicit the obverse role of extension's prized "bottoms up" planning approach, namely in seeing extension officials as political actors who are seeking to maintain state and county support in the midst of marked changes in their clientele/electorate. Indeed, from this perspective, the continued increase in state support in the face of declines in extension's traditional clientele can be seen as an indicator of the political skills of state extension directors. As the federal contribution has declined, first in relative terms and now possibly in absolute amounts as well, extension, perforce, has had to become more attuned to the priorities of the state groups that form its new coalition of support. Although in some states this may be compatible with strengthened ties between research and extension programs--the fundamental justification for having extension as part of the university system--and increased attention to a technology transfer capability in agriculture, neither of these outcomes is necessary.

TECHNOLOGY TRANSFER AS A FEDERAL POLICY ISSUE

OMB's view of extension is rooted in a straightforward and conventional view of the process of technological change in which the Cooperative Extension Service serves to adopt generic scientific and technical findings to site-specific production settings, to disseminate these findings, and to educate farmers as to the efficient use of these new approaches. Indeed, the administration's proposed support for "applied" R&D and technology transfer activities through the activities of "agricultural extension" is itself an exception to its more general policy of confining its support to basic research and leaving the private sector to support the development, commercialization, and transfer of the technologies made possible by basic research (Keyworth 1984).

Technology transfer/research utilization became a discrete programmatic interest beginning in the late 1960s, following the sharp increases in federal support for science and technology in the domestic (nondefense/nonspace) agencies.[13] There were widespread and overlapping beliefs that (1) public investment in R&D in domestic agencies was justifiable only in terms of the tangible, programmatically discernible benefits that could be derived from mission-oriented research; (2) considerable research-based knowledge already in hand, if effectively transferred, could aid in realization of the Great Society's domestic goals in improved housing, transportation, environmental protection, education, and more; and (3) that the same combination of targeted scientific, technological, and management skills and approaches that had produced a successful manned space program could be successfully applied to the country's domestic agenda (Hough 1975; NSF 1975; Anuskiewiez 1973).

Technology transfer became a subsystem within the federal government as many agencies developed or expanded their technology transfer/research utilization/research dissemination programs. New interagency units such as the Federal Council for Science and Technology and the Federal Laboratory Consortium emerged. Federal agencies also increased their support of research on "diffusion" processes and on the dynamics of the transfer process. There was great interest in sorting through the relative influences of "technology-push" or "user-pull" strategies. There was also considerable agency interest in developing a program rationale and sets of performance indicators. A search for a paradigm ensued (Havelock and Lingwood 1973).

Everett Rogers' work was influential in generating this paradigm. In Diffusion of Innovations, for example, Rogers described the agricultural extension model as "an integrated system for the innovation-development process" (Rogers 1983, 159). Rogers' research laid out in both functional and organizational terms how the U.S. agricultural system of research and cooperative extension provided for the non-linear, reflexive set of interactions between the several stages of research, development, adaptation, demonstration, and diffusion that were emerging in contemporaneous research as the hallmarks of "successful" technology transfer programs. In the literature on technology transfer that burgeoned in the 1960s and 1970s, extension served as the real world exemplar and criterion for organizational arrangements that linked the multiple phases of a research-diffusion and dissemination system.[14]

Basking in the glory shared with the state agricultural experiment stations and their counterparts in the U.S. Department of Agriculture, extension had little reason to introduce the qualifications concerning breadth of mission and decentralized priority setting that are now seen as politically needed ripostes to stylized and narrow characterizations of itself as a technology transfer organization. Extension was also relatively unscathed by the criticism of publicly funded agricultural research that began in the early 1970s with the release of the National Academy of Sciences (Pound) report, and that only recently has shown signs of abating (OTA 1986; Feller et al., forthcoming).

Lost to many participants in these events, federal personnel as well as many academic researchers, were the distinctions between the agricultural extension model and the agricultural extension system, and between agricultural extension and cooperative extension. To those officials concerned mainly with developing a technology transfer program for their agency or in designing effective arrangements for "coupling," "linking," or "interfacing" research and technology transfer either within or among organizations, this depiction of agricultural extension, however limited by some broader measure of understanding, was sufficient. Few agencies had the need for detailed consideration of intra- and interorganizational relationships between the constituent units in the stylized agricultural technology delivery system. The agricultural technology transfer system was (and, indeed, still is) widely cited, but little studied.

EXTENSION'S SELF-ASSESSMENT: TECHNOLOGY TRANSFER/NONFORMAL EDUCATION

Extension's historic contribution to increased agricultural productivity is not in question: its current contribution is.[15] Several technological and economic changes have created an environment that has called into question the relevance of the extension system, particularly its emphasis on the county agent in the transfer of agricultural technology. Among these changes are the increased technical complexity of agricultural production, the increased demand for research-based information by agricultural producers, the trend toward a bimodal distribution of farms with production concentrated in a small number of farms, the availability to farmers of multiple information sources, including the development by input suppliers of farm information management systems, the increased educational levels of farmers, and the bypassing of the county agent by producers who seek direct contact with extension specialists or experiment station researchers.

These trends have been noted by many, both within and without extension.[16] In the new policy environment extension has had to simultaneously seek to correct (in its view) the narrow conceptualization of its missions and to demonstrate its effectiveness as a technology transfer system. But the matter does not end there. Acceptance of the proposition that extension is more than a technology transfer organization does not fully explain (1) in what way(s) it is a technology transfer organization (so as to be able to respond to OMB and similar stylized interpretations of its activities related to this mission), or (2) extension's own seeming difficulties in resolving perceptions of itself as a technology transfer organization or as a nonformal education system.[17]

It is at this point that mission, strategy, effectiveness, and vocabulary begin to part; the complex cluster of extension's composite activities become fragmented into discrete pieces; different emphases in orientation, activities, and personnel policies become evident among state extension programs; and extension is compelled to grapple--rhetorically, substantively, politically--with questions of alternative organizational styles and competencies. These questions relate to (1) the functional relevance not simply of the "extension" system, but of the relative contributions of extension specialists and county agents; and (2) the emphasis given in different states to the different skills and activities of extension professionals.

Complicating any analysis here is extension's diversity--across states, across counties within a state, between clusters of states, and between the states and the federal government. With this caveat, extension's choices relate to the <u>relative</u> weight assigned to three functional roles of specialists and agents: (1) researcher, adapter, demonstrator, and disseminator; (2) capacity-builder; (3) information disseminator, educator, and facilitator.[18]

The first role corresponds to OMB's delineation of cooperative extension. It is extension's historic role; it is a role it continues to perform; it is, indeed, extension's central role in some states. Portrayal of extension, as done by OMB, as an adapter and disseminator of new research-based knowledge does not include several of extension's roles relating to technology transfer. These include "capacity-building" (e.g., organization of artificial insemination cooperatives, irrigation districts), authenticating information, the wholesaling of information, the direct provision of services (e.g., soil testing, forage testing, entomological and pathological identification), and serving as a link between users and research and teaching faculty (Young 1982, 6).

More broadly, extension's contribution to increased agricultural productivity through technological change involves different tasks and skills at different stages in the life cycle of a technology. In a stylized manner, these tasks can be viewed as follows: In the phase of initial or early use, extension serves to foster diffusion. This may be done through a combination of adaptive research, organizational innovation to provide the infrastructure necessary for the use of specific innovation/demonstration trials, and information dissemination. When a technology is in wider but not general use, extension's role shifts more toward communicating the experiences of adopters to non-adopters. Finally, when there is a relatively standard technology in use, extension's role shifts more toward that of a consultant or troubleshooter, with the agent typically addressing either through his own competencies or through the information he can obtain from other sources the specific production problems encountered by producers in a particular growing season.[19]

This orientation is evident in selected state systems. For example, the University of California/Cooperative Extension's Long-Range Planning Statement (which serves as the model for the OTA's description of extension) presents extension as a link between the agricultural experiment station and research and extension users. Among the functions identified for extension are development of specific applications of new knowledge, field testing, development of educational strategies, and dissemination. The rationale for these assignments rests upon the view that research findings cannot quickly and effectively be put to use simply because these are available. Rather, "Applications of research are not always obvious, either to the researcher or to the potential adopter, even when the information has been published or otherwise disseminated. For this reason, effective extension activity must go far beyond a public service information role" (University of California 1982, 2).[20]

Extension's need to renew its emphasis on a research orientation is also found within the extension community. In a recent examination of extension activities, Pigg has observed that Seaman Knapp and his first field agents "taught only farming methods that had been fully tested by practical farmers. Today, extension agents often try to transfer information that comes directly from research labs with little practical experience; this procedure rarely encourages widespread adoption" (Pigg 1983). Dillman's account of his discussion with a county agent over the future role of extension presents the same message. According to this agent, for years farmers had bypassed extension in their search for information. The agent noted, however, "that there was much more to making a decision than simply having the information. The important act was not the securing of information from a particular place, it was the interaction that interpreted the information in a local context" (Dillman 1985).

This focus on the technical aspects of extension work is not self-evident. For in its historically evolving emphasis on its "education" role, extension has deemphasized the technical skills needed for its personnel to perform the technology transfer functions envisioned for it not only by OMB but also by users and its own members. This pattern appears, for example, in historical accounts of the evolution of extension's activities.[21] It is evident in the differences in emphasis between extension's 1980 Evaluation report and Extension in the '80s. Indeed, if the technology transfer community may at times be said to have truncated extension's activities to fit a narrow mold, it is also striking to see how extension at times has descriptively lopped off its technology transfer activities and cast itself increasingly in terms of "extension education."

This shift is illustrated by Charles Beer, former deputy administrator for agriculture, Cooperative Extension Service, in his presentation entitled, "Extension Programs--Technology Transfer and More." In this presentation, Beer counters the accusation that growers have not adopted recommendations based on new research results because of extension specialists and agents' ineffectiveness in transferring new technology. He suggests that this charge is based on a simplistic view of the technology transfer process:

> To delve more deeply into these questions is to wrestle with the issue of extension education and technology transfer. What is it and how is it accomplished? Transfer means to "convey"--technology in this case--from one person or group to another. In this context the definition does not necessarily include understanding. Education, on the other hand, includes training-development of knowledge and skills that should lead to understanding (Beer 1983).[22]

More significantly, this substitution of education for technology transfer appears in mission statements and job descriptions that set forth the activities of extension personnel. The Pennsylvania Extension Orientation Handbook at Pennsylvania State University, for example, expresses the philosophy of extension as follows: "The job of the Cooperative Extension Service is informal education--to help people help themselves" (Pennsylvania Extension Orientation Handbook, p. 4). Each of the next four paragraphs in this statement of philosophy employs the word education. Only in the fifth paragraph does the concept of the transfer and use of knowledge to solve problems appear. Similarly, in vacancy announcements, the larger number of program responsibilities relate to developing and conducting educational programs. Interpreting, disseminating, and applying the latest research in the field of agriculture and related science appears as only one of several responsibilities.

Within segments of the extension community, the tenor of the descriptions of extension found in Extension in the '80s and other publicity releases that usually refer to extension "as an agency that provides information" are matters of concern. As stated by Henry Wadsworth, director of the Cooperative Extension Service, Purdue University, and chairman of the Extension Committee on Organization and Policy, in his comments on the report, Extension in the '80s:

> In my viewpoint, the report needed to place more emphasis on the need to allocate greater resources to applied research and demonstrations. Much of current research is directed toward understanding basic concepts. Frequently there is a considerable gap in moving from laboratory results to field applications. Who will do the work to make such a transition possible in order to realize the full potential of a basic work? The state extension services could assume

> more of this responsibility if resources were available to do so (Wadsworth 1984, 3-4).

It is not enough to say that a "good" extension system combines each of these attributes--"high tech" and "high touch" in current megatrends parlance (Gibson 1984)--both in terms of a division of labor between extension specialists and county agents and in the multiple tasks performed by each, or that "In practice ANR (Agricultural Natural Resources) extension activities frequently involve aspects of all three of these functions (disseminating information, teaching skills, and providing services)" (USDA 1980, 40). In practice, the demands of diverse constituencies on specialist and agent alike threaten, where they have not already attenuated, their abilities to effectively perform technology transfer activities. In practice, users, whatever may be their historic loyalties to extension, are opting for alternative sources of information and advice. In practice, the stylized set of relationships between specialists and agents is strained by opposing influences that serve to strengthen linkages between specialists and researchers while agents assume the role of convenor or facilitator at meetings where others provide the technical information required by producers. And in all this, the county agent's effort to preserve technical proficiency is hampered by requirements to serve multiple constituencies and to operate within a set of organizational norms, activities, and symbols that in many jurisdictions rationalize a weaning away from demonstration research activities and dissemination of research-based information.[23]

Somewhere along the way extension shifted its own world view from the system portrayed in NSF and other depictions of a technology transfer system to a system of informal or non-formal education. A philosophical and programmatic umbrella that had to be broad enough to protect the horizontal span of extension programs began to substitute for the vertical, research-based concept of technology transfer formerly ascendant in extension's agricultural programs.[24]

The possibility is that without such a commitment, agricultural producers will regard CES as increasingly less relevant and cease to turn to it for assistance. As an organization, cooperative extension may continue to survive and indeed flourish as it continues to reposition itself to serve a new, more diversified constituency retaining modest ties to its traditional clientele. Its role in the agricultural technology transfer system will atrophy, however, and its claims to public support will be challenged.

CONCLUSION

Cooperative Extension's missions are broad, its clientele are changing, and its methods of program delivery are diverse. Extension's flexibility in the face of changing clientele is a prized attribute, both in terms of organizational self-assessment and in adaptation to changing political realities at state and local levels of government. This flexibility, however, has had its price. Operationally, it has constituted openness rather than reallocation as new roles for existing clientele and new clientele have been added without commensurate reductions in existing programs, even in the face of stagnant real resources. One consequence has been that extension has been under recurrent criticism for its inability to articulate a coherent set of priorities, either by clientele or by program area. More directly relevant to this paper, its broadening range of activities is widely held to have adversely affected its effectiveness as a technology transfer organization, particularly in its ability to upgrade the technical skills of its personnel and its internal protection, relatively, of county agents over extension specialists.[25]

No single metric exists to gauge the quantitative and qualitative dimensions of the direction or pace of change in extension. It is possible to point to state and county extension services that are held in high regard both for providing valued services in technology transfer and for their openness to change. It is also possible to identify extension officials who are keenly aware of the many changes occurring in the structure of U. S. agriculture. Several state extension services are systematically seeking to ensure that they continue to modernize quickly as both technology and user characteristics change. They are reorganizing their modes of organization and program delivery so as to change at a pace commensurate with their intended clientele. Other state services are going through a process of self-examination in an effort to identify basic structural changes in the number and characteristics of their intended users and of the type and levels of skills they will need in order to be functionally relevant to them. Other state extension services, however, either in the main or at significant levels of decision-making, seem to be responding to these changes by using extension's historic grassroots political base to ward off proposals for change or by using the introduction of a computer-based network as the symbol (and substance) of their technological contemporaneousness.

Extension is confronting difficulties in functionally justifying the maintenance of its current organizational arrangements, most notably effective linkages between experiment station researchers, extension specialists, and county agents in a research-based, technology transfer system. It has begun to substitute for this orientation those missions and activities that can be justified within the framework of an educational system (Wildavsky 1979). This shift toward emphasis on education, information dissemination, and organization-building mutually reinforces extension's service to a broader and more diverse constituency in program areas that do not exhibit the same links between research and extension activities found in agriculture. This new constituency, in turn, is an essential element in extension's current ability to maintain public support, particularly within state and county governments.

Given this setting, there are uncertain dynamics generated by the proposed budget cuts and OMB's view of extension. Extension, certainly agricultural extension, is a technology transfer organization in almost every sense in which the term is defined, and in empirically based surveys of what county agricultural agents do, but its activities in this role extend beyond those contained in OMB's depiction. Moreover, extension's "grass-roots" characteristics, its encouragement of direct involvement of clientele in the process of planning and carrying out programs, and its strong local identity, as expressed by Warner and Christenson, produce a feeling of goodwill among its constituents. The support that flows from this goodwill, in their view, "would not be specified but would exist as a diffuse obligation or commitment to the agency and its programs. In all likelihood, this positive sentiment would lay dormant until the organization is threatened. In which case, support could be mobilized for specific purposes" (Warner and Christenson 1984, 87).

It is this dormant support that appears to be quickening. The quandary is that this quickening need not lead to a strengthening of extension's capacities as a technology transfer organization. Marshalling this diverse group to ward off budget cuts involves an implied contract to return to them the diverse set of services they demand. Extension's resolution of priorities among constituents, program areas, and emphasis on functional competencies may go in quite different ways depending on the nature of constituent support, by size and saliency, among jurisdictions.

Submerged in the current debate is any statement of extension's activities grounded in new economic conditions that are relevant to its efficiency as an agricultural technology system; in equity considerations relating to the distribution among producers of the benefits of publicly supported agricultural research; or to the delivery of other public services.

Historically, extension has seen itself as a lead agency in moving research findings from the land-grant university to farmers. Today, farmers receive new ideas from several sources, including researchers, extension specialists, private firms, and private consultants. Extension's role as a technology transfer agency in agriculture may have given way to a system in which it serves to help screen technically and economically "useful" from "nonuseful" innovations after a technology has been placed in use either through the initiation of industry sales efforts or through direct contacts between producers and land-grant or other researchers. Moreover, firms sell discrete technologies, not the integration of these technologies, into technically and economically efficient production units. Extension's contribution may be to take the hardware and farm practice components of a new technology and, as one county agent expressed extension's role, put the pieces together (Madden 1985).

By reaching small-scale producers with information on the availability and impacts of new techniques earlier, as would be the case under profit-oriented, private sector marketing strategies, extension may also serve to promote a more equitable distribution of the benefits of publicly funded research. There also may be social benefits from extension activities in agriculture such as making producers aware of the externalities associated with the use of specific techniques.

In all, regardless of the eventual level of appropriations for FY87 or immediate outyears, there are risks that the debate initiated by the president's budget messages serves either (1) to transform extension into the system projected for it by OMB, truncating its delivery of public services that meet resource allocation and equity criteria both in agriculture and its other program areas, or (2) to lead extension to defend itself in a manner that weakens its none too robust internal capacity, political willingness, and budgetary condition to undertake the corrective measures necessary for it to serve as an efficient provider of traditional and/or new services.

NOTES

1. A survey of diffusion research authored by the National Science Foundation expresses this view: "The classic version of a technology transfer system is the agricultural extension programs.... The agricultural extension system is perhaps the most 'complete' of any of the existing technology transfer programs" (NSF 1983, 162; Claar, Dahl, and Watts, n.d.).

2. For example, "If the Extension Service is to be a socially oriented organization with broad educational and behavioral modification objectives, then changes may have to be made to its basic funding formulas and organizational structure. On the other hand, if its mission is to be limited to more traditional focuses, then the scope of its programming may have to be reduced" (GAO 1981, 21).

3. "A key mission of the Federal Extension Service should be to facilitate technology transfer between USDA research agencies and the State extension services as well as between states. If this function is not adequately performed, research agencies become motivated to develop their own outreach programs. The need then is for increased integration of the research and extension function--not greater fragmentation" (OTA 1986).

4. Considerable variation exists in the organization of CES within the land-grant university system. For descriptions of these arrangements, see Feller et al., 1984b.

5. These shifts have occurred largely as a result of attribution as extension's personnel complement has declined from approximately 18,000 to below 17,000 between 1979 and 1985. The adjustment process has followed two main forms. First, vacancies in the three program areas other than agriculture have not always been filled. This has been most noticeable with vacancies occurring in 4-H, where the pattern has been to add responsibility for 4-H to extension personnel in agriculture or home economics. Second, budget stringencies have compelled a withdrawal from the traditional extension objective of having a

complement of agents to provide the full range of extension programs in each county and acceptance of the need for multi-county agent arrangements.

6. "By its very charter, Cooperative Extension was established as an entity that would modify its programs and outreach in response to such factors as new knowledge, changes in its clientele's needs, and alterations in the socioeconomic landscape. And, over the years, Cooperative Extension *has* changed in accordance with changing surroundings" (*Extension in the '80s*, 1983, 3) (italics in original).

The principal mandates of Cooperative Extension are set out as follows:

> ... to aid in diffusing among the people of the United States useful and practical information relating to Agriculture, uses of solar energy with respect to agriculture, home economics, and rural energy and to encourage the application of the same" (Smith-Lever Act, 1914).
>
> ... the establishment and maintenance of permanent and effective agricultural industry including ... the development of the rural home and rural life, and the maximum contribution of agriculture to the welfare of the consumer and the maintenance of maximum employment and national prosperity (Bankhead-Jones Act, 1935).

The Federal Extension Service summarizes these mandates as follows: "The basic mission of Cooperative Extension is to improve American agriculture and strengthen American families and communities through the dissemination and application of research-generated knowledge and leadership techniques" (USDA 1983, 1).

7. "... Extension functions through locally based offices that attempt to be responsive to a variety of clientele with differing needs. In agriculture, for example, services are extended to large commercial farmers, to limited resource farmers, and to backyard gardeners. In home economics, traditional homemakers, women employed outside of the home, and single parents are served. Youth programs serve teen leaders, farm boys and girls, and urban children; while community development lends assistance to local officials, low-income community groups, and tourist agencies. With these varied programs and clientele has come a diversification of subject matter" (Warner and Christenson 1984, 9).

8. OMB's Major Policy Initiatives describe home economics extension programs as consisting of "cooking, sewing, upholstery and quilting" (OMB 1986, 66), whereas the American Home Economics Association (AHEA) has emphasized that extension programs relate to teenage pregnancy, teenage suicide, nutrition, child care, and rural family stress (AHEA 1986).

9. New York Times, March 21, 1986. Similar identification of the proposed cuts with agricultural extension are widespread. Tom Wicker, New York Times columnist, reacted to the proposed cuts passionately:

> This nation's farmers have been the most resourceful and productive in history, and another such Federal agency--the Agricultural Extension Service, which teaches farmers the latest and best agricultural knowledge--has been a major reason.
>
> Now Ronald Reagan wants to kill the Extension Service to save money; if the service is needed, his aides say, let the states pay for it.
>
> What effrontery! (New York Times, December 20, 1985, A35).

10. Between 1966 and 1976, the number of congressional districts defined as predominately rural declined from 181 to 86, or less than 20 percent of the total.

11. "Available evidence suggests that the progress of the agricultural research community in establishing priorities is more advanced than that of the extension community. The agricultural research community has been widely studied and critically evaluated within and without the system in a series of projects extending back to the mid-1960s. In light of these analyses, the agricultural research system has adjusted the distribution of its resources in recognition of potential advances evolving from biotechnology and information technology.

"Similar progress is not apparent in extension. Extension administrators suggest that this is because most of the extension planning occurs at the local level through advisory committees. Yet such a system does not obviate the need for setting national plans and priorities" (OTA 1986, 71).

12. The National Agricultural Research and Extension Users Advisory Board, which was established by the 1977 Farm Bill, has consistently sought a more technologically oriented focus to extension's activities: "Greater

attention must be given to training local extension agents in emerging biological and computer technologies to shorten the transfer time between discovery and adoption of strategic research. There is a need to reestablish an effective body of extension agents who are closely linked to the agricultural departments and experiment stations of our land-grant universities where fundamental advances in agricultural science and technologies will occur. Without such linking, we believe the capacity of the CES to transfer agricultural research and education will diminish" (National Agricultural Research and Extension Users Advisory Board 1983, 32-33).

13. The accent here is on the federal interest in applications of domestic R&D. Historically, social utility has always been present in federal sponsorship of scientific endeavors (Dupree 1957).

14. Illustrative of the examples of scholarly citation of cooperative extension as the exemplary public sector technology transfer system is the following:

> A primary example is the cooperative (Federal and State) Agricultural Extension Service through which information about innovations is brought to farmers. This program provides a two-way channel between the agricultural research and the farmer (Baker et al. 1963; Knoblauch et al. 1962). This kind of effective feedback mechanism does not exist in NASA's Technology Utilization Program, in which technologies developed in the course of NASA's mission spill over into other sectors of the economy (Kelly and Kranzberg, 1978, 40).

15. "Basic organizational issues must be addressed by the Extension Service. The premise on which extension was developed was that of research scientist conveying the knowledge of discoveries to the extension specialist who, in turn, supplied information to the county agent who then taught the farmer. Over time, this concept has gradually but persistently broken down as agricultural technology has become more complex and insufficient resources have been devoted to staff development.... As these changes have occurred, the role of the county agent has become increasingly unclear. Appreciation for and use of county agents as educators and technology transfer agents has declined. As a result of these changes, a basic structural reevaluation of the organization of the extension function

of the agricultural research system is needed" (OTA 1986, 279).

16. "The county agent is cited as a fundamental strength of the Extension system. But the agent's role is changing.

"Certain problems now addressed by Extension require more specialization. Many commercial farmers want contacts with the county agents but often go directly to the campus specialists for help. County agents continue to perform an important referral function" (USDA 1980, xii).

"Many interviewees report that while the theoretical relationship between research and extension is clear (i.e., Extension translates and demonstrates new research findings for users and simultaneously reports users' research needs to scientists), in reality, the relationship is not well articulated. Some Extension personnel worry about a growing separation of researchers and Extension workers" (Lipman-Blumen and Schram 1984, viii-16; see also Feller 1984; Busch and Lacy 1983; Cochrane 1979; and Wolek 1985).

17. A tangential debate exists here concerning the multiple definitions of technology transfer and the variable congruence between specific definitions and extension's missions. A summary compendium of the conceptual exercises on technology transfer was produced for use by the House of Representatives Committee, as it sought itself to examine what the federal government was doing in the way of domestic technology transfer. The report notes that "technology transfer programs can have at least four different types of objectives--i.e., to improve research utilization, to develop organizational capabilities, to serve spinoff functions from other R&D activities, and to provide technical assistance" (U.S. Congress 1978, 2).

Three of the four functions cited for technology transfer programs are clearly within the intent of extension's mission statement. Only the "spinoff" function, an activity most generally associated with defense and space-related research, would seem to strain the gambit of extension's activities.

18. This classification is close to that advanced by the Extension Service's 1980 Evaluation report. The report categorizes the functional responsibilities for agricultural and natural resources as follows:

> 1) Collecting, interpreting, and disseminating information and knowledge through an information system linking farmers and other clientele with the research and knowledge base of the land-grant

universities, the U.S. Department of Agriculture, and other government agencies;

2) Teaching skills and principles and providing other assistance to facilitate developing clientele (individual and group) capacities for problem-solving; and

3) Providing services to clientele including identifying and diagnosing problems, formulating or recommending alternative solutions to clientele problems, referring or giving other aid enabling clientele to identify and utilize additional (public or private) sources of assistance (USDA 1980, 40).

19. These multiple roles are evident in the history of several technologies. In the early history of the use of artificial insemination on dairy cattle, cooperative extension played a variety of roles that included: helping to organize the local farmer cooperatives that turned into the early bull studs associations; helping farmers and researchers take proposals for needed research in the technology of semen collection and preservation to state legislatures for special funding; and in some instances actually using their offices as clearing-houses for calls to the artificial insemination technician. Extension's role of organizer, promoter, and conduit for research needs and research findings has evolved as the technology of artificial insemination and the structure of the breeding industry have changed. The advent of frozen semen began the consolidation of local bull studs. In 1950 there were almost 100 bull studs in the United States; by 1980 there were fewer than twelve studs of commercial significance. Extension is now less involved with the actual operation of these studs and cooperatives and has focused more on helping individual dairymen assess their management priorities; these include ways of using artificial insemination, selecting superior sires, and assessing the applicability of emerging technologies such as embryo transplants (Moore 1984).

20. This commitment to research is based in part on the diversity of commodities produced in the region, the variety of microclimates, and the technical sophistication of growers or the managers they hire. But it is also an explicit strategy intended to maintain the functional relevance of the county agent in the linked relationships between researchers, extension specialists, private industry vendors, and growers (University of California 1982, 2).

21. "In the early years of extension, the main emphasis was on information and technology transfer for all types of programs--agriculture, home economics and youth activities. These are still important concerns but as the organization matured, attention was given to individual capacity building and leadership development" (USDA 1980, 31). Extension representatives may respond with considerable accuracy that the system has always described itself in the language of an educational system and that it has been the "innovation-technology" transfer community that has led itself and others astray. Sanders and Maunder, for example, use the term extension education to describe the introduction in extension lectures in agricultural science by Cambridge University as early as 1873 (Sanders and Maunder 1966, 3-12).

22. Formally establishing a terminological concordance between the technology transfer and education is less important than establishing an analytical correspondence between the two concepts. Economists have generally established this connection by seeing extension and schooling--informal and formal education--as ways in which abilities are created to make improved allocative decisions. Boyce and Evenson have written:

"Fundamentally this approach provides a basis for understanding not only the economic contribution of decision making or information processing skills, but of the information itself which may be supplied by extension systems. It recognizes that farmers cannot make strictly optimal decisions under conditions of uncertainty and the introduction of the products of the research system creates uncertainty. There is then an important role for agencies which attempt to provide information and to summarize that information in easily understood form. Extension programs, both private and public, are primarily engaged in this information supply activity" (Boyce and Evenson 1975, 107-8).

23. Patterson and McCubbin's study of Minnesota agents found that of a possible list of twenty-two stressors and strains identified by extension agents as part of their job, each of the six top-ranking accessors was related to "overload and feeling pulled in too many directions." In particular, the expectation that agents would be too knowledgeable among many diverse program areas was found to create both a quantitative overload in the sense that "more program areas needed to be researched and planned for than there is time for," and a qualitative overload in the sense that "it may stretch the agent's

ability beyond where their expertise lies and where they feel they can be effective" (Patterson and McCubbin 1985).

24. Extension's dilemma in simultaneously articulating its role to OMB and its own members is suggested in a 1978 study of the characteristics of county extension agent work. The study was undertaken in response to legal challenges involving state performance appraisal systems and is principally directed at establishing valid grounds for determining employment conditions for agents and not the issues of attention here. Still, it is interesting to observe that the duality noted here in the characterization of the agent's work is evident as well in the study's findings.

In terms of the rankings of relative time spent in duty areas, agricultural agents ranked "respond to client requests for specific information" and "respond to client requests for technical assistance" as their first two duties, and "conduct programs" as the third. This ranking points to the agricultural agent's role as a technical consultant and then as an educator.

The study also asked agents to rank the relative criticality of major functional duty areas. For agricultural agents, the same ranking as above--responding to client requests for specific information, responding to client requests for technical assistance, and conducting programs--emerges. For the total sample of agents, which includes home economics/family living, 4-H/youth development, other, and those agents with mixed program area responsibilities, a different ranking occurred. Here the three most critical tasks were responding to requests for specific information, and developing and maintaining public relations. It is this latter ranking that was presented in the study's summary of its findings (Brumback, Hahn, and Edwards 1978, tables 10 and 11, pp. 25 and 27).

25. Between 1975 and 1984, the number of county agents declined from 11,357 to 11,140, while the number of specialists declined from 4,224 to 3,581. OTA termed this latter decline "particularly alarming since the specialist staff has the largest level of training and is the best equipped to educate both county agents and farmers on evolving agricultural technologies... (W)hile the need is for an increased emphasis on specialist staff, just the opposite is occurring" (OTA 1986, 269-70).

REFERENCES

American Home Economics Association (AHEA). "Proposed Termination of Federal Funding for the Extension Service." AHEA Newsletter, February 5, 1986.

Anuskiewicz, Todd. Federal Technology Transfer. Washington, D.C.: George Washington University, 1973.

Beer, Charles. "Extension Programs--Technology Transfer and More." Paper presented at symposium, Soil and Water Conservation in the '80s, Increasing Challenges with Diminishing Resources. Meeting of the American Society of Agronomy, Washington, D.C., August 1983.

Bonnen, James T. "Observations on the Changing Nature of National Agricultural Policy Decision Processes, 1946-76." In Farmers, Bureaucrats and Middlemen: Perspectives on American Agriculture, edited by Trudy H. Peterson, 309-27. Washington, D.C.: Howard University Press, 1980.

Boyce, James, and Robert Evenson. Agricultural Research and Extension Programs. New York: Agricultural Development Council, 1975.

Brumback, Gary, Clifford Hahn, and Dorothy Edwards. Reaching and Teaching People: A Nationwide Job Analysis of County Extension Agents' Work. Washington, D.C.: American Institute for Research, 1978.

Busch, Lawrence, and William Lacy. Science, Agriculture and the Politics of Research. Boulder, Colo.: Westview Press, 1983.

Claar, J., D. Dahl, and Lowell Watts. The Cooperative Extension Service: An Adaptable Model for Developing Countries. Series No. 1. University of Illinois: Interpaks, n.d.

Cochrane, Willard W. The Development of American Agriculture. Minneapolis: University of Minnesota Press, 1979.

Cournant, Paul, Edward Gramlich, and Daniel Rubinfeld. "The Stimulative Effects of Intergovernmental Grants or Why Money Sticks Where It Hits." In Fiscal Federalism and Grants-in-Aid, COUPE Papers on Public Economics, vol. 1, 5-22. Edited by Peter Mieszkowski and William Oakland. Washington, D.C.: Urban Institute, 1979.

Dillman, Don. "The Social Impacts of Information Technologies in Rural North America." Rural Sociology 50(1)(Spring 1985): 1-26.

Dupree, A. Hunter. Science in the Federal Government. New York: Harper and Row, Torchbook Edition, 1957.

Evenson, Robert, Paul Waggoner, and Vernon Ruttan. "Economic Benefits from Research: An Example from Agriculture." Science 205(14)(September 1979): 1101-7.

Extension in the '80s. A Report of the Joint USDA-NASULGC Committee on the Future of Cooperative Extension. Cooperative Extension Service, University of Wisconsin, Madison 1983.

Feller, Irwin. "Reconsideration of the Agricultural Technology Transfer Model." Journal of Technology Transfer 8(Spring 1984): 47-56.

Feller, Irwin, D. Lynne Kaltreider, Patrick Madden, Dan Moore, and Laura Sims, eds. The Agricultural Technology Delivery System, 6 vols. Institute for Policy Research and Evaluation, Pennsylvania State University, University Park, 1984a.

________. Surveys of Organizations and Their Linkages, Vol. 2 of The Agricultural Technology Delivery System. Institute for Policy Research and Evaluation, Pennsylvania State University, University Park, 1984b.

Feller, Irwin, Patrick Madden, D. Lynne Kaltreider, Dan Moore, and Laura Sims. "The New Agricultural R&D Policy Agenda." Research Policy, forthcoming.

General Accounting Office (GAO). Cooperative Extension Service's Mission and Federal Role Need Congressional Clarification. CED-81-119. Washington, D.C.: GPO, 1981.

Gibson, Terry. "High Tech and High Touch--New Roles and Responsibilities for Extension." Paper presented at annual meeting of the American Association of Adult and Continuing Education, Louisville, Kentucky, 1984.

Guttman, Joel. "Interest Groups and the Demand for Agricultural Research." Journal of Political Economy 80(1979): 467-84.

Hadwiger, Don. The Politics of Agricultural Research. Lincoln: University of Nebraska Press, 1982.

Havelock, Ronald, and David Lingwood. "R&D Utilization Strategies and Functions: An Analytical Comparison of Four Systems." Report to the Manpower Administration, U.S. Department of Labor. Ann Arbor, Mich.: Center for Research on Utilization of Scientific Knowledge, 1973.

Hobbs, Daryl. "Review of Warner and Christenson's The Cooperative Extension Service." Rural Sociology 50(2)(Summer 1985): 275-81.

Hough, Granville. Technology Diffusion. Mt. Airy, Md.: Lemond Systems, Inc., 1975.

Huffman, Wallace. "Allocative Efficiency: The Role of Human Capital." Quarterly Journal of Economics 91(1984): 59-80.

Huffman, Wallace, and John Miranowski. "An Economic Analysis of Expenditures on Agricultural Experiment Station Research." American Journal of Agricultural Economics 63(1981): 104-18.

Johnson, Bruce E. "From Analyst to Negotiator: The OMB's New Role." Journal of Policy Analysis and Management 3(1984): 501-15.

Kelly, Patrick, and Melvin Kranzberg, eds. Technological Innovation: A Critical Review of Current Knowledge. San Francisco: San Francisco Press, 1978.

Keyworth, George II. R&D in FY 1985: Budgets, Policies, Outlooks. In Colloquium Proceedings, March 29-30, 1984, edited by Mary Morrison, 19-30. Washington, D.C.: American Association for the Advancement of Science, 1984.

Lipman-Blumen, Jean, and Susan Schram. The Paradox of Success. Washington, D.C.: USDA-Science and Education, 1984.

Madden, J. Patrick. "The Roles of the Land Grant University in the Development and Transfer of Mechnical Tomato Harvester Technology." Paper presented at annual meeting of the American Association for the Advancement of Science, Los Angeles, 1985.

Marshall, H. Peter, and James C. Summers. Strengthening the Research Base for Extension, Final Report. USDA Extension Service cooperating with West Virginia Cooperative Extension Service and Missouri Cooperative Extension Service, 1985.

McDonald, Kim. "Reagan Plan to Cut Farm-Extension Aid Assailed by Colleges." Chronicle of Higher Education, 26(1)(March 5, 1986): 1, 12.

Moore, Dan. "Artificial Insemination in Dairy and Beef Cattle: Technological Developments and Organizational Change." In vol. 4 of The Agricultural Technology Delivery System, edited by Feller, et al. Institute for Policy Research and Evaluation, Pennsylvania State University, University Park, 1984.

National Agricultural Research and Extension Users Advisory Board. Appraisal of the Proposed Budget for Food and Agricultural Sciences. Report to the President and Congress. February 1983, pp. 32-33.

National Association of State Universities and Land Grant Colleges (NASULGC). "Campuses React to Proposed Budget Slash for Cooperative Extension." The Green Sheet, March 31, 1986.

National Science Foundation (NSF). Division of Industrial Science and Technological Innovation. The Process of Technological Innovation--Reviewing the Literature. Washington, D.C.: National Science Foundation, 1983.

National Science Foundation (NSF). Office of National R&D Assessment. Federal Technology Transfer, An Analysis of Current Program Characteristics and Practices. A report prepared for the Committee on Domestic Technology Transfer, Federal Council for Science and Technology, Washington, D.C., 1975.

Office of Management and Budget (OMB). Executive Office of the President. Budget of the United States, Fiscal Year 1986. Washington, D.C.: GPO, 1985.

________. Major Policy Iniatives, Fiscal Year 1987. Washington, D.C.: GPO, 1986.

Office of Technology Assessment (OTA). U.S. Congress. An Assessment of the United States Food and Agricultural Research System. Washington, D.C.: OTA, 1981.

________. Technology, Public Policy and the Changing Structure of American Agriculture. Washington, D.C.: GPO, 1986.

Paarlberg, Don. "Agriculture Loses Its Uniqueness." American Journal of Agricultural Economics 60(5)(December 1978): 769-76.

Patterson, Joan, and Hamilton McCubbin. Minnesota County Extension Agents: Stress, Coping and Adaptation. St. Paul: University of Minnesota, 1985.

Pennsylvania Extension Orientation Handbook. University Park: Pennyslvania State University, Cooperative Extension Service, n.d.

Pigg, Kenneth. "Shades of Seaman Knapp." Journal of Extension 21(July/August 1983): 4-8.

Rogers, Everett. Diffusion of Innovations, 3rd ed. New York: Free Press, 1983.

Rose-Ackerman, Susan, and Robert Evenson. "The Political Economy of Agricultural Research and Extension: Grants, Votes, and Reapportionment." American Journal of Agricultural Economics 67(February 1985): 1-14.

Ruttan, Vernon. Agricultural Research Policy. Minneapolis: University of Minnesota Press, 1982.

St. Pierre, Tena. "Addressing Work and Family Issues among Extension Personnel." Journal of Home Economics 76(Winter 1984): 42-47.

Sanders, H. C., and A. H. Maunder. "Why an Extension Service Today?" In The Cooperative Extension Service, edited by H. C. Sanders. Englewood Cliffs, N.J.: Prentice-Hall, 1966.

Schuh, G. Edward. "Revitalizing Land Grant Universities." Choices (Second Quarter 1986), pp. 6-10.

Sims, Laura, D. Lynne Kaltreider, J. Lynne Brown, and Patricia Adams. National Assessment of Extension's Food and Nutrition Programs. Institute for Policy Research and Evaluation in cooperation with the Cooperative Extension Service, Pennsylvania State University, University Park, 1986.

U.S. Congress. House. Subcommittee on Science Research and Technology. Domestic Technology Transfer: Issues and Options, Vol. 1, 95th Cong., 2d sess., November 1978.

U.S. Department of Agriculture (USDA). Science and Education Administration. Evaluation of Economic and Social Consequences of Cooperative Extension Programs. Washington, D.C.: GPO, January 1980.

________. Federal Extension Service: Challenge and Change. Washington, D.C.: GPO, 1983.

University of California. Cooperative Extension Service. Long-Range Planning Statement, Summary, revised December 1982.

Wadsworth, Henry. Statement before the Subcommittee on Department Operations Research and Foreign Agriculture, Committee on Agriculture, U.S. House of Representatives, June 12, 1984.

Warner, Paul T., and James A. Christenson. The Cooperative Extension Service: A National Assessment. Boulder and London: Westview Press, 1984.

Wildavsky, Aaron. "Strategic Retreat on Objectives." In Speaking Truth to Power. Boston: Little, Brown and Co., 1979.

Wolek, Francis. "Transferring Federal Technology in Agriculture." Journal of Technology Transfer (Spring 1985): 57-70.

Young, J. Orville. "Future Directions for Cooperative Extension in Washington." Paper presented at Cooperative Extension Faculty Inservice Education Week, Pullman, Washington, March 17, 1982.

7
Mobilizing Support for Agricultural Research at the Minnesota Agricultural Experiment Station

Richard J. Sauer and Carl E. Pray

A major problem of government agricultural research systems is insufficient resources. Despite studies that show very high rates of return to agricultural research, real investments in research by many states in the United States, the U.S. Department of Agriculture, and many nations around the world are stagnant or declining. To reverse this trend, state and national research systems have to mobilize political support for research budgets. This paper is a case study of the way in which the Minnesota Agricultural Experiment Station (MAES) develops political and financial support for its research program. After reviewing the institutional development and economic impact of the MAES, we review past and present methods of mobilizing support.

INSTITUTIONAL GROWTH OF THE MINNESOTA SYSTEM[1]

By the middle of the nineteenth century, farmers in Minnesota were interested in scientific ways to improve their agricultural practices. Many farmers conducted individual experiments using different types of seed, fertilizer, crops, feed, and livestock. These experiments were generally on their own property using trial and error methods. Group research efforts by the state horticultural society established "agricultural experiment stations" on the farms of some of their members. Periodicals for farmers were started to disseminate information and to share the results of experiments by farmers and by the society. In an effort to reach the large number of farmers who did not subscribe to the magazines, the Minnesota

Agricultural Society sponsored the first agricultural fair in 1855.

By 1858, there was sufficient support for scientific agriculture that the state legislature authorized the creation of an agricultural college but did not provide any funding. In 1869 Minnesota established an agricultural college and farm as part of the University of Minnesota. Its financial support came from the U.S. government through the Morrill Act of 1862. The agricultural college, like others in the United States in the nineteenth century, was unstable. It had a small and changing faculty and few full-time bachelor degree students until 1899.

Table 7.1 shows the growth and distribution of the agricultural staff in the different disciplines at the college. Staff size increased from ten in 1900 to 467 in 1970. There was a major shift in the distribution of staff in favor of the social sciences. In 1910, social science staff made up about 2 percent of the total staff. This figure increased to 28 percent by 1970.

Staff resources and the number of affiliated institutions also increased rapidly. One indication of growth was the development of the branch station network. Within sixty years every major agricultural region in Minnesota had an experiment station. The Crookston and Grand Rapids stations were established before 1900. Between 1900 and World War I the Forest Experiment Station at Cloquet and the stations at Morris and Waseca were set up. After World War II the university took over a large tract of army land just south of the Twin Cities at Rosemount and established a satellite station for the Twin Cities campus. The Lamberton station, the last station created, was established in 1957 in the southwest part of the state.

The public sector played an important role in financing the Minnesota agricultural research system. Initial funding to establish an agricultural college came from the federal government in 1962. The federal government's contribution has continued to grow, but at a slower rate than the state's contribution (see table 7.2). By 1900 the state government was providing more money than the federal government. This has continued to be the case. A third source of funding that has increased is that from the private sector, although it is difficult to disaggregate this data from table 7.2. Some resources have not been included in table 7.2 because they are nonmonetary. One example is the land for the Northwest Experiment Station given by James J. Hill of the Northern

TABLE 7.1
Percentage Distribution of Staff, 1900-1970

Department	1900	1910	1920	1930	1940	1950	1960	1970
Staff Size	10	47	87	106	129	173	311	467
Crop Related	30.00	29.79	40.22	24.53	31.01	28.32	23.47	17.77
Agriculture-Agronomy	20.00	10.64	13.79	6.60	9.30	7.51	7.40	7.07
Entomology		10.64	10.34	7.55	8.53	10.40	6.75	4.71
Plant Pathology		8.51	16.09	10.38	13.18	10.40	9.32	6.00
Animal Related	40.00	34.04	18.39	20.76	19.38	30.06	28.62	27.41
Animal Poultry Husbandry	10.00	19.15	5.75	7.55	7.75	6.36	6.11	7.07
Dairy Husbandry	20.00	4.26	5.75	5.66	6.20	7.51	2.57	
Veterinary Medicine	10.00	10.64	6.90	7.55	5.43	16.18	19.94	20.34
Total: Farm-Related	70.00	63.83	58.61	45.29	50.39	58.38	52.09	45.18
Physical Sciences	30.00	34.04	36.80	41.51	35.66	30.64	25.40	26.77
Horticulture			8.05	5.66	9.30	6.36	5.46	5.14
Forestry			5.75	7.55	4.65	7.51	8.36	7.71
Soils			4.60	3.77	3.88	3.47	4.18	6.42
Agricultural Biochemistry			9.20	16.04	9.30	5.20	2.89	3.64
Agricultural Engineering			9.20	8.49	8.53	8.09	4.50	3.85
Social Sciences-Services	--	2.13	4.60	13.20	13.96	10.98	22.51	28.05
Agricultural Economics	--	2.13	4.60	6.60	8.53	6.94	8.36	11.99
Rural Sociology	--	--	--	0.94	0.78	1.16	0.96	1.50
Home Economics	--	--	--	5.66	4.65	2.89	13.19	11.99
Agricultural Journalism	--	--	--	--	--	--	--	2.57
Total: Non-Farm Related	30.00	36.17	41.40	54.71	49.62	41.62	47.91	54.82

Source: Joseph C. Fitzharris, "Public Institutions and Agricultural Growth: The Minnesota Experience," University of Minnesota, Department of Agricultural and Applied Economics Staff Paper P76-22, June 1976.

TABLE 7.2
Funds Available in the Fiscal Year, 1890-1982/83[a]
(in thousands of current dollars)

Year	Federal	State			Endowment and Fellowships	Fees, Sales, Private Industry	Total Funds
		Special	General University	Total			
1890	15.0	--	3.0	3.0	--	2.8	20.8
1900	15.0	n.a.	32.6	32.6	--	9.3	56.9
1910	28.0	n.a.	40.2	40.2	--	6.2	74.4
1920	30.0	32.7	257.2	257.2	--	84.1	371.3
1930	90.0	40.3	210.1	250.4	53.5	102.8	496.8
1940	146.8	50.3	333.8	384.1	10.1	62.8	612.8
1950	207.5	248.0	918.2	1,166.2	108.2	306.8	1,898.7
1960	729.4	707.5	1,733.3	2,440.8	502.7	708.4	4,381.4
1970	1,451.8	n.a.	6,055.0	6,055.0	1,137.5	1,007.3	9,890.6
1982/83	8,188.9	10,026.4	11,260.9	21,287.3	--	5,652.4	35,129.1

Source: Updated and adapted from Joseph C. Fitzharris, "Public Institutions and Agricultural Growth: The Minnesota Experience," University of Minnesota, Department of Agricultural and Applied Economics Staff Paper P76-22, June 1976.

[a]For the years 1890-1940, the data is from the Annual Report of the Director of the Office of Experiment Stations; it can be cross-checked in the Station's Annual Report for 1900-1940. For 1950-1960, see the Station's Annual Report; data for 1970 is from Minnesota Science, 26 (Fall 1970).

Pacific Railroad. Other contributions by industry, cooperatives, and commodity groups are included in this endowment and fellowships category.

At present, the state government is responsible for the major share of the MAES budget, followed by the federal government and the private sector. In 1982/83 the breakdown was roughly 62 percent from the state, 12 percent from unrestricted federal funds, and 18 percent from federal competitive grant funds and private companies, organizations, or foundations. Fees and earnings from the experiment stations themselves provide 6 to 8 percent of the MAES budget. The private sector contribution was primarily from relatively small agribusiness firms and farmers who were organized into commodity groups. Large multinational firms fund little research at the University of Minnesota.

During the early years Forestry and Home Economics functioned as units within the College of Agriculture. When they became separate colleges under an Institute of Agriculture, Forestry and Home Economics, the Experiment Station continued to provide research funding for the new units. By the mid-1970s the organization of the Institute had stabilized along the lines illustrated in Figure 7.1.

THE IMPACT OF EXPERIMENT STATION RESEARCH

Most economists measure the impact of experiment station research in terms of the increase in agricultural productivity due to this research. This paper will also concentrate on that contribution, but it recognizes that experiment station research does more than simply increase productivity. Social science research often has goals that go beyond increased productivity. Improved income distribution is one common goal. Research on human nutrition has improved the health of consumers. Resources in the traditional agricultural sciences have been devoted to pollution and ecological problems and to such issues as animal welfare. This research may not directly increase productivity but it does yield benefits to many important groups.

During the early years, most productivity change resulted from farm management research, which consisted of identifying the best farming practices and then transferring those practices to other farms in the state. Yield trials of different crop varieties used by farmers led to planting recommendations, and the testing of

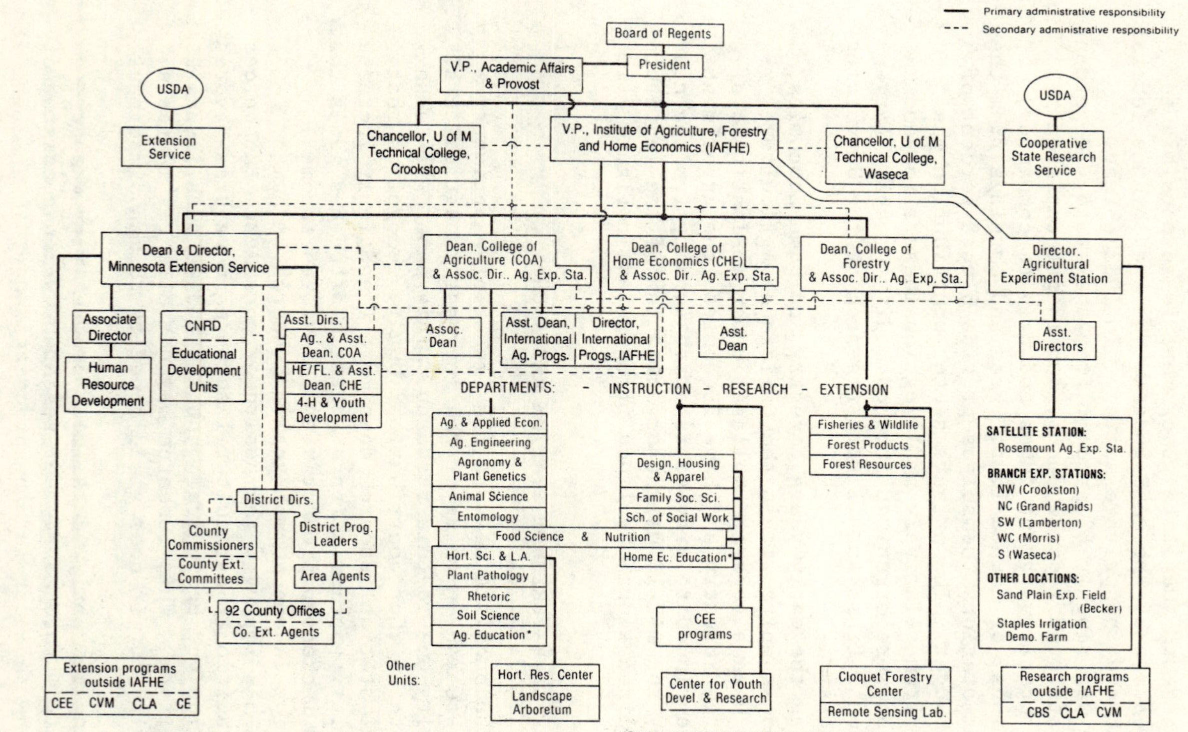

Figure 7.1. Organization chart: Institute of Agriculture, Forestry and Home Economics, (IAFHE), University of Minnesota

different feeds at the university led to recommendations for adoption.

Much of the early scientific research could be classified as maintenance research (Peterson and Fitzharris 1977, 70). Its purpose was not so much to increase productivity as to prevent it from declining. As such, it was frequently difficult to measure using aggregate productivity statistics. An example of maintenance research that had a very large economic impact was the research on rust control. In 1916 this research led to the barberry eradication campaigns in which the winter host of rust, the barberry bush, was destroyed. As plant breeding became more sophisticated, breeding for rust resistance became the chief method of controlling rust. Likewise, much of the early research on animals focused on maintaining their health, not on improving their output. Early veterinary research helped to eradicate several poultry diseases.

Another important line of work has been crop adaptation. Peterson and Fitzharris describe the process:

> Efforts to move crops northward, adapting them to shorter growing seasons and colder climates, began in the 1890s. The initial work involved trial experiments and the selection of the best varieties. Considerable success was achieved in moving corn northward and in selecting wheat varieties better adapted to the shorter growing season of the northern two-thirds of the state. After the turn of the century, breeding and crossbreeding experiments were initiated. Breeding efforts were even more successful than trial experiments in producing varieties adapted to the rigorous climate and soil conditions in Minnesota. Much of this work has been cooperative, involving the neighboring state experiment stations, the Minnesota branch stations, and various bureaus in the USDA (Peterson and Fitzharris 1977, 71).

H. K. Hayes was one of the early leaders in hybrid corn research. However, crop breeding did not make a major impact on average yield per acre until the widespread adoption of hybrid corn in the 1930s. For the first time, this adoption was accompanied by the use of large amounts of chemical fertilizer.[2] At about the same time, wheat yields also started to increase. The introduction of soybeans and their improvement through plant breeding after 1940 had a major impact on the state's economy. Since World War II there has been a gradual improvement in yields of most crops through breeding, cultural practices, chemicals, and other factors. Perhaps the most important breakthrough in recent years was the development of the Era wheat variety, which was released to farmers in 1970 and gave farmers an increase of five bushels per acre over other varieties available at that time.

Important contributions in the animal sciences have been made that have improved both health and productivity.[3] T. L. Haecker's early work developed the feeding standards for beef and milk cattle. University scientists improved techniques for making silage and included it as a regular part of dairy rations. Animal disease research significantly improved the productivity of Minnesota animals. Poultry diseases were greatly reduced. In the 1930s, pharmaceutical treatments for swine disease were developed. Diagnostic tests refined by the university helped to eliminate bovine tuberculosis. University research on genetics and breeding techniques have improved swine production, and recent work on antibiotics in pork feed has significantly improved the health of hogs.

The contributions of MAES scientists show up in estimates of the productivity of Minnesota research. Productivity measured in value of output per worker or value of output per acre has risen steadily throughout the last 100 years; output per acre has generally risen since 1900 (figure 7.2). Output per worker has increased since 1900 with a slight decline between 1930 and 1940 and then rapid growth after 1940. These productivity figures give preliminary evidence that there was substantial technological change in Minnesota agriculture during this period. Increases in output per acre indicate that changes in biological and/or chemical technology were taking place. Increases in output per worker indicate that there were mechanical innovations that allowed one worker to cultivate more land.

Recent studies by economists have confirmed the impressionistic evidence that research has been a good investment of government money. A study of soybean research indicates a rate of return to Minnesota soybean research above 50 percent. Research by Bredahl and Peterson indicates that the marginal product of every dollar spent on Minnesota research was $13.15 for cash grains and $75.55 for livestock (Bredahl and Peterson 1976).

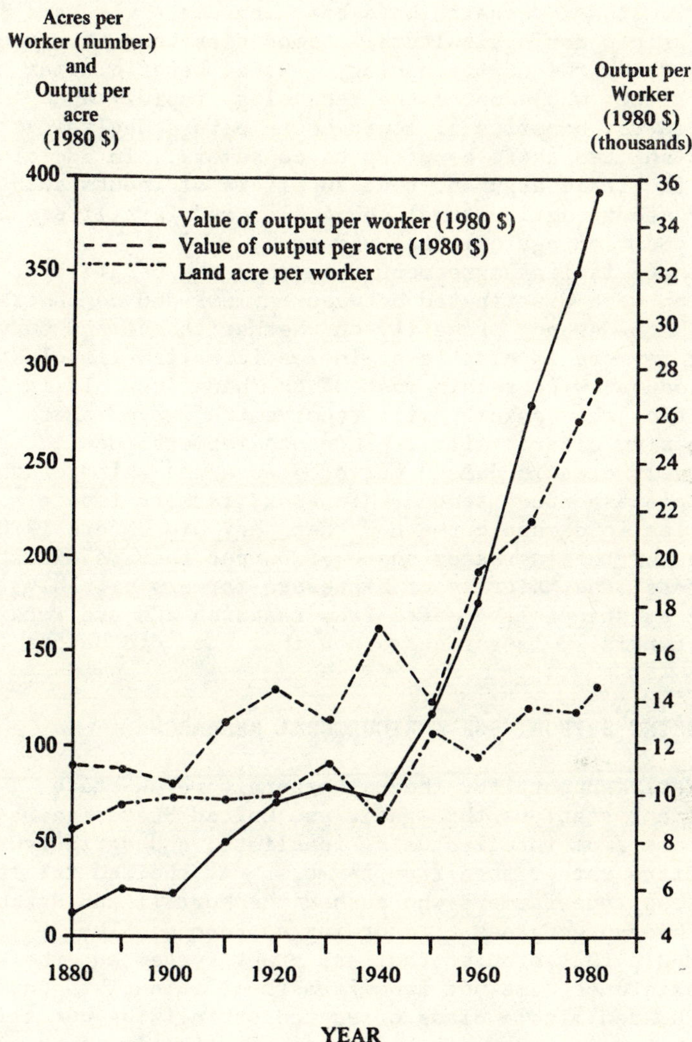

Figure 7.2. Minnesota agriculture, 1880-1980: Labor and land productivity. Updated and adapted from Willis Peterson and Joseph C. Fitzharris, "Organization and Productivity of the Federal-State Research System in the United States," in _Resource Allocation and Productivity in National and International Agricultural Research_, ed. Thomas M. Arndt, Dana G. Dalrymple, and Vernon W. Ruttan (Minneapolis: University of Minnesota Press, 1977).

The consensus of scholars is that major beneficiaries of agricultural research have been consumers who have paid lower prices for agricultural commodities than they would have without the new technology. Other beneficiaries have been farmers who adopted the technology rapidly and captured the benefits of lower costs before declining prices shifted these benefits to consumers. In addition there is little argument that suppliers of inputs such as hybrid seeds, agricultural chemicals, and fertilizers also benefited from agricultural research.

There is less agreement about how the benefits of research were distributed between farmers and consumers. The debate depends primarily on whether the demand curve facing farmers is elastic or inelastic. If it is elastic, the producer will retain most of the benefits. If it is inelastic, the consumer will receive the larger share. Two points seem clear: first, Minnesota farmers face a relatively elastic demand curve because one-third of their output is exported; second, Minnesota farmers face a much more elastic demand curve now than they did before 1970 because of the increased share of output that is exported. Therefore, the majority of Minnesota farmers probably retain substantial benefits from research and are retaining a substantially larger share now than they did before 1970.

MOBILIZING SUPPORT FOR AGRICULTURAL RESEARCH

Early support for federal expenditure on state experiment stations throughout the United States seems to have come from intellectuals, idealists, and agricultural scientists rather than from farmers. At the federal level, it was not the farmers who pushed the Morrill and Hatch acts through Congress. According to Bonnen, "The leadership that brought the land grant system and the USDA into existence came not from farmers or scientists but from the educated middle class of an industrializing but still largely agrarian society. These professionals, lawyers, ministers, doctors and journalists, shared a faith in the 'idea of progress' and saw access to education as the road to individual opportunity as well as the primary assurance of a vital democratic social order" (Bonnen 1987).

Major farmer organizations like the Grange did not support agricultural research or the agricultural colleges until after 1900. Given the absence of major breakthroughs that benefited farmers, this lack of interest by farmers is not surprising. Two groups that supported agricultural research were the Minnesota Horticultural Society and the Minnesota Agricultural Society. These groups consisted of a small group of urban and rural citizens who were interested in the scientific advancement of agriculture. Another early source of financial and political support in Minnesota was the Great Northern Railway, which provided land for experimental farms and either free or inexpensive transportation for experiment station staff and equipment. The railway was interested in the rapid development of agriculture in western Minnesota in particular.

As the Minnesota Experiment Station began to produce useful results after 1900, it was able to build political support among farmers. Farmers became an increasingly powerful force in politics in the state, and their support helped to ensure the continuation of research funds from the state. As scientists tried to spread the results of their research to farmers, they actively helped to develop farmer organizations. These organizations in turn supported the experiment station both politically and to a lesser extent financially. In some cases scientists were actively involved in establishing these organizations. More often, however, they offered technical and managerial expertise, which helped to strengthen groups or institutions that already existed.

The earliest institutions that the university and experiment station helped to organize were the farmers institutes. "In 1886 Edward D. Porter, professor of agriculture at the University of Minnesota, formulated the plan of going out among the farmers and holding meetings in the hope that they might manifest sufficient interest in the agricultural course at the university to send their sons to attend it. He persuaded the agricultural committee of the board of regents to appropriate $1,000 for farmers' institutes, and 31 were held that year, largely in connection with county fairs" (True 1928). A few years later the farmers institutes were set up with an independent board that had two members from the university. These institutes generally consisted of one- or two-day meetings in which university specialists and leading farmers talked to farmers about ways to improve their practices.

Farmers clubs were started in 1908 under the influence of the farmers institutes. These clubs were more permanent institutions than the institutes and from 1910 were promoted by the extension division of the agriculture department of the university. By 1914 there were over 300 farmers clubs in Minnesota.

Extension work by county agents began in 1912. The first county agents worked in the Red River Valley and were financed by the Great Northern Railway, millers, bankers, elevator operators, implement dealers, and farmers (True 1928, 76). These contributions were soon supplemented with money from the federal government and the farmers institutes. True's account of the early progress of county extension indicates that the university soon established considerable control over extension's activities.

> The university, through the division of agricultural extension, assumed leadership in this movement and, in cooperation with the Office of Farm Management, appointed a State leader of county agents. When the State legislature passed the act of April 19, 1913, appropriating $25,000 for aiding the appointment of county agents in 1913, and $35,000 to be used in 1914, the law gave control of these and county funds for this purpose to the dean of the agricultural department of the university. Each county might receive not to exceed $1,000 a year, provided it contributed at least an equal amount. County commissioners were given authority to appropriate not to exceed $1,000 for county-agent work, and their approval was a necessary preliminary to the appointment of county agents, who must be satisfactory to the dean (True 1928, 92).

By the time the Smith-Lever Act was passed in 1914 there were extension agents in over twenty counties in Minnesota. These agents were supported by a variety of sources, many of which were business based rather than farmer based. There were a considerable number of farmers clubs in Minnesota, which provided a means to spread new technical information. The business connections of the early county agent movement, however, kept many farmers and the farmers clubs from linking up with the county agents (True 1928, 93).

The Smith-Lever Act provided federal funds for extension to each state. It included $10,000 annually plus an additional amount based on rural population. The provision of this additional amount was conditional on the state, county, or local authorities providing matching funds. This provided the university with an important incentive for organizing local farm organizations because a university-controlled extension system would receive both state and federal money on a matching basis.

In response to Smith-Lever the university spent considerable effort in organizing farmers. Between 1917 and 1918 the number of Minnesota counties with an extension agent expanded from sixteen to eighty-five. During World War I a movement developed to standardize the form of local organization (which was not specified by Smith-Lever) into county farm bureaus in the northern and western states, including Minnesota. The county extension agents became the primary organizers of the farm bureaus. By 1921 almost all counties of Minnesota had Farm Bureaus (True 1925, 162). These county Farm Bureaus organized themselves into a state organization after the war and joined the national organization--the American Farm Bureau Federation--in 1920.

The Farm Bureaus had three major activities: education, economics, and political action. Their educational role was both to educate farmers about the latest technology and to educate the nonfarm sector about the importance of agriculture. In their economic role they encouraged the development of cooperatives for purchasing inputs and marketing their output. Their political role was as a lobbyist for farmers' interests at both the state and national level. The Farm Bureaus provided grass roots political support for the program and the budget of Minnesota's agricultural college and experiment station. The relationship between the university and the Farm Bureau was close, not only at the county level but also at the state level, with the head of the Farm Bureau serving for a number of years as a regent of the university.

The decline of the influence of the Farm Bureau in Minnesota was signaled by the 1953 extension law that separated extension and the Farm Bureau (Block 1960, 206). Before 1953, Minnesota law, unlike the laws in many other states, formally linked the Farm Bureau and the extension service. The law that funded extension required that the Farm Bureau approve the county extension agent and work with the agent and the university to prepare the annual program and budget. The law also required a minimum number of Farm Bureau members in the county in order to get state money for extension. This ensured that the agent would please the Farm Bureau members and recruit new members when possible.

The university also assisted in the development of other types of institutions. It played a key role in the development of cooperatives, which in return provided political and financial support to the university. During the 1890s, T. L. Haecker, the first faculty member to focus on dairy production, was a major force in the spread of cooperative creameries. In addition to his outstanding scientific achievements, he helped creameries develop a program of quality cream production and provided advice on cooperative management.

University faculty also played an important role in the development of other special interest groups. University professors helped establish the Minnesota Crop Improvement Association. This association multiplies new varieties developed by the university and maintains the certification process that helps the farmer get pure seed. Members of the association include small seed companies and farmers who multiply and sell seed. Faculty and extension personnel have helped to develop commodity research and promotion councils that now finance research through a check-off system. The extension service has played an important role in organizing producer organizations at the county level. These commodity organizations finance research at the university and provide political support for the budget when necessary.

At the federal level, national commodity groups such as the National Association of Wheat Growers seem to be the most important source of political support for research. The American Farm Bureau Federation has also been supportive. In contrast, the National Farmers Union, the National Farmers Organization, and the American Agriculture Movement are more likely to be "suspicious and critical rather than supportive of research" (Hadwiger 1982, 92). Trade associations that lobby for industries like agricultural chemicals seem to have expertise and interest in agricultural research but less political influence than the commodity organizations.

At the state level, commodity groups seem to be the most active supporters of research. It is through these organizations that farmers express themselves most effectively. Processors and industry personnel are sometimes involved in commodity organizations (e.g. the sugarbeet and poultry organizations), but more often they consist solely of farmers. Another important group that actively supports the university includes the Minnesota Crop Improvement Association and the smaller seed companies that do not have their own research programs and thus depend on the university for the varieties they sell.

From the late 1960s to the early 1980s the Agricultural Experiment Station, jointly with the Agricultural Extension Service, mobilized political support from farmers, agribusiness leaders, and homemakers for its proposed program and state budget request in eight to twelve legislative support meetings held annually around the state. These meetings brought together farmers, leaders of commodity organizations, representatives of agribusiness, and members of the legislature who were identified by extension agents and branch station personnel. Fifty to 100 or more people attended each meeting. At each meeting the director of the experiment station and director of the extension service reviewed the program and achievements of the past year, presented their proposed new initiatives for which increased state funding was requested, and then discussed ways of developing the political support for the request. This gave clientele who attended a chance to criticize and make constructive suggestions about the research and extension program and then to develop a strategy for building the political support necessary to secure the appropriations increases. After these meetings the directors invited the prominent participants to St. Paul to lobby for the budget. They also kept these participants informed about the progress of the request during the legislative session and about what they could do to help in a timely way.

The above process, combined with the directors focusing on the members of key legislative subcommittees, worked well. During this period, when state fiscal resources were increasing rapidly, the recurring state appropriation for the Agricultural Experiment Station grew to become the largest in the twelve-state north central region and the sixth largest in the United States.

The Institute of Agriculture, Forestry and Home Economics and all of the branch experiment stations have advisory boards that represent a wide range of interests and help the station develop priorities and generate political support for the institute. The institute's board was organized in 1947. Branch station boards started in 1982. In addition, some departments or special research programs have advisory groups. When a department like agronomy needs a new building, it works with the industry or commodity groups with which it is most closely associated to develop support in the legislature. Discussions with faculty members indicate that poultry producers and seed companies were very effective in lobbying for new research and teaching facilities in recent years.

A number of these same organizations provide financial support as well as political support to the university.

The commodity check-off groups, about ten of which provide money to support research at the experiment station, are a growing source of support. Because of the present surpluses of many commodities, some of these organizations are debating whether they should invest money in research that will increase productivity or whether they should invest money in market development and utilization. This issue was debated by the soybean producers organization during the last two years (as of 1983). The majority of farmers in this organization believe that research is necessary to keep U.S. soybean producers competitive in world markets. The soybean producers provide about $200,000 a year to the university for research. In addition, when the university soybean breeder retired recently, they provided money to hire his replacement a year early so that the two could work together for a year. Other organizations provide similar assistance.

Corporations and cooperatives that supply inputs to farmers also support research and hire faculty members as consultants. Corporations and cooperatives in food processing and marketing also finance research at the experiment station. The Department of Food Science and Nutrition has an advisory board from industry and receives a substantial amount of money from them. Private gifts to the university by private individuals are a relatively new source of funds.

By the early 1980s it became increasingly obvious that efforts were needed to further organize and mobilize political support for state appropriations for the Minnesota Agricultural Experiment Station. Several factors led to this conclusion. First, the agricultural constituency was shrinking in numbers and rural legislators were now in the minority: 55 percent of the legislators now represented the seven-county Twin Cities metropolitan area. Second, state revenue growth had slowed dramatically and there were frequent state budget shortfalls, resulting in greater competition for the state's dollars. Third, there were and are increasing concerns that the Agricultural Experiment Station was not focusing an adequate share of its state resources in certain areas, areas such as the reduction of purchased capital inputs, the improvement of profitability to small to moderate-sized family farms, water quality, soil conservation, and even organic farming. Fourth, the MAES could no longer assume that the present base level of resources would be continued automatically from one biennium to the next; its director was being increasingly challenged to explain and defend the use of present resources rather than focus just on justifying the requested increases.

In 1984, it was decided to alter the statewide political lobbying network in order to involve more key citizens to reach more legislators on their home turf before the legislative session began. At the previous large meetings described earlier, participation by legislators was modest. In collaboration with the institute's advisory council, a plan was developed to hold a small meeting in each of the state's sixty-seven legislative districts (each district has one senator and two representatives for a statewide total of 201 legislators). Each meeting was to be hosted by a local volunteer who was a member of one of the institute's advisory boards or committees, and/or an alumnus of one of the institute's programs, or a citizen who in some way had benefited from the research or extension programs. The district volunteer invited ten to twelve other leading citizens of the district (including farmers, agribusiness people, bankers, and local elected officials) and the senator and two representatives from the district. The meetings were often held over breakfast or coffee in a local restaurant, town hall, or even in the volunteer's home.

The meetings had a three-fold purpose: (1) to create an increased awareness and understanding of the institute's programs; (2) to listen to clientele, seeking their input and constructive criticisms on how programs could be improved; and (3) to review the legislative request for additional funding.

Sixty-six such meetings were actually held. The station director attended all of them, sometimes participating in three or four in a single day. Approximately 150 legislators participated; thus, three-fourths of the legislators received direct input from constituents before the legislative session began.

The district volunteers and some of the other attendees then served as a continuing network of citizens who were kept informed and activated periodically during the 1985 legislative session. The result: the Experiment Station and the Extension Service received one of their largest budget increases ever for the 1985-87 biennium.

The district meeting approach was used again in 1985-86, in a reduced way, focusing largely on key legislative districts (districts represented by legislators who served on the agricultural and appropriations committees). In the 1986 legislative session, amidst a $300 million state shortfall in revenues, the Experiment Station received further increases.

DEVELOPING GREATER SUPPORT FOR THE SYSTEM

The evidence from studies on the returns to research is that greater investment in agricultural research in Minnesota would increase the income and well-being of farmers and consumers. The benefits are sufficiently high to justify the allocation of additional resources for agricultural research.

How should the university mobilize political support for larger agricultural research budgets? The university first needs to answer three questions. First, who are the groups who benefit from research but are not at present providing political support for the Experiment Station? Second, if there are beneficiaries who are not now mobilized, how can the university gain their support? Third, can the present supporters provide more support?

Turning to the first question about potential beneficiaries currently not providing political support, there is evidence that several groups do not provide active political support. Consumers are the one group who economists agree have benefited but who do not effectively support agricultural research. The MAES is aware of this but has been frustrated in attempts to generate support from this group. Part of the problem is that the benefits to an individual consumer are small and consumers as a group are not well organized (Olson 1971). Minnesota residents receive a small fraction of the total benefits from research. In addition, popular books like <u>Hard Tomatoes, Hard Times</u> (Hightower 1973), along with the efforts of farmer commodity groups to increase prices, have created the general impression among consumer activists that the agricultural research system primarily benefits agribusiness and may actually be working against the consumer's nutritional interests.

The most difficult problem the MAES faces in trying to mobilize consumer support is how to present this message to consumers without alienating its major supporters--farmers and their commodity organizations. If consumers are told that agricultural research had led to lower prices for food, this may antagonize farmers who want higher prices for their output. The only possible way out of this dilemma is to educate consumers gradually about the benefits they have received. At the same time, it must be emphasized to farmers that reducing the costs of production through new technology may reduce prices but it also increases the amount of grain that can be sold in foreign markets.

A second latent group that offers potential support is the conservation and ecology movement group. Scientists at the Minnesota Agricultural Experiment Station and College of Forestry have completed a number of studies on ecological and resource conservation problems. Virtually all agricultural research now tries to consider the impact of research on the environment and on soil conservation. Organizing the support of the conservation groups has not been easy. If it is essential for the university to work more closely with large private corporations in the future to get sufficient funding, it may be even more difficult to develop support among both conservation and consumer groups who see corporations as the enemy.

A third group whose support has not been effectively mobilized is large-scale agribusiness--the Cargills and Pioneers. Rather, the university has relied on support from small-scale agribusiness and co-ops. Closer ties with large-scale agribusiness may not be possible without weakening support from the Experiment Station's traditional clientele. The financial resources and political influence of large-scale business and the fact that they have benefited from research suggest that it would be wise for the university to continue trying to tap these resources.

Several groups might support research in the future. First, the Experiment Station is only starting to recognize the importance of the part-time farmers. Their numbers are growing while the number of full-time farmers continues to decline. Part-time farmers undoubtedly have received some benefits from research and their political importance should grow with their numbers. A second group includes farm management companies and cooperatives who provide farm management advice. Their numbers and importance are growing rapidly (Sundquist, Menz, and Neumeyer 1982). Most of them depend on technology developed by the university. They are using computer programs developed in the Department of Plant Pathology for estimating the economic loss from pests to advise farmers on when to use pesticides. It may be that the equivalent of the Minnesota Crop Improvement Association is needed for the distribution of this expertise. At the least, the university needs to develop political support from these companies.

Efforts are currently under way to develop additional, more effective political support from traditional clientele groups. The MAES is seeking to enlist a minimum of three volunteers from each legislative district so that each volunteer has the specific assignment of targeting input to a single legislator. Meetings held in the fall of 1986 focused on training these volunteers and the volunteers then making legislator contacts independently (without university personnel present). The volunteers were then to provide feedback to university administrators and continue to serve as an active network throughout the 1987 legislative session. There have also been discussions with the Institute Advisory Council and the Extension Citizens Advisory Committee about organizing a completely independent support organization funded by private donations and led by the volunteers themselves.

LESSONS FROM MINNESOTA HISTORY

Government research institutions must produce technology, service, and knowledge that are useful to the people who pay the bill. If these institutions are not productive, no amount of publicity or organization will provide long-term funding. If an institution is productive, then there may be some applicable lesson from the MAES experience. The Minnesota experience reviewed here, and the U.S. experience generally, suggests a number of principles that can guide efforts to generate more political support:

1. Research institutions have to convince society that research is important. Thus, they must do research on the needs of the general society, not just on the needs of narrow interest groups who are politically powerful and well organized. The institutions must publicize the way in which research has helped society meet more general goals.
2. Research institutions frequently have to organize groups to provide political and financial support. If there are potential beneficiaries who are not well organized, it may be necessary to help them organize.
3. Research institutions should identify and work with important, well-organized groups of beneficiaries that can provide political and financial support. In Minnesota the small-scale businessman provided financial support to the early extension system. At present, groups such as small seed producers provide political support to the Experiment Station.
4. Leaders of farmer and other groups should be involved in the process of setting research priorities.

This is essential in order to get their support for the program and to make sure that the future research results will be useful to them. Client advisory boards have been useful in setting priorities in Minnesota.

5. Decentralization of the research system has been important for developing appropriate technology, spreading technology, and generating local support in the U.S. research system.

6. Support for research depends on an effective extension service that can spread results and educate farmers about the importance of government expenditure on research. Therefore, research must cultivate close ties with extension. It seems to be advantageous to have them in the same institution.

7. More investment should be made in social science research in LDC research institutions. Social science research is important because it can help identify the impact of past and future research, it can help set research priorities, and it can help develop support for research. Social science research plays an important role in improving the welfare of rural people directly by designing improved institutional arrangements and by helping farmers and other rural institutions operate more efficiently and equitably. Finally, it can assist in the design and management of price and regulatory policies that have an important impact on rural welfare.

NOTES

1. This section is based on Fitzharris 1976.
2. For a detailed breakdown of the impact on yield per acre of various technological improvements in corn production, see Cardwell 1982, pp. 984-90.
3. This paragraph is based on Cardwell 1975.

REFERENCES

Block, William J. Separation of the Farm Bureau and the Extension Service. Urbana: University of Illinois Press, 1960.

Bonnen, James T. "U.S. Agricultural Development: Transforming Human Capital, Technology and Institutions." In U.S.-Mexico Relations: Agriculture and Rural Development, edited by Bruce F. Johnston et al., 267-300. Stanford, Calif.: Stanford University Press, 1987.

Bredahl, Maury, and Willis Peterson. "The Productivity and Allocation of Research: U.S. Agricultural Experiment Stations." American Journal of Agricultural Economics 58(November 1976): 684-92.

Cardwell, V.B. "Care for the Animals" (pp. 27-28) and "20th Century Animal Science: More Food Per Animal" (pp. 40-42). Minnesota Science, Special Centennial Issue, 31(1) Spring 1975. (A publication of the Minnesota Agricultural Experiment Station, University of Minnesota, St. Paul.)

________. "Fifty Years of Minnesota Corn Production: Sources of Yield Increase." Agronomy Journal 74(November-December 1982): 984-90.

Fitzharris, Joseph. C. "Public Institutions and Agricultural Growth: The Minnesota Experience." Department of Agricultural and Applied Economics Staff Paper P76-22, University of Minnesota, June 1976.

Hadwiger, Don. The Politics of Agricultural Research. Lincoln: University of Nebraska Press, 1982.

Hightower, Jim. Hard Tomatoes, Hard Times. Cambridge, Mass: Schenkman Publishing Company, 1973.

Olson, Mancur. The Logic of Collective Action: Public Goods and the Theory of Groups. Cambridge, Mass.: Harvard University Press, 1971.

Peterson, Willis, and Joseph C. Fitzharris. "Organization and Productivity of the Federal-State Research System in the United States." In Resource Allocation and Productivity in National and International Agricultural Research, edited by Thomas M. Arndt, Dana G. Dalrymple, and Vernon W. Ruttan. Minneapolis: University of Minnesota Press, 1977.

Sundquist, W. Burt, Kenneth M. Menz, and Catherine F. Neumeyer. A Technology Assessment of Commercial Corn Production in the United States. Agricultural Experiment Station Bulletin 546, University of Minnesota, 1982.

True, Alfred C. A History of Agricultural Extension Work in the United States 1785-1923. USDA Misc. Publication No. 15. Washington, D.C.: Government Printing Office, 1928.

8
Anticipating Advances in Crop Technology*

Gary H. Heichel

The knowledge that fragile physical and chemical resources may pose significant constraints to the future productivity of U.S. cropland enhances the need to understand which crop technologies might lead to tomorrow's yield improvements. Furthermore, the current profitability crisis afflicting U.S. agriculture and the noncompetitiveness of many U.S. agricultural commodities on world markets heighten the need to develop research policy initiatives that will improve the efficient direction of technology development in public research institutions to enhance agricultural productivity.

Predictions of future productivity have usually been made from historical trends of yields (Evans 1980), by analyzing the contributions of weather and technology to crop yields (Thompson 1975), by analyzing the separate contributions of genetics and of management to productivity of specific crops (Duvick 1977; Jensen 1978), by analyzing the responses of crop yield to a historical series of increments in fertilization (National Academy of Sciences 1975), from measurements of the rate of growth of per-hectare yields for specific crops (Farrell 1981), and from technology assessments (Sundquist et al. 1982; Heichel 1984). Evidence is available to support the interpretation of either a gradual deceleration of productivity or, conversely, continually increasing productivity.

A feature common to all these analyses of future productivity is retrospective analysis of past yields or scientific breakthroughs in an effort to predict the future. Little attention in public institutions has been focused on tracking advances in specific crop technologies to facilitate identification of scientific problems that need solution before a new or improved crop technology is fully developed and delivered to the end-user. My thesis is that the time has come for a shift from predictions of productivity advances based on retrospection to predictions based on scientifically realistic expectation or anticipation. I will use the chronological development of scientific advances in a specific crop technology, biological nitrogen fixation, to exemplify the principle, although the methodological analysis could be applicable to any crop technology.

A DELIVERY SYSTEM FOR IMPROVING CROP PRODUCTIVITY

Faced with limited financial and human resources, research managers and administrators must decide the relative proportions of funds to allocate to applied research (activities producing a steady yield of solutions to problems for specific constituents or society-at-large), and basic research (activities advancing the frontiers of scientific knowledge while addressing mission or problem-oriented goals). In contrast to applied research, basic research is autocatalytic in the sense that knowledge (capital) is reinvested at a compound interest rate so that over time growth occurs in the framework of knowledge upon which solutions to applied problems are based (Heichel 1985).

The delivery system for crop technologies developed in public institutions generally conforms to a five-tier pyramidal model (figure 8.1). Traditional scientific disciplines (tier 1), which may or may not conform to institutional management units, form the base. Notice that "genetic engineering" and related biotechnologies are not categorized as a separate new discipline since they are a family of new research techniques or tools potentially applicable to all of the existing biological disciplines involved in crop technology research. Since genetic engineering and other biotechnologies amplify the effectiveness of research in the traditional disciplines, the capability for this type of research activity must be built into existing institutional management units or multidisciplinary research teams rather than into separate management units. This tactic keeps the biotechnologist at the cutting edge of knowledge in the parent discipline and

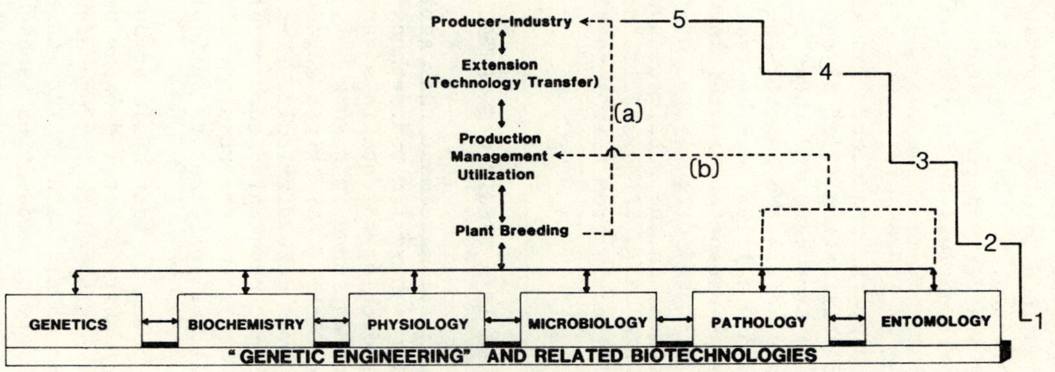

Figure 8.1. A five-tiered pyramidal model for the utilization of basic information from scientific disciplines in plant breeding and the development of new crop technology for producers and industry (adapted from Heichel 1984).

assures constant exposure to critical problems in the parent disciplines that need solution (Heichel 1985).

The science of plant breeding (tier 2) involves the recombination of desirable genes and application of new knowledge to the development of improved crop cultivars or germplasm. There is little doubt that traditional plant breeding will continue to play a pivotal role in delivering new crop technologies, whether the desirable genes are "engineered" or are derived from genetic selection in exotic plant introductions or in plant-breeding populations.

Development of a new cultivar is still several steps removed from use by the producer. Research on production, management, and utilization (tier 3) and the transfer of this knowledge to producers and industry (tier 4) are usually necessary before the genetic and economic potentials of a new cultivar can be realized in the field (tier 5). General exceptions to this pyramidal model occur when germplasm released directly to industry without cultivar development in the public sector moves directly from tier 2 to tier 5 (a, figure 8.1), or when pest control practices move directly from tier 1 to tier 3 (b, fig. 8.1).

TRACKING ADVANCES IN A SPECIFIC CROP TECHNOLOGY

There is general agreement that administrators and managers of science have limited methods for objectively tracking and predicting scientific advances <u>within</u> specific disciplines such as those in figure 8.1. Objectively tracking and predicting scientific advances <u>across</u> several disciplines is even more difficult. Yet this can be done for specific crop technologies such as biological nitrogen fixation (fig. 8.2), and the results illustrate several points potentially significant to agricultural science policy.

The first point is that basic advances in a crop technology may require several decades before the applied technology is transferred to the producer. In the case of biological nitrogen fixation, thirty-three years elapsed after the discovery of nodulation genes before an improvement in nitrogen fixation technology (a unique crop cultivar in this case) was available to the producer.

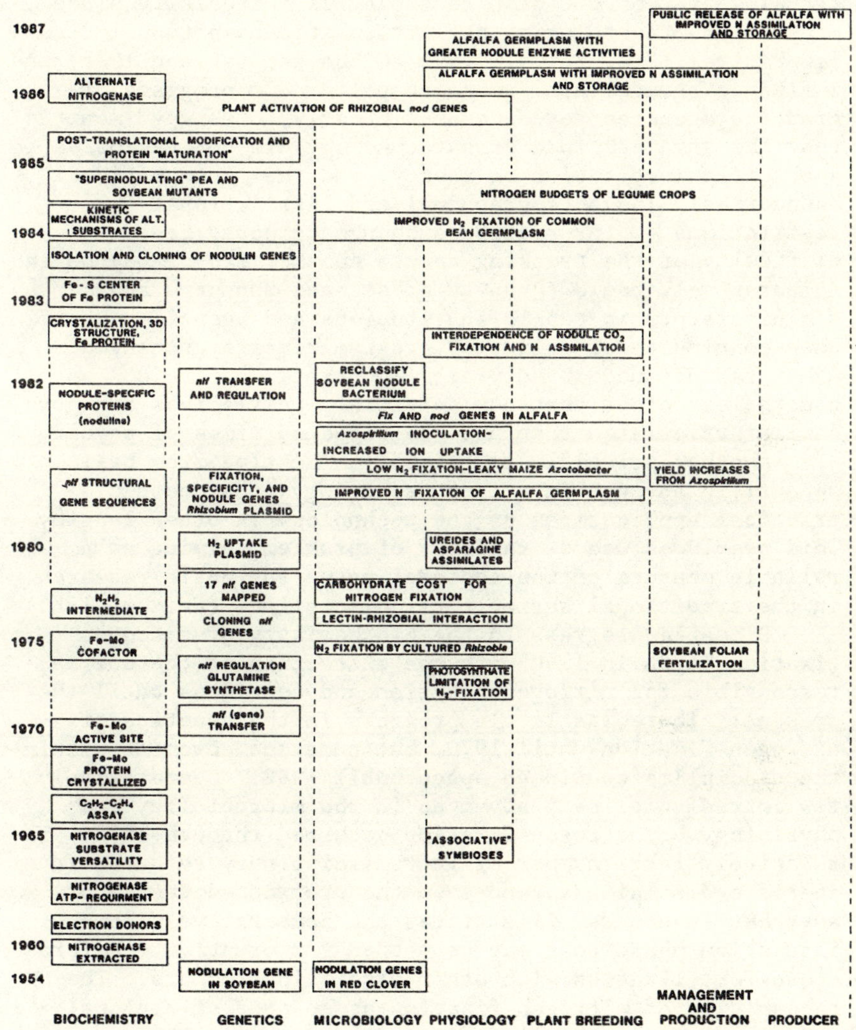

Figure 8.2. Chronological development, intra-, and interdisciplinary sources of advances in nitrogen fixation research and technology development since 1954. The designation of the discovery of nodulation genes in red clover as an advance in microbiology is equivocal since the breakthrough was made by a microbiologist who independently gained sufficient knowledge of plant genetics to conduct the critical experiments. (Adapted from Hardy et al. (1983) and Heichel (1985)).

It is difficult to understand why public institutions have not yet assumed a leadership role in developing such tracking systems for crop technologies, especially since many of the technologies are articulated as national imperatives (Brown et al. 1975; Gibbs and Carlson 1985) and influence agricultural research policy and programs of granting agencies for decades. Clearly, one impediment is that the advances in a crop technology are not exclusive to a specific institution or agency, and they span national boundaries. These geographical and institutional restrictions on information exchange compound the difficulty of the tracking task. Another factor may be the comparatively rapid turnover of science managers and administrators in public institutions and agencies, even when continuity of policy or program remains unchanged. Other factors undoubtedly are important, but it is nonetheless clear that new scientific initiatives are more easily articulated than are the expected times of payoff.

Another principle is evident. The time from basic scientific advance in a crop technology to successful practical applications of the technology is often lengthy. This precludes use of rapidity of problem solving as a reliable measure of the societal value for basic research in the traditional agricultural disciplines (Heichel 1985).

Dramatic progress in the biochemistry of nitrogen fixation began in 1960 with the extraction of the enzyme responsible for nitrogen fixation and continues until the present. There was little progress in the genetics of nitrogen fixation until 1970, but numerous advances within the discipline continued apace until 1982. Comparatively few intradisciplinary advances in the microbiology and physiology of nitrogen fixation occurred through 1974. The relatively large number of intradisciplinary advances in specific disciplines, and even the protracted dearth of advances in others, illustrates the comparatively long incubation phase that may be needed for creativity before cross-fertilization with other disciplines occurs. The recognition of nitrogen fixation as an area of national priority for extramural funding (Brown et al. 1975) undoubtedly stimulated many recent intra- and interdisciplinary advances.

Since 1975, much new knowledge has derived from interdisciplinary efforts, first between microbiologists and physiologists, then between microbiologists, physiologists, and plant breeders, and almost simultaneously between groups of biochemists and geneticists, groups of geneticists, microbiologists, and physiologists, and groups of physiologists, plant breeders, and management or production scientists. The key advances of the first twenty years of nitrogen fixation research were intradisciplinary, while most of the key advances in the last thirteen years have resulted from interdisciplinary collaboration. In 1987, the peak of the research pyramid will be reached by the first of presumably many successive producer-related achievements that will provide strategies to enhance the managerial skills and economic viability of producers, preserve the land base while increasing its productivity, increase the efficient use of climatic resources, and enhance the genetic endowments of plant and microbial germplasm.

Several conclusions from the tracking scheme in figure 8.2 are justified. If the mission of research on a specific crop technology, e.g., biological nitrogen fixation, is to deliver an application of societal importance, the initial application will soon be achieved. The direction of research activity indicates that many more applications are likely in the future. Even policy makers without scientific background in a specific crop technology can examine figure 8.2 and readily accept the probability of successful application within the next five years. Finally, the interdisciplinary advances among biochemists, geneticists, microbiologists, and physiologists since 1984 have resulted from application of the tools of "genetic engineering" and related biotechnologies that were first successfully applied to plants in the late 1970s. This reinforces the argument advanced earlier (Heichel 1985) that the capabilities for this research should be built into existing institutional units for maximum effectiveness.

SUMMARY

In contrast to the more traditional approach of retrospective yield analysis to predict future gains in crop productivity, development of tracking systems for specific crop technologies such as the example in figure 8.2 offer the potential of identifying the need for interdisciplinary teamwork instead of maintaining intradisciplinary parochialism, and offer a predictive model for the time when new or modified crop technologies will achieve application. This "tracking capability," or an equally effective system of monitoring scientific advances, is imperative if the new biotechnologies are going to amplify effectively the inherent power of traditional disciplines to solve pressing problems of productivity, profitability, marketability, and environmental stewardship.

NOTES

*Joint contribution of USDA-ARS and the Minnesota Agricultural Experiment Station. Scientific Journal Series No. 15,150.

REFERENCES

Brown, A. W. A., T. C. Byerly, M. Gibbs, and A. San Pietro, eds. Crop Productivity--Research Imperatives. Proceedings of an international conference sponsored by Michigan State University Agricultural Experiment Station and the Charles F. Kettering Foundation, Boyne Highlands, Michigan, Oct. 20-24, 1975, p. 399.

Duvick, D. N. "Genetic Rates of Gain in Hybrid Maize Yields During the Past 40 Years." Maydica 22 (1977): 187-96.

Evans, L. T. "The Natural History of Crop Yield." American Scientist 68(1980): 388-97.

Farrell, K. R. Productivity in U.S. Agriculture. ESS Staff Report AGESS-810422, Washington, D.C., 1981, p. 29.

Gibbs, M., and C. Carlson, eds. Crop Productivity--Research Imperatives Revisited. Proceedings of an international conference held at Boyne Highlands, Michigan, Oct. 13-18, 1985, and Airlie House, Virginia, Dec. 11-13, 1985, p. 304.

Hardy, R. W. F., P. G. Heytler, and R. M. Rainbird. "Status of Novel Nitrogen Inputs." In Better Crops for Food, edited by J. Nugent and M. O'Connor, pp. 28-48. London: Pitman Publishers, 1983.

Heichel, G. H. "Anticipating Productivity Responses to Crop Technologies." In Future Agricultural Technology and Resource Conservation, edited by B. C. English, J. A. Matezold, B. R. Holding, and E. O. Heady. Ames: Iowa State University Press, 1984.

________. "Molecular Genetics, Biotechnology, and Experiment Station Research." In Proceedings, Modules I and II, Agricultural Research Policy Seminar, edited by F. Hoefer, C. Pray, and V. W. Ruttan. Minneapolis: University of Minnesota Agricultural Extension Service, St. Paul, 1985.

Jensen, N. F. "Limits to Growth in World Food Production." Science 201 (1978): 317-20.

National Academy of Sciences. Agricultural Production Efficiency. National Research Council, Washington, D. C., 1975.

Sundquist, W. B., K. M. Menz, and C. F. Neumeyer. _A Technology Assessment of Commercial Corn Production in the United States_. University of Minnesota Agricultural Experiment Station Bulletin 546, 1982, p. 158.

Thompson, L. M. "Weather Variability, Climate Change, and Grain Production." _Science_ 188 (1975): 535-41.

PART 3

RESEARCH ORGANIZATION AND REFORM IN DEVELOPING COUNTRIES

INTRODUCTION

Carl E. Pray

This section deals with three major issues that concern research administrators in developing countries. The first issue is how to mobilize the resources needed for an effective research program. The second issue is how to allocate research resources efficiently between different research topics. The third issue is to determine what types of institutions generate stable funding and efficient research. The final paper in this section deals with how research institutions can be evaluated.

In many countries in the Third World, government research institutions are facing serious funding problems. Despite the growth of research expenditures in most developing countries during the 1960s and 1970s, documented by Judd, Boyce, and Evenson (chapter 1), research expenditure per dollar of agricultural output was quite low in many countries. Even during the 1970s research expenditure declined in some countries in Latin America. Since 1980 real expenditure in many of the countries has declined as budgetary pressure on governments grew and some of the worst food problems were solved temporarily. A number of studies indicate that investment in agricultural research is generating rates of high returns. By reducing funds for agricultural research, these countries are depriving themselves of an important source of growth.

The paper by Eliseu Alves deals with the issue of how to mobilize the political support needed to fund a public agricultural research system. He draws on his experience as director of the Brazilian research system, EMBRAPA. EMBRAPA was able to mobilize general support by choosing research topics that were important to the Brazilian people and then publicizing their successes on these topics.

EMBRAPA also invested resources in economists who could speak the language of the technocrats in government who were important in deciding how much money EMBRAPA would receive. These tactics allowed the EMBRAPA budget to grow during a period when many other research programs in Latin America were losing support.

A second major issue facing all research administrators is how to allocate research resources between competing research programs. The Binswanger-Pingali and Harwood papers suggest that many countries have not allocated research resources efficiently. The Binswanger-Pingali paper starts with the induced innovation framework of Hayami and Ruttan that suggests that research priorities should be based in part on resource scarcity. In the African case, Binswanger suggests that in the past too many research resources have been devoted to yield-increasing technologies rather than to labor-saving technologies.

Harwood indicates that if research systems want to assist resource-poor farmers they need to work on low-input technologies that do not strain poor farmers' financial resources. He argues that by using a farming systems approach researchers can develop technologies that will increase production without upsetting the ecological balance present in traditional agriculture. This will increase farmers' incomes without requiring as many expensive industrial inputs--such as chemical fertilizers or pesticides.

The funding and planning problems of developing country research programs raise the question of whether there is some optimal organizational structure for research institutions. This optimal structure generates stable funding and an efficient research structure. The absence of stable funding has given rise to considerable experimentation with various institutional structures for agricultural research in the Caribbean, Latin America, and Africa. Eduardo Trigo approaches this problem by first developing a taxonomy of research institutions. He then examines these institutions in the national context and concludes that there is no ideal research structure for all countries. He concludes that the research institutions must be carefully shaped to fit the local political structure of the country.

The paper by Idachaba is a case study on Nigeria that brings together the resource mobilization, research planning, and institutional development issues in one country. He indicates how institutional constraints, political instability, and the availability of oil have led to excessive centralization of funding and management. A series of reorganizations has kept the system disorganized and has prevented the Nigerian research system from making a breakthrough in food production.

An important issue for people outside the agricultural research system, such as someone from a Ministry of Science and Technology or from donor agencies or an international organization, is how to evaluate the effectiveness of a research system. The paper by Elliott represents a very useful step in developing a methodology for such an evaluation. It provides a way of collecting and organizing the important events in this history of a research system that any evaluator of such a system would need.

9
Agricultural Research Organization in the Developing World: Diversity and Evolution*

Eduardo J. Trigo

Successful agricultural research systems require mutually reinforcing interactions among three groups of variables: the policy environment, the system's organizational structure, and a set of basic operational processes. These operational processes include the setting of objectives and priorities, resource acquisition, and development--including the development of a critical mass of experienced scientists, program development, and the establishment of adequate scientific linkages. The linkages monitor and evaluate program implementation and assure the flow of information between research and extension workers, farmers, policymakers, and the public. Within this three-sided perspective the system's organizational structure provides the framework that links research and the broader social, political, and economic environment, and conditions the implementation of the system's basic operational processes and thus the actual research activities.

Recognizing the importance of the organizational structure, however, does not imply that any particular format is better than the alternatives in all circumstances. Information from agricultural research (and other fields of activity) shows that there is no one optimum method of organizing a system; a country's agricultural conditions, history, economic characteristics, and socio-political traditions play a key role in shaping the optimum organizational structure. Even within a country, the most effective way to organize research activities will change through time as social, economic, and political conditions change.

This paper attempts to summarize the ways in which national agricultural research systems in the developing world are organized and examines some of the aspects that may have affected their characteristics and evolution. It is divided into six sections. The first section considers the nature of the basic organizational options currently found at the level of the national agricultural research system and their extent in the different developing regions. The next three sections consider some of the main organizational trends in Asia, Latin America, and Africa. The fifth section tries to point to some of the commonalities and differentiating elements in these trends. Finally, in the last section the main aspects are summarized, and areas for future work are highlighted.

THE BASIC ORGANIZATIONAL OPTIONS

The organizational structure comprises the durable organizational arrangements through which responsibilities and authority are distributed, and the reporting relationships among the different organizational components. These relationships correspond to the patterns for division of labor--single versus multicommodity, basic versus applied research, research and extension--and coordination among the different units responsible for research. The organizational structure also includes the channels for interaction with the system's environment, which reflect the system's guidance and input mechanisms.

The analysis may begin either at the level of the overall system or at the level of the individual organization. Specific descriptive variables at the system level are the types and numbers of organizations that perform research (degree of decentralization); their mandates (scope of work); their governance and resource acquisition mechanisms and the degree of control they allow over decision-making in regard to operational policies and resource management matters (degree of autonomy); and the patterns they follow in working with each other and with other relevant nonresearch organizations (planning/ coordination and resource allocation mechanisms).

At the level of the individual organization, governance and resource acquisition mechanisms are the main differentiating characteristics. Using these as typological variables, agricultural research organizations can be summarized in five basic organizational types.[1]

The Ministry Model. Research is organized in one or more line departments within the bureaucratic structure of a ministry.[2] The basic feature of this format is that the unit responsible for research has a low degree of control over decision-making, particularly in matters concerning resource management. Funding usually flows from allocations within the national budget through the ministry treasury, and administrative policies and procedures are subordinated to those of the ministry. Mandates, both in product and functional terms, are highly variable. Research and extension functions are usually located in separate units. There is no predominant base with respect to the product scope.

The Autonomous or Semiautonomous Institute. Research responsibilities are placed within an administratively independent organization. The basic characteristic of this format is a high level of control over decision-making with respect to program and administrative policy and resource allocation matters, which is usually exercised through an independent board of directors or governors. At the funding level, the autonomy allows the existence of an independent treasury, which increases research management control over fund administration. Funding flows as a special budget line within the national budget, and in some cases funds are directly tied to specific sources of revenue (a cess on sales of given crops, export revenues, etc.). As with the ministry model, functional and product mandates are variable. The first experiences with autonomous research institutions were with single commodities. More recently, however, the broad-mandate national research institute type has become quite widespread.[3]

The University Model. Research is closely integrated with education within a university context. Extension activities may or may not be part of the same structure. However, the crucial feature of this approach is the integration of applied research activities oriented to technology generation within the educational environment. Because of this very characteristic of the university structure, this model has a high degree of autonomy and decentralization. Funding flows through a variety of mechanisms from both public--national, state, or provincial--and private sources.[4]

The Agricultural Research Council (ARC). The agricultural research council model represents a variant of the autonomous research organization, emphasizing the coordination function rather than the direct implementation of research activities. Several different organizational arrangements are usually included under the general concept of the ARC. Autonomy and a high level of control over program policy matters, through an independent board of directors or governors, are the key distinguishing features of the council model. However, specific functions assigned to them range from those of merely a review and advisory role to responsibility for the consolidation of budgets for all government-sponsored research, funding specific research projects, and even directly implementing research. From the point of view of mandate, the ARC almost invariably has a wide national scope of work and concentrates solely on research activities.

Private Sector Research Organizations. The basic characteristics of research organizations operating in the private sector domain are highly specific and concentrated mandates with program policy subordinate to that of the parent organization. There are two basic variations of private sector involvement in agricultural research: (1) research departments of the firms producing technological inputs such as seeds, agrochemicals, fertilizers, farm machinery, and veterinary products; and (2) crop-specific research associated with agricultural producer associations. Autonomy in program and administrative matters tends to be low in the first type, particularly at the applied-adaptive end of the research scale, where research efforts are usually directly integrated into the firms' overall production and marketing strategies. In producer associations, the second type, there is a greater similarity to the autonomous commodity institutes facing comparable conditions.

The types of research organizations described above should be considered in terms of "ideal types"; they are very seldom found in isolation as pure forms. At the national level, it is usual for different types of research organizations to coexist. In such cases, the number and type of different organizations that conduct research, and the coordination patterns and mechanisms among them, become the important differentiating features among systems. Two basic types of system can be envisaged: single-organization systems, where most research activities are carried out within one organization; and multiorganizational systems, where a variety of different organizations perform research activities. The first type is generally directed from a ministry of an autonomous research institute with a broad mandate. In the multiorganizational situation, the most important differentiating element is the existence or not of formal coordination mechanisms. Agricultural research councils are characteristic of the multiorganizational framework with formal coordination mechanisms.

The current organizational formats of National Agricultural Research Systems (NARS) for the majority of the countries in Asia and the South Pacific, West Asia and North Africa, Africa South of the Sahara, Latin America and the Caribbean are summarized in table 9.1.[5] One highlight shown by the table is that no organizational format can be said to be predominant throughout the developing world; on the contrary, much "variability" exists both within and across geographical regions. A cross-regional analysis highlights two aspects: (1) the concentration of the model with formal coordination of research activity (ARC) in the Asian countries; and, (2) the concentration of the autonomous or semiautonomous national institute model in Latin America. The ministry model (without considering how many ministries are involved, and allowing for some autonomous--mainly commodity-specific--research activities) seems to be present in all three regions; however, it is more common in Africa. In Asia, the South Pacific, Latin America, and the Caribbean there seems to be an association between the size of the country and the prevailing model; the smaller countries tend to carry out research within ministerial structures.

In the next section we discuss the characteristics of these main organizational forms by region and their evolution over the last twenty to twenty-five years.

THE ASIAN AGRICULTURAL RESEARCH COUNCILS (ARC'S)

The ARC model has emerged as one of the main features of agricultural research organization in the Asian continent over the last twenty to twenty-five years (Moseman 1971; Drilon 1977).

Historically, ARCs have emerged in response to situations characterized, on the one hand, by a complex network of institutions with overlapping mandates, lack of skilled personnel and scientific critical mass in key organizations, unstable funding levels unrelated to organizational needs, neglect of important research areas, and inadequate responsiveness to national needs as determined by policymakers; and on the other hand, by an agricultural or food situation severe enough to induce the

TABLE 9.1
Organizational Structure of the National Agricultural Research Systems in 81 Countries of the Developing World

	1*	2	3	4		1	2	3	4
ASIA AND THE SOUTH PACIFIC					AFRICA SOUTH OF THE SAHARA (continued)				
Bangladesh				X					
Burma	X				Mozambique			X	
India				X	Niger	X			
Indonesia	X				Nigeria				X
Malaysia		X			Rwanda				X
Nepal	X				Kenya	X			
Pakistan				X	Ghana			X	
Philippines				X	Senegal			X	
South Korea	X				Sierra Leone	X			
Sri Lanka			X		Sudan		X		
Thailand	X				Swaziland	X			
Fiji	X				Tanzania			X	
Papua New Guinea	X				Zaire			X	
Solomons	X				Zambia	X			
Tonga	X				Zimbabwe	X			
Western Samoa	X				Somalia	X			
AFRICA SOUTH OF THE SAHARA					LATIN AMERICA AND THE CARIBBEAN				
Benin	X				Argentina		X		
Botswana	X				Bolivia		X		
Burkina Faso			X		Brazil				X
Burundi				X	Chile		X		
Cameroon	X				Colombia		X		
Cape Verde		X			Costa Rica	X			
Central Africa				X	Ecuador		X		
Chad	X				El Salvador	X			
Ethiopia		X			Guatemala		X		
Gambia	X				Honduras		X		
Ivory Coast			X		Mexico		X		
Lesotho	X				Nicaragua	X			
Madagascar		X			Panama		X		
Malawi	X				Paraguay	X			
Mali	X				Peru		X		
Mauritania			X		Guyana	X			
Mauritius			X		Belize				X

(Continued)

TABLE 9.1 (Cont.)

	1*	2	3	4		1	2	3	4
LATIN AMERICA AND THE CARIBBEAN (continued)					WEST ASIA AND NORTH AFRICA				
					Algeria			X	
Peru		X			Cyprus	X			
Uruguay	X				Egypt			X	
Venezuela		X			Morocco			X	
Barbados	X				Syria	X			
Cuba				X	Tunisia			X	
Dominican Republic		X			Turkey	X			
Jamaica	X								
Haiti	X								
Trinidad & Tobago	X								

*Key: Types of NARS

1. Research carried out predominantly by ministries (one or more; there may be one or more autonomous efforts, restricted to specific crops).
2. Research carried out predominantly by an autonomous or semi-autonomous agency with a broad mandate, both in commodity and territorial terms (there may also be one or more single-crop efforts and some research at universities).
3. Research is carried out by several different entities: ministries, autonomous and/or semiautonomous agencies, universities, without the existence of a central coordinating authority.
4. Research is carried out in a multi-organizational situation with a central coordinating body (Agricultural Research Council).

SOURCE: Elaborated by the author on the basis of primary and secondary information available at ISNAR.

government to attempt to bring agricultural research under control (Ruttan 1982, chap. 4).

The particular characteristics and powers vested in the ARCs vary, but as indicated in the previous section, coordination and planning functions constitute the foundation of the research council idea. Specific functions may be:

- review and advisory role in regard to the program and projects of other organizations
- responsibility for developing a long-term research plan
- preparation of a consolidated research budget

for approval by the government
- financing, monitoring, and evaluation of research projects of national interest out of own funds
- final decision on the allocation of all agricultural research funds among executing agencies
- responsibility for coordinating training for agricultural research
- responsibility for coordinating external technical and scientific assistance in agricultural research
- responsibility for coordinating external financial assistance in agricultural research

In terms of legal status, ARCs are autonomous organizations, with full powers to set administrative policies and procedures. The highest authority is the board of directors/trustees, whose members are chosen, by legal requirement, often according to their role as appropriate representatives of particular institutions or interest groups. They usually operate with an executive office/secretariat that includes permanent technical staff and is complemented by ad hoc members from other organizations in the system mobilized for specific tasks.

Following the creation of the Indian Council of Agricultural Research (ICAR) in 1964, a number of councils were created: in Pakistan, the Pakistan Agricultural Research Council (PARC, 1964); in the Philippines, the Philippine Council for Agricultural and Resources Research Development (PCARRD, 1972); and in Bangladesh, the Bangladesh Agricultural Research Council (BARC, 1973).

In addition to these, the Malaysian Agricultural Research and Development Institute (MARDI) and the Indonesian Agency for Agricultural Research and Development (AARD) are frequently mentioned as having the ARC's basic characteristics. They differ substantially from the "model," however, since their central mandate is to implement research activities, and their coordination function is quite limited. For example, AARD exercises no control or coordination over what happens in research outside the Ministry of Agriculture in the National Science Department Board and the Ministry of Research and Technology. Furthermore, the degree of autonomy of AARD is limited, and it does not escape the ministerial structure in administrative and personnel policies (ISNAR 1981b). MARDI is an autonomous body with a growing board that has participation from both the private and public sectors.

But, its functions do not include the coordination of research activities outside the program it implements directly (Hasim 1981).

Each of the aforementioned ARCs (ICAR, PARC, BARC, and PCARRD) constitutes the legal apex of the national agricultural research systems in their respective countries. However, they have varying degrees of formal and de facto power and involvement in research activities per se.

Beyond this there is a tendency to move away from being a body with merely coordinating and advisory powers to one with greater directional, executive control over the actual implementation of the research program. The force behind this trend appears to be the increasing conviction that without at least partial control over funding and the capacity to actually implement certain strategic components of the research program, the coordination function cannot be properly performed.

This trend is clearly present in the Indian case, where the very creation of ICAR in its modern concept in 1964 corresponded to the desire to transform its predecessor organization, the Imperial (later Indian) Council of Agricultural Research, established in 1929, into a more effective coordinating mechanism. In its pre-1964 conception, ICAR did not operate or control any research facilities and was restricted mostly to making ad hoc grants to the various institutes, ministries, and other research organizations. Under those conditions, ICAR coordination functions were severely restricted. The changes introduced in 1964 included the transfer of control of the commodity research institutes and the central research institutes previously under the Department of Agriculture or the Department of Food to ICAR. An additional institutional innovation was the creation of the Coordinated Crop Improvement Programs as the basic instrument for coordinating the research activities in the country's priority crops at the state level (Ruttan 1982, chap. 4; Jain 1984).

In its new--and present--format, ICAR brings together two functions. At one extreme ICAR has a self-contained "agricultural research institute" and implements its own programs through its own research infrastructure. At the other, ICAR is intended to mobilize the entire Indian research capacity and acts as the main linkage between the Ministry of Agriculture, the body responsible to Parliament for the agricultural development effort, and the research community of the states and the agricultural university system. Within this context, the autonomous nature of ICAR has allowed the creation of separate conditions of service for its personnel and the flexible management style necessary for successful research. Accountability is assured through its special relationship with the Ministry of Agriculture and the composition of the board.

The pattern of development of the other councils mentioned has been similar to that of ICAR. However, the degree of control they exercise over their respective countries' research activities varies.

The council with broader powers in these terms seems to be PCARRD (originally the Philippine Council for Agricultural Research--PCAR) in the Philippines. According to its constitution, PCARRD functions cover a wide field that includes, among others, the development of objectives and definition of goals for research, the development of a national agriculture and resources program, the establishment of priorities, the development and implementation of a fund-generating strategy, and programming. It also allocates all government revenues earmarked for research and controls the incentive mechanisms for researchers and, since 1977, relationships with international funding agencies and technical assistance organizations. The establishment, support, and management of a national network of centers of excellence for the various research programs in crops, livestock, forestry, fisheries, soils and water, mineral resources, and socioeconomic research related to agriculture and natural resources are also functions formally assigned to PCARRD (Drilon and Librero 1981).

To implement its coordination function, PCARRD has the power to review all research proposals in agriculture and natural resources, and to recommend research proposals to the Ministry of the Budget for funding. This power was recently bolstered by a policy of the Ministry of the Budget that only research proposals recommended by PCARRD would be eligible for government funding.

The functions of PARC in Pakistan and BARC in Bangladesh are somewhat more restricted in terms of actual control over the research infrastructure and stay within the coordinating role. However, over the last few years both have gradually increased their powers (Drilon 1977). In 1978, following a catastrophic wheat crop failure (caused by yellow rust), PARC was reorganized into an autonomous body with representation from various provincial and national sectors, and with a subcommittee of the council designated as the Executive Board. The strengthening continued throughout 1981, when a World Bank credit was made available for the development of PARC headquarters, as well as for the expansion and completion of the National Agricultural Research Center (NARC) facilities. The Pakistan Agricultural Research Council Ordinance of 1981 acknowledged the administrative and institutional advances made by PARC so far, with what could be construed as an enlargement of the mandate. Fully autonomous PARC employees were then placed outside civil service regulations.

In Bangladesh a number of decrees, starting in 1976 and 1979, placed practically all research activities legally under BARC. However, a number of the research institutes retained control over their own sources of funding and their administrative councils (such as in the case of BARI, the Bangladesh Agricultural Research Institute).

The trend toward the existing ARCs has continued. In addition, a number of countries are moving toward the creation of similar structures. One example is that of Sri Lanka, where plans and specific proposals are advanced and already at the project preparation stage. Here the intention is to create a coordinating body to facilitate priority setting and coordination among the commodity institutes, units within ministries, and universities currently involved in research activities (ISNAR 1984c).

The Sri Lankan experience represents an interesting summary of the ARC idea and evolution. The reorganization presently being discussed arises out of a preoccupation with the state of dispersal in agricultural research activities and the difficulty of integrating the present research effort, particularly in those areas that fall between the jurisdiction of different ministries. The situation is similar to that encountered in neighboring countries when they initially established their ARCs; the response is also similar, favoring coordination and planning functions rather than direct control over research infrastructures and funding. What remains to be seen is whether the Sri Lankan coordinating body will stay as it is or will move toward an increase in control and executive powers.

THE LATIN AMERICAN NATIONAL AGRICULTURAL RESEARCH INSTITUTES

National Agricultural Research Systems in Latin America and the Caribbean clearly fall within two main forms of organizational structure: the ministry model and the autonomous or semi-autonomous research institute with broad national mandate (Trigo, Piñeiro, and Ardila 1982). These two models cover, in practice, the entire region (table 9.1). There seems to be a correlation between country size and the type of system: all the larger countries have national research institutes, while the ministry structure usually appears in the smaller countries of South America, Central America, and the Caribbean Islands. However, it is necessary to highlight a number of national institutes in countries such as Panama and Honduras, which clearly fall within the small-country category. Moreover, in a number of other countries, such as the Dominican Republic and Guyana, there have been recent developments toward the creation of national institutes (ISNAR 1982a, 1983).

An important feature of the Latin American experience, however, is that these two forms of organization cannot be seen as alternatives since, almost without exception, the creation of the national institute has followed and replaced a structure of research based in the ministry of agriculture.

The early agricultural research efforts in most Latin American countries developed on an ad hoc basis under a number of different, and often unstable, institutional arrangements. The initial experiment stations were usually developed as isolated efforts linked, in some instances, to ministries of agriculture or to their predecessors in the administrative structure (such as in the case of Pergamino and other experiment stations in Argentina); to agriculture schools (such as Palmira in Colombia); or to agricultural producer organizations (such as La Platina in Chile and Cañete in Peru). During the 1940s and the early 1950s these initial undertakings were streamlined, and essentially all research activities, with the sole exception of some export crop cases, such as coffee in Colombia, were centralized as line activities of varying hierarchy within the ministries of agriculture. This was the predominant institutional model in the mid-1950s (Elgueta 1967; Marzocca 1967).

This form of research organization soon came under attack. The criticisms stemmed mainly from the ministries' essentially bureaucratic nature. Some of the most commonly expressed deficiencies were the lack of stable budgetary support; poor expression of the problems and priorities of the producers; lack of coordination of efforts; inadequate communication between researchers, on the one hand, and technical assistance and extension workers on the other; and finally, absence of any coordination between organizations generating technology and others responsible for implementing different components of agricultural policy, prices, credits, services, and others (Trigo, Piñero, and Sabato 1983).

The national agricultural research institutes resulted from these preoccupations. The general model is common to them all, entailing the legal and administrative character of an autonomous or semi-autonomous public entity with a broad mandate covering a wide range of products, regions, and types of farming situations. The basic objectives were to solve the problems created by the bureaucratic environment of the ministries; to allow for an improvement in the funding situation and conditions of service for research personnel; and at the same time to maintain research in the public domain, closely linked to agricultural development policy. Organizationally, the model adopted in most cases was one that combined centralized decision-making, with respect to priority setting and resource allocation, and operational decentralization through a network of experiment stations and commodity discipline programs.

The efforts to create the national research institutes had large support from technical and donor assistance, and particularly that originating from what came to be known as Point IV of the U.S. Foreign Aid Policy. This assistance included crucial support for human and infrastructural development. Perhaps more important, however, was its role as a key element in the development of the national research institute model as a Latin American expression of the U.S. experiment station system.

From this process emerged the following institutions: the National Institute of Agricultural Technology (INTA) of Argentina in 1957; the National Institute of Agricultural Research (INIAP) of Ecuador in 1959; the complex CONIA-FONAIAP in Venezuela between 1959 and 1961; the National Institute of Agricultural Research (INIA) in Mexico in 1960; the Agricultural Research and Promotional Service (SIPA) in Peru, which after successive modifications became the National Institute of Agricultural Research Promotion (INIPA) in 1984; the Colombian Agricultural Research

Institute (ICA) in 1963; and the Agricultural Research Institute (INIA) in Chile in 1964. This trend continued into the seventies with the creation of the Bolivian Institute of Agricultural Technology (IBTA); the Institute of Science and Agricultural Technology (ICTA) in Guatemala; the Agricultural Research and Development Institute (IDIAP) in Panama in 1975; and the National Institute of Agricultural Technology (INTA) in Nicaragua (back under direct control of the Ministry of Agriculture since 1980).

All of these institutions share the organizational characteristics mentioned above. However, variation exists with respect to some specific aspects covering their governance structure, mandates, and/or sources of funding.

In regard to the governance structure, all of the institutes are organizations with a legal status of their own, reporting in most cases to the ministry of agriculture or its equivalent. A differentiating characteristic, however, is the existence or not of a board of directors or trustees responsible for policy guidance and management control. Of the above-mentioned institutes, INTA of Argentina, ICA of Colombia, ICTA of Guatemala, and INIA of Chile have boards; the remaining institutions do not have such a body and the directors general or the chief executive officers report directly to the ministries of agriculture.

Another difference relates to the scope of the mandate. The institute model has tended to bring research and extension together. However, in some instances, such as INIAP in Ecuador, IDIAP in Panama, and INIA in Mexico, the two functions have been kept separate, with extension remaining a ministerial function. Education was generally kept separate from research and extension. However, in a number of cases--Argentina, Colombia, Peru, Uruguay, Mexico, and Brazil--due to the need to develop a minimum critical mass of human resources, ad hoc attempts were made to develop postgraduate training infrastructures in conjunction with universities. With the exception of Brazil and Mexico, most of them have been short-lived and unstable and have not become integral parts of the institutional model.

Funding is also a differentiating factor. The original concept was to seek as much funding autonomy as possible. While this was seldom achieved as a permanent feature, autonomy as regards financial management has allowed the institutes to attract substantial amounts of donor assistance. However, only INTA of Argentina has had special funding mechanism treatment, receiving its resources through a 2 percent tax on agricultural exports. Usually, funds flow from direct allocations in the national budgets, with the result that, although some benefits have been derived from greater control and flexibility in budget management, funding instability continues to be a serious limiting factor in many countries (Trigo and Piñeiro 1983).

The development of the Brazilian agricultural research system has followed a somewhat different pattern. Chronologically speaking, Brazil is the only major country in the region where the sixties brought no major change. More significant, however, is a difference with respect to the institutional model followed to create the Brazilian Corporation of Agricultural Research (EMBRAPA). Established in 1973, EMBRAPA is an institutional development similar to the research institutes in the other Latin American countries; its objective is to set the national basis for linking Brazil to the international system and making research an active instrument of agricultural development policy. As in the cases of INTA in Argentina, ICA in Colombia, and other institutes, it was not an isolated event. It resulted in and remains an integral part of a broader effort to influence agricultural development.

The organizational format is, however, different. EMBRAPA combines two separate sets of functions. On the one hand, there is the mandate to carry out research, for which it has a substantial research capacity of its own in the national commodity centers. On the other, it has the function of leading and coordinating, as far as objectives and priorities are concerned, a multi-organizational model involving separate levels of administration in the public sector (federal and state) as well as in the private sector. In this context EMBRAPA is probably closer to the concept of the agricultural research councils than to the rest of the national research institutes in Latin America.

POST-COLONIAL AFRICA: IS THERE A PREVAILING ORGANIZATIONAL TREND?

By examining the information in table 9.1, one may be tempted to associate the current situation in Africa with the ministerial model of agricultural research organizations. This association is probably correct but should be made carefully, and with a number of qualifications, especially in reference to the subsequent evolutionary trends that may be involved.

The first consideration relates to the colonial heritage. Colonial strategies in Africa varied widely, not only depending on the colonial power involved, but also within any given colonial heritage. Nonetheless, it is pertinent to attempt a summary of the main phases that have marked the evolution of agricultural research organization since the colonial era, especially if the African experience is to be included (Eicher and Baker 1983; Cooper 1970).

The second consideration is that in a number of countries the national research institutions are in the early stages of development, often just beginning to develop their human resource base. Consequently, any attempt to generalize trends on the basis of the current situation should be treated with extreme care (FAO 1984; ISNAR 1982b).

The main differentiating element among the colonial experiences (British, French, Belgian) in regard to agricultural research is the way in which research in the colonies and research in the metropolis were linked, and the type of relationship maintained after independence. The first affects the starting point of today's structures; the second affects the nature of the changes that have taken place and the level of resources that have been available to national research since independence.

Under British colonial rule each colony was perceived as a distinct entity, to be ruled and developed in accordance with its particular characteristics. This acted against the centralization of research, and in some cases--particularly in food crops--also against the regionalization of research activities, although regional efforts were present in East and West Africa in the post-World War II period.[6] In line with this approach, general responsibility for research came under the aegis of a department of agriculture in each colony, although a number of commodity-specific efforts were developed outside the ministries.

At the time of independence there was a dual structure in situ, where research in the food crops in departments of agriculture coexisted with a number of autonomous, or quasi-autonomous, efforts servicing specific export crops, where planters or external commercial interests were significant. Since independence, the modifications in the power structure and a very dynamic, and often chaotic, social, political, and economic environment constitute the basic framework for the evolution of the research structures. The main features are the "nationalization" of the structure, with a rapid fading of colonial presence and the substitution of expatriate researchers for national research personnel, and a shift of research emphasis from export to domestic food crops.

Specific changes in agricultural research organization followed these tendencies in the context of acute shortages of trained manpower and the need to protect some important export crops as sources of external revenue. This sometimes prompted post-independence administrators to leave untouched the organizational arrangements in those commodities. The general trend, however, was to maintain the preeminence of the ministry or ministries vis-à-vis other types of organization and in recent times to develop a central coordinating capacity, either by combining the different ministerial units involved in research under one roof, as in the case of Kenya, with the Tanzanian Livestock Research Organization (Taliro), or Tanzania, with the Tanzanian Agricultural Research Organization (Taro), or by formally assigning the coordination role to a special unit or a ministry of research and scientific development (or similar), as in the case in Nigeria.

Experience in former French colonies was significantly different. Before independence, agricultural research was highly centralized and closely linked to the metropolis through the GERDAT institutes, which had an applied orientation and a worldwide mission covering not only Africa but also the French colonies in other parts of the world.[7] The budgets of these institutions, with headquarters in France, were met largely by French taxpayers. Staffed by expatriates, the stations abroad were outreach establishments of the specialized institutes. No consideration was given to creating an independent research capacity in the colonies, either individually or regionally.

The end of French colonial rule in 1960 did not immediately change the characteristics of the French agricultural research presence in the former colonies, with which France maintained close economic, political, and cultural ties. In most instances the activities of the various French agricultural research organizations in the former colonies continued under formal cooperation agreements with the national governments.

The most important feature of the post-independence research system in terms of organizational structure was the growth of an indigenous agricultural research and agricultural administrative capacity largely staffed, funded, and controlled by French organizations and nationals. As a consequence of increased national participation, there was also a shift from export crops to food crops in the overall focus of the research system. This process was greatly affected by the political evolution of the relationship with France and by the resource situation in each of the countries. The particular array and distribution of responsibilities between ministries, agencies, and institutes in each case resulted from shifts in power distribution during the successive alternations of military and civilian rule. Although no clear evolutionary pattern can be identified, it is possible to mention some tendencies. These refer to the creation of the ministries of scientific and technical research (Senegal, Ivory Coast, Cameroon, Central African Republic, Mali) in the 1970s and the development of horizontal linkages among the research institutes working in a country, to substitute for the vertical links that existed between the individual institute and its parent in France, which continued into the post-independence period.

For how long these dual structures, with heavy participation of the former colonial institutes, will last is difficult to say. Three essential issues are: (1) the nature of the privileged relationships between the countries and France; (2) the evolution of the research capacities in the local institutions created since independence, particularly with respect to the availability of research staff with proper levels of training; and (3) the willingness of a national government to bear the costs of its national research effort.

In the former Belgian colonies the situation was rather different. Again in this case the colonial strategy with respect to agricultural research played a key role in determining the present situation. The research efforts initiated under Belgian rule were based in the Institut National pour l'Etude Agronomique du Congo Belge (INEAC), which had stations throughout the Belgian Congo, Rwanda, and Burundi. Created in 1933, it was funded primarily by Belgian funds but was highly decentralized in program development and implementation. At the time of independence, or soon thereafter, this infrastructure was transferred to the full and separate control of those independent states, and constituted the basis of the national agricultural research systems in those countries. The salient feature of the evolution since then has been the inability to use the vast infrastructure inherited (e.g., Zaire, Rwanda). Political problems and lack of resources--human and financial--to substitute for the

Belgian support as it was withdrawn have been the main deficiencies (ISNAR 1981a, 1982c; 1984a).

To summarize, the postcolonial structure of agricultural research in Africa appears to be characterized by the existence of a vast array of organizations, which mostly correspond to what was in place at the time of independence. The "nationalization" of those research structures has undoubtedly been the main task of the last twenty to twenty-five years. This process has taken place against a background of different colonial heritages, which has affected the types of institution established in the newly independent countries and the decolonization strategies, which influenced the nature and pace of the nationalization. The agencies, ministries, universities, etc., are still confronted with many of the same problems prevalent in Asia and Latin America when the processes that led to the national institutes and ARCs were started: namely, too few human resources, unstable funding, and duplication. In recent years efforts have concentrated on the development of an appropriate resource base. At the organizational level the ministry model seems to be widespread, but it would be premature to talk about a well-established trend toward a "dominant" model as in the other regions.

COMMONALITIES AND DIFFERENCES AMONG THE PREVAILING FORMS OF ORGANIZATIONAL STRUCTURE

The issues discussed in the previous sections highlight evolutionary patterns of interaction between the research institutions and their environment and how at any point in time the existing structures reflect the influence of a complex set of forces. They also provide a good basis from which to discuss the idea that there is no single "best" way to organize agricultural research and that any particular format is not equally effective in all situations. Without going into a detailed discussion, it is relatively easy to accept that agricultural research in Asia and Latin America over the last twenty to twenty-five years has been highly effective and has contributed significantly to the improvement of agricultural production and productivity. It suffices to point out the successes evidenced by India's present maintenance of a buffer stock of around 25 million tons of cereals, the significant improvements in rice production throughout Asia and Latin America, the near doubling of grain production in Argentina since the early 1970s, and the Brazilian experience with wheat and soybeans. Although a one-to-one relationship is not argued, it is not difficult to associate these successes with changes in the organizational structures that allowed research to address the problems of the farmers. Since the organizational approaches were quite different, it seems relevant to ask, therefore, "What were the factors that prompted the evolution of the systems?" and "What were the differentiating factors?" Bearing these questions in mind, we will now examine how the environmental changes took place and briefly discuss some of the factors that may have affected the particular shape of the institutions that were created.

The Demand for Research and Institutional Change

The process of institutional change is clearly affected by political, social, and economic forces (Ruttan 1978; Alves 1984). For purposes of this paper, a detailed examination of how these function is not pertinent. However, it is postulated that for effective institutional change to occur, a clear need must exist and decision-makers must see the necessity of structural change for meeting that need. If "effective" change is to happen, there must be political support and commitment to assuming the costs--political and otherwise--associated with that change. The changes in Latin America and Asia since the late 1950s-early 1960s are interesting examples of the dynamics of these processes. They allow us to raise a number of hypotheses about the situation in Africa and its likely evolution. The important aspect is that, although the countries in the regions differ substantially in resources and in cultural and political traditions, the processes that led to the establishment of the national research institutes and the agricultural research councils have striking similarities.

The emergence of the national institutes and the ARCs, and the cases of MARDI and AARD, resulted from situations in which technology and consequently research were seen by the relevant political systems as a key to solving their problems. In both regions the need was made obvious by the poor performance of the agricultural sector and its inability to satisfy the national requirements for food and to provide exportable surpluses. In some Latin American countries, such as Mexico, Colombia, Peru, and Ecuador, national production was rising at a rate well below the increase in demand resulting from population growth and urbanization. In others, such as Argentina and Uruguay, the stagnation of the agricultural sector generated balance-of-payment problems, which augured the appearance of even more serious difficulties as the industrial processes began to gain headway. In still other countries, such as Brazil, the agricultural sector was inextricably linked to both foreign trade and domestic demand problems (Piñeiro and Trigo 1983).

In Asia, most countries were confronted by both sets of problems, as they were highly dependent on food imports, which represented a major drain on foreign exchange and a substantial constraint on the overall growth of the economies. In some years it was not possible even to meet domestic requirements through imports, since it was difficult to purchase the grain, irrespective of price, and because of the logistical problems in transporting the food to where it was needed. There was also a political dimension: the poor agricultural performance was a major contributing factor to political instability. In Indonesia, the "rice crisis" of the second half of the 1960s contributed to the fall of the Sukarno regime. In other countries there was an increasing realization of the dangers of depending on other countries for food supplies. India and Pakistan both experienced difficulties with US PL 480 foodgrain shipments during the 1960s, when the United States stopped food aid or threatened to do so to force these countries to make certain political decisions. In 1974 the food aid to Bangladesh was delayed in a shortage year, and the Bangladeshis perceived this as an attempt by the United States to force them to break their trading relations with Cuba (Pray 1983).

At the international level there was a growing conviction that these problems could be solved through new technology. Furthermore, by that time it was clear that the soil, climate, and nature of the dominant crops were amenable to major technological breakthroughs, but institutions capable of producing and disseminating them were needed (Schultz 1964). The existing structures did not meet the requirements. In some cases there was a network of overlapping institutions; in others the existing structure was too dependent on volatile political factors. In almost all circumstances there were insufficient human and material resources.

These conditions set the stage for the domestic demand for research and the reorganization of the existing structures. Foreign assistance played a key role in facilitating these changes in several important ways: first, by helping link the production and productivity problems with research and conceptualizing the need for institutional change; second, by providing foreign scientists and administrators to help identify appropriate institutional forms and adapt them to local needs; finally, by providing support for the implementation of the new structures. USAID, the Ford and Rockefeller Foundations, together with a number of American universities, participated actively in these processes. In more recent times the FAO and the World Bank, and in Latin America, IDB and IICA, were other important sources of ideas and support.

In Africa, there were two important factors in the institutional changes that have taken place over the last twenty to twenty-five years. The first was the local situation and the demand for agricultural research. At the political level, there was no local demand for research until recently. The changes that took place resulted not from the decision to strengthen research institutions, but as part of the overall nationalization of the public administration that followed independence. The tendency in many countries was toward policies that discriminated against the agricultural sector; consequently there was no role for research. It is only in the past few years that some local initiatives have begun to appear.

The second difference was in the role of donor assistance in the region. As stressed above, external agencies played a crucial role in both the conception and the implementation of the institutional changes that took place in Asia and Latin America. In Africa they were also actively involved, but their role was different. Donor assistance focused mainly on specific projects rather than on long-term institution-building programs. Furthermore, there was a high level of direct involvement in the implementation of the projects and of research activities proper, often within ad hoc structures and not as part of the local research organization. In a few instances more recently, donors have begun to emphasize institutional characteristics in their assistance efforts. An additional important difference was that while for the other regions there was--rightly or wrongly--the conviction that the problem was technological and that technologies were available, in the African case there was no general agreement on technology's role (Mellor 1985).

Country Characteristics and Choice of Organizational Format

The debate as to how essential organizational questions, such as the degree of decentralization, were dealt with in different situations provides important additional insights into the relationships between environment and organizational structure.

The centralization-decentralization issue lies at the very center of the discussion about agricultural research organization. Agricultural research has a need for decentralization, not because decentralization is inherently superior from an organizational point of view, but because it is responsive to the nature of the problems that the research systems address (Ruttan 1982, chap. 4; Bonnen 1984). Agricultural production is location specific, and agricultural technologies need to reflect this location specificity. However, diversity of agroecological environments is not the sole source of variability that must be considered; technology also has a social variable.

For research to be successful, its products must have not only an effective biophysical adaptive capacity, but also the ability to reflect accurately the diverse socioeconomic, political, and cultural constraints facing the farmers who make the adoption decisions (ISNAR 1984b). This characteristic of agricultural production calls for a physical infrastructure and for decision-making processes capable of reaching all relevant environments and accurately reflecting the needs of the different clientele. Both of these attributes appear to be better achieved through a decentralized organizational structure. Nevertheless, it is important to recognize that this need for decentralization has a counterbalance in the need to achieve program coherence, and to relate research to the other components of the agricultural development strategy. Furthermore, decentralized systems are more management intensive than centralized structures (Paul 1983).

As stressed in the previous section, the conditions of demand in each case were similar: poor agricultural performance, together with the recognition that agricultural research was essential to altering the situation. The state of the existing agricultural research systems was also similar: weak institutions with inappropriate human and financial resources. Under these conditions the prevailing trend was toward a centralized structure, but the capacity to mobilize research in terms of a given agricultural development was lacking and human and managerial resources were scarce. Hence, high priority was given to minimizing duplication of effort and to reducing the number of decision-making levels. The different nature of the structural responses to these common problems can be explained by the characteristics of the existing research infrastructures and the politico-administrative styles of the countries.

In Latin America the national institutes followed an already established centralization trend. At the outset, agricultural research was not a central government responsibility, although it became so in the 1930s and 1940s. This centralization of responsibility arose from the unified nature of the political organization in most of the countries and the financial weakness of the regions or provinces that prevented them from taking any substantial initiatives in this area. In the mid-1950s the existing research capacity was centralized in the ministries of agriculture. The national institutes followed as a natural development, and the needs for operational decentralization were handled through their internal organization strategy, which emphasized program development and decision-making at the regional and local levels.

The influence of background and political system is further highlighted in the case of Brazil. As previously mentioned, very little happened in Brazil during the 1960s. The problems confronted were similar to those of the other countries in the region, and it was exposed to the same ideas that prompted the creation of the national institutes. However, Brazil had a stronger federal organization, which made it difficult to move in the same direction. A major political change had to take place before EMBRAPA could come into existence, and even then centralization was limited, as some of the existing state research systems remained outside the control of EMBRAPA (i.e., São Paulo) (Ruttan 1982, chap. 4; Pray 1983).

By contrast, in Asia (especially India), where the council model originated, there was a highly decentralized system in place. This decentralization occurred when the Indian Department of Agriculture was placed under the aegis of provincial governments, and was furthered by the proliferation of research programs in the 1950s and early 1960s. The strengthening of the functions of ICAR was a response to the need to coordinate and to optimize the use of available research resources. It would have been unrealistic to have attempted to substitute a new institution of the type of the national institutes for the existing structure (Pastore and Alves 1977).

The dynamics of the Pakistan and Philippines experiences are similar, although the trends toward centralization were greatly facilitated by political changes toward a more centralized form of administration.

The size of the country and the diversity of the agricultural sector are also relevant to the centralization issue. It is difficult to envisage a single organization able to manage the entire research effort in countries the size of Brazil or India.

CONCLUDING REMARKS

This paper was developed out of the proposition that in organizational format matters, there is no one optimal way of organizing agricultural research systems, and not all formats are equally effective. Without attempting to put forward a formal testable hypothesis, it was stressed that "optimality" results from political and technical fit within a given environment. An optimal format is one that gets the job done.

The previous sections reviewed the ways in which agricultural research systems in the developing world are organized and attempted to find commonalities and differences that could help to advance the understanding of relationships between organization and environment. In doing so, a great diversity in the ways in which agricultural research is organized was identified. At the same time it would not be difficult to associate success stories with each of the four main types of systems presented in table 9.1, thus giving some validity to the proposition that there is no one best way to organize. It was also found that each of the formats reviewed results from evolutionary adjustments to changing environments where the pre-existing structures were not seen as effective ways of mobilizing the needed resources and delivering the products expected from research. This observation may explain the proposition that not all of the formats are equally effective.

When comparing the evolution of the organization "models" in Asia and Latin America it was found that the efforts that led to the development and consolidation of the ARCs and the national institutes resulted from a confluence of forces and interests that created a favorable policy environment for research and institutional change.

The contrast with the Asian and Latin American experiences may be of value when discussing how to meet the challenge in Africa, particularly in relation to the time scale and the set of concomitant actions that should accompany efforts to strengthen agricultural research.

Two aspects seem to be important. First, there is the time scale involved in the institutional development process. The present state of development of research institutions in Asia and Latin America is the result of more than twenty to twenty-five years of continued support evolution. Most postcolonial African experiences are much more recent. Second, donor assistance in Asia and Latin America was channeled mostly into institution-building programs; in Africa the predominant trend has been to support individual projects, often directed at solving very specific problems rather than at creating new capacity.

A second issue concerning the evolution of the systems is how they have coped with new developments. During the last ten to fifteen years conditions in the countries have changed substantially, and in many cases as the result of the very success of the new forms of organizing research. One of those changes, not discussed here, was the increasing role and importance of private agricultural research activities. The analysis of the implications of this phenomenon in terms of the organizational structure, and the role of certain formats such as the ARCs or the national institutes, remains an important area for investigation and discussion.

Finally, some specific organizational dimensions were touched upon, particularly the degree of centralization-decentralization. Available evidence points to certain general patterns related to a country's stage of development, its political system and size, and the type of organizational format chosen. However, more detailed information is required before the nature of the parameters of the optimal environment for each different type of organization can be examined.

NOTES

*The author wishes to acknowledge the contribution to the concepts expressed in this paper made by Joseph Chang through his work on the governance of national agricultural research systems in developing countries while he was a research fellow at ISNAR, in 1984.

1. In describing the different formats, no effort is made to provide a fully comprehensive typology. Each organizational type is presented to emphasize what ISNAR considers to be its main differentiating feature in terms of its impact on the performance of the essential management processes and the effectiveness of the research activity.

2. Usually the ministry of agriculture and/or livestock. However, there are situations where other ministries are also involved: the most frequent cases are the ministries of education (or higher education) and science and technology.

3. An autonomous agricultural research organization meets the following criteria:

a) it has legal personality and its own board of directors/trustees that oversees the execution of its mandate;
b) it has independence in the management of its budget, and it does not have to go through the financial service of a ministry, even where it may formally report to the ministry;
c) it controls its internal organization, as well as sets its own criteria for hiring, firing, and conditions of service (which may depart from civil service norms);
d) it has formal reporting obligations to some public body (e.g., president, prime minister, ministry, research council, etc.) from which it is otherwise legally and operationally independent.

A semiautonomous agricultural research organization is an organization that has legal existence apart from that of a line division of a ministry, but does not meet all the criteria necessary for definition as autonomous.

4. Examples of this type of institutional model are the U.S. land-grant universities, and the agricultural universities of India and the Netherlands.

5. Private sector research activities are not included in the table because of lack of information.

6. For example, the East African Agricultural and Forestry Research Organization (EAAFRO), which operated until the mid-1970s under the auspices of the East African Federation.

7. Le Centre Technique Forestier Tropical (CTFT); L'Institut d'Elevage et de Medecine Veterinaire des Pays Tropicaux (IEMVT); L'Institu Francais de Recherches Fruitieres Outre-Mer (IFAC); L'Institut de Recherches Agronomiques Tropicales et des Cultures Vivrieres (IRAT); L'Institut Francais du Cafe et du Cacao et autres Plantes Stimulantes (IFCC); L'Institut de Recherches sur le Caoutchou en Afrique (IRCA); L'Institut de Recherches du Coton et des Textiles Exotiques (IRCT); L'Institut de Recherches pour les Huiles et Oleagineux (IRHO).

REFERENCES

Alves, E. "Major Issues in Resource Allocation." In The Planning and Management of Agricultural Research, a World Bank and ISNAR symposium, edited by D. Elz. Washington, D.C.: World Bank, 1984.

Bonnen, J. T. "A Century of Science in Agriculture: What Have We Learned?" Department of Agricultural Economics Staff Paper No. 86-79, Michigan State University, East Lansing, March 1984.

Cooper, St. G. C. Agricultural Research in Tropical Africa. Nairobi, Kenya: East African Literature Bureau, 1970.

Drilon, J. D. Agricultural Research Systems in Asia. College, Laguna, Philippines: Southeast Asia Regional Center for Graduate Study and Research in Agriculture (SEARCA), 1977.

Drilon, J. D., and A. R. Librero. "Defining Research Priorities for Agricultural and Natural Resources in the Philippines." In Resource Allocation to Agricultural Research, edited by D. Daniels and B. Nestel. Ottawa, Canada: International Development Research Center (IDRC), 1981.

Eicher, C. and D. Baker. Research on Agricultural Development in Sub-Saharan Africa: A Critical Survey. East Lansing, Michigan: Michigan State University, 1983.

Elgueta, M. "Evolución de la Investigación Agricola en América Latina." In Las Ciencias Agricolas en América Latina, edited by Instituto Interamericano de Ciencias Agrícolas (IICA) and Asociación Latinoamericano de Fitotécnica. San José, Costa Rica: Editorial Trejos, 1967, pp. 125-41.

Food and Agriculture Organization (FAO). "Agricultural Research Organization and Management in Africa." In Advancing Agricultural Production in Africa, proceedings of the CAB (Commonwealth Agricultural Bureau) First Scientific Conference, Rome, Italy, 1981. London: CAB, 1984.

Hasim, M. Y. "The Agricultural Research System in Malaysia." In Resource Allocation to Agricultural Research, edited by D. Daniels and B. Nestel. Ottawa, Canada: International Development Research Center (IDRC), 1981.

International Service for National Agricultural Research (ISNAR). Rapport d'une ISNAR/IITA aupres de l'Institut de Recherche Agronomique et Zootechnique de la Communaute Economique des Pays des Grands Lacs. Burundi, Rwanda, Zaire: ISNAR, July 1981a.

________. The Agency for Agricultural Research and Development of Indonesia. The Netherlands: ISNAR, October 1981b.

________. The Agricultural Research System of Guyana. Report of the ISNAR Review Mission to Guyana. The Netherlands: ISNAR, March 1982a.

________. Strategies to Meet Demands for Rural Social Scientists in Africa. The Netherlands: ISNAR, May 1982b.

________. Le Systeme National de Recherche Agricole au Rwanda. Report of the Mission to the Government of Rwanda. The Netherlands: ISNAR, December 1982c.

________. El Sistema de Investigacion Agropecuaria en la Republica Dominicana. Report to the Government of the Dominican Republic. The Netherlands: ISNAR, July 1983.

________. Improvement of Agricultural Research Management in Cameroon. Report to the Ministry of Higher Education and Scientific Research of Cameroon. The Netherlands: ISNAR, June 1984a.

________. Considerations for the Development of Agricultural Research Capacities in Support of Agricultural Development. The Netherlands: ISNAR, 1984b.

________. The Agricultural Research System in Sri Lanka. Report to the Government of Sri Lanka. Report to the Government of Sri Lanka. The Netherlands: ISNAR, June 1984c. (Out of print.)

Jain, H. K. "India's Coordinated Crop Improvement Project--Organization and Impact." Indian Farming, July 1984.

Marzocca, A. "Los Pioneros." In Las Ciencias Agrícolas en Américan Latina, edited by Instituto Interamericano de Ciencias Agrícolas (IICA) and Asociación Latinoamericano de Fitotécnia. San José, Costa Rica: Editorial Trejos, 1967, pp. 27-66.

Mellor, J. W. The Changing World Food Situation. IFPRI Food Policy Statement. Washington, D.C.: International Food Policy Research Institute (IFPRI), January 1985.

Moseman, A. National Agricultural Research Systems in Asia. New York: Agricultural Development Council (ADC), 1971.

Pastore, J., and E. Alves. "Reforming the Brazilian Agricultural Reearch System." In Resource Allocation and Productivity in National and International Agricultural Research, edited by T. Arndt, D. Dalrymple, and V. Ruttan. Minneapolis: University of Minnesota Press, 1977.

Paul, S. Strategic Management of Development Programs, International Labor Organization (ILO), Management Development Series No. 19. Geneva: ILO, 1983.

Piñeiro, and E. Trigo. "Towards an Interpretation of Technological Change." In Technical Change and Social Conflict in Agriculture: Latin American Perspectives, edited by M. Piñeiro and E. Trigo. Boulder, Colo.: Westview Press, 1983.

Pray, Carl E. "The Institutional Development of National Agricultural Research Systems in South and Southeast Asia." Paper presented at the meeting of the Association for Asian Studies, San Francisco, March 25, 1983. Photocopy. Department of Agricultural and Applied Economics, University of Minnesota, St. Paul.

Ruttan, V. "Induced Institutional Change." In Induced Innovation: Technology, Institutions and Development, edited by H. Binswanger and V. Ruttan. Baltimore: Johns Hopkins Press, 1978.

________. Agricultural Research Policy. Minneapolis: University of Minnesota Press, 1982.

Schultz, T. W. Transforming Traditional Agriculture. New Haven: Yale University Press, 1964.

Trigo, E., and M. Piñeiro. "Funding Agricultural Research." In Selected Issues in Agricultural Research in Latin America, proceedings of a conference for Latin American research directors, sponsored by IFARD, IICA, and ISNAR in cooperation with the Spanish government. The Netherlands: ISNAR, August 1983.

Trigo, E., M. Piñeiro, and J. Ardila. Organizacion de la Investigación Agropecuaria en América Latina. San José, Costa Rica: Interamerican Institute for Cooperation in Agriculture (IICA), 1982.

Trigo, E., M. Piñeiro, and J. Sabato. "Technology as a Social Issue: Agricultural Research Organization." In Technical Change and Social Conflict in Agriculture: Latin American Perspectives, edited by M. Piñeiro and E. Trigo. Boulder, Colo.: Westview Press, 1983.

10
The Evolution of Farming Systems and Agricultural Technology in Sub-Saharan Africa*

Hans P. Binswanger and Prabhu L. Pingali

In this paper we present a simple conceptual framework to evaluate the benefits one should expect from investment in agricultural research under different factor endowment situations. The framework has important implications for the allocation of research resources. It draws on both the induced innovation framework developed by Hayami and Ruttan (1971, 1985) and Binswanger and Ruttan (1978), and on the work on farmer-generated changes in farming systems and technology by Ester Boserup (1965, 1981). Boserup's work is particularly relevant today for understanding the pattern of agricultural development in sparsely settled areas of Sub-Saharan Africa. But even elsewhere the relevance of Boserup's work lies in the continued complementarity of farmer-generated innovations with science- and industry-generated innovations. Her work has been badly neglected, perhaps because she focused solely on population density as the main driving force of agricultural intensification.

In this paper we briefly discuss other important determinants of agricultural intensification such as transport infrastructure, external final demand, and the endogenous distribution of population across different agroclimatic zones. We treat the overall rate of population growth of a country as exogenous for this enquiry. Nevertheless and in accordance with Boserup, we recognize that the rate of population growth has important endogenous components. These endogenous components are not, however, relevant to the topic at hand, as the rate of growth itself is exogenous to the farmer. We then discuss the consequences of agricultural intensification.

In Boserup's (1965) analysis, intensification has eight principal effects: (1) it reduces the fallow period; (2) it increases investment in land; (3) it encourages the shift from hand hoe cultivation to animal traction; (4) it encourages soil fertility maintenance via manuring; (5) it reduces the average cost of infrastructure; (6) it permits more specialization in production activities; (7) it induces a change from general to specific land rights; (8) it reduces the per capita availability of common property resources (forest, bush and/or grass fallows, communal pastures). To these effects we add agroclimatic and soil considerations, which were almost totally neglected by her, but were emphasized in the seminal work of Hans Ruthenberg, Farming Systems in the Tropics (1971). Our own particular addition is a discussion of the systematic changes in land use patterns and soil quality preferences associated with intensification.

The discussion in this paper is illustrated by two-way frequency charts using data from fifty-six villages in ten countries of Sub-Saharan Africa and India. This data was collected in 1983-84 using the group interview technique as part of a larger research effort on agricultural mechanization and the evolution of farming systems in Sub-Saharan Africa. Although not substitutable for rigorous empirical testing, these frequency tables provide preliminary support for the proposition presented in the paper.

DETERMINANTS OF THE INTENSITY OF LAND USE

Population Density

The existence of a positive correlation between the intensity of land use and population density has been shown by Boserup (1965, 1981). She argues from the premise that during the neolithic period forests covered a much larger part of the land surface than today. The replacement of forests by bush and grassland was caused by (among other things) a reduction in fallow periods due to increasing population densities.

> The invasion of forest and bush by grass is more likely to happen when an increasing population of long-fallow cultivators cultivate the land with more and more frequent intervals (Boserup 1965, 20).

Table 10.1 presents the relationship between population density and the intensity of the agricultural system. At very sparse population densities, up to perhaps four persons per square kilometer, the prevailing form of farming is the forest fallow system. A plot of forest land is cleared and cultivated for one or two years and then allowed to lie fallow for twenty to twenty-five years. This period of fallow is sufficient to allow forest regrowth. An increase in population density will result in a reduction in the period of fallow and eventually the forest land will degenerate to bush savannah. Bush fallow is characterized by cultivation of a plot of land for two to six years followed by six to ten years of fallow. The period of fallow is too short to allow forest regrowth. Increasing population densities are associated with longer periods of continuous cultivation and shorter fallow periods. Eventually the fallow period becomes too short for anything but grass growth. The transition to grass fallow occurs at population densities of around sixteen to sixty-four persons per square kilometer. Further increases in population result in the movement to annual and multi-cropping, the most intensive systems of cultivation. All along the course of this transition farmers make investments in land. Initially, they are confined to land clearing and destumping. Drainage investments, leveling, erosion control investments, and eventually irrigation investments follow. The capacity to double or triple crop usually requires very large investments.

Since the turn of this century we have observed a substantial increase in the natural rate of population growth across the world, mainly due to a sharp decline in the death rates caused by rapid advances in public health services. At the worldwide level, and at the level of a specific country, the decline in arable land per capita must be attributed primarily to this general increase in population. Within a country and within regions, however, population concentrations vary by soil fertility, altitude, and market accessibility. These intra-country variations are briefly discussed below using examples primarily from Sub-Saharan Africa. Table 10.2 provides the major causes and consequences of population concentrations.

Soil Fertility

The marginal productivity of labor is relatively higher on more fertile soils and hence one would expect

TABLE 10.1
Food Supply Systems in the Tropics

Food Supply System[d]	Farming Intensity[a] (R-Value)[b]	Population Density Group[b] Persons/km[b]	Climatic Zone[c]	Tools Used
G. Gathering	0	0-4		
FF. Forest fallow	0-10	0-4	Humid	axe, matchete, & digging stick
BF. Bush fallow	10-40	4-64	Humid and semi-humid	axe, matchete, digging stick, and hoe
SF. Short fallow	40-80	16-64	Semi-humid, semi-arid, and high altitude	hoes, animal traction
AC. Annual cropping	80-120	64-256	Semi-humid, semi-arid, and high altitude	animal traction, and tractors

[a] R = # of years of cultivation *100/# of years of cultivation + # of years of fallow.
Source: Ruthenberg 1980, p. 16.

[b] Source: Boserup 1981, pp. 19 and 23.

[c] Source: Ruthenberg 1980.

[d] Description of food supply systems:

Gathering: wild plants, roots, fruits, nuts
Forest-fallow: one or two crops followed by 15-25 years of fallow
Bush-fallow: one or more crops followed by 8-10 years of fallow
Short-fallow: one or two crops followed by one or two years of fallow, also known as grass fallow
Annual cropping: one crop each year
Multi-cropping: two or more crops in the same field each year

Note 1: The above food supply systems are not mutually exclusive. It is quite possible for two or more of the systems to exist concurrently (e.g., cultivation in concentric rings of various lengths of fallow, as in Senegal).

Note 2: The above population density figures are only approximations; the exact numbers depend on location specific soil fertility and agroclimatic conditions.

TABLE 10.2
Causes and Consequences of Population Concentration

	Causes		Direct Consequence	Implications
Natural population growth:	improved public health and lack of emigration			Reduction in fallow periods: movement from shifting to permanent cultivation
Soil fertility:	immigration to capture the benefits of higher returns to labor input			Mechanization: Plowing: where agro-climatic and soil conditions make it profitable
Transport facilities:[a]	immigration to capture the benefits of reduced transport costs			Transport: where markets exist for food and other crops
Urban demand:[a]	immigration to capture the benefits of market proximity		Reduction in available area per capita	Milling: in response to higher opportunity cost of time for female household members
Health:	avoidance of malaria and tsetse fly	immigration to cooler highlands		
Historic:	tribal war/ slave trade	immigration to inaccessible highlands		Land investments: for soil fertility, drainage, terracing, etc. Increase in the marginal lands brought under cultivation
Land laws, rights:	restrictions on the right to open new land			Land rights: from general use rights to specific land rights

[a]In the case of improved transport facilities and urban demand one may observe an expansion in the area under cultivation in the absence of immigrations.

immigration from less endowed areas leading to reductions in cultivable areas per capita. A few examples of fertile areas that are relatively densely populated and intensively cultivated are Ada District, Ethiopia; Nyanza Province, Kenya; and the southern province of Zambia. High altitude areas are similarly densely populated due to immigration from the lowlands because of lower disease incidence (notably malaria and sleeping sickness). Population concentrations on the Ethiopian and Kenyan highlands are popular examples of this phenomenon.

Given suitable soil conditions, areas with better access to markets, either through transport networks or those in the proximity of urban centers, will be more intensively cultivated. Intensification occurs because of two reasons:

1. Higher prices and elastic demand for exportables imply that marginal utility of effort increases, hence farmers in the region will begin cultivating larger areas; and
2. Higher returns to labor encourage immigration into the area from neighboring regions with higher transport costs.

Intensive groundnut production in Senegal, maize production in Kenya and Zambia, and cotton production in Uganda have all followed the installation of the railway and have been mainly concentrated in areas close to the railway line. Similarly, agricultural production around Kano, Lagos, Nairobi, Kampala, and other urban centers is extremely intensive compared to other parts of these countries. It should be noted that agricultural intensification in response to improved market access could occur even under low population densities because of individual farmers expanding their area under marketed crops. The consequences of intensification in these circumstances do not differ from those in areas with high population densities.

Soil fertility and transport infrastructure interact in two ways. First, a large region of high-quality land provides incentives for the construction of infrastructure to exploit that potential. Second, where roads or railway lines are built for nonagricultural purposes, such as to the copper mines of Zambia, they will cross both fertile and infertile areas. The impetus for intensification, however, is felt only in the fertile areas, as the infertile areas are unable to compete in output markets because of their higher costs of production.

Other Causes

Finally, it should be noted that inter- and intra-country variations in population densities, especially in Sub-Saharan Africa, have historically been caused by tribal warfare and slave trade, resulting in population concentrations in relatively inaccessible highlands. Population concentration on the high plateau of Rwanda and Burundi was in response to the incursion of slave traders and for health reasons. Similar migrations from the lowlands to the Mandara Mountains in Cameroon, the Jos Plateau in Nigeria, and the Rift Valley in Kenya and Tanzania have been based on the desire for personal security. Subsequent natural population growth has made many of these areas the most densely populated parts of Africa.

The above discussion leads to the broad generalization that for given agroclimatic conditions, increases in population density and/or improvements in market access will gradually move the agricultural system from forest fallow to annual cultivation and eventually to multi-cropping. Empirical support for these two determinants of agricultural intensification is provided by means of two-way frequency charts using our data set from Sub-Saharan Africa.

Figure 10.1 presents the positive relationship between population density and farming intensity. Forest and bush fallow systems are predominant under sparse population densities (less than 15 persons/km^2); nine of the ten sparsely populated areas in our sample practice these systems of farming, and the remaining case is under short fallow. In the medium density group (16-50 persons/km^2) one begins to observe the transition to more intensive systems of farming. Here the majority of the cases (9 of 14) fall in the short fallow and emerging annual cultivation categories; there are no forest fallow cases at this density level, and bush fallow tapers off to 3 out of 14 cases. In the dense population groups (51-100 persons/km^2), annual and multi-crop cultivation are well established with 12 of 18 cases in these systems and 3 cases emerging toward annual cultivation. There are 3 remaining short fallow cases presumably at the lower end of the population density group. Nine of the 10 cases in the very dense population group (greater than 100 persons/km^2) practice annual or multi-crop cultivation, with the remaining case emerging toward annual cultivation.

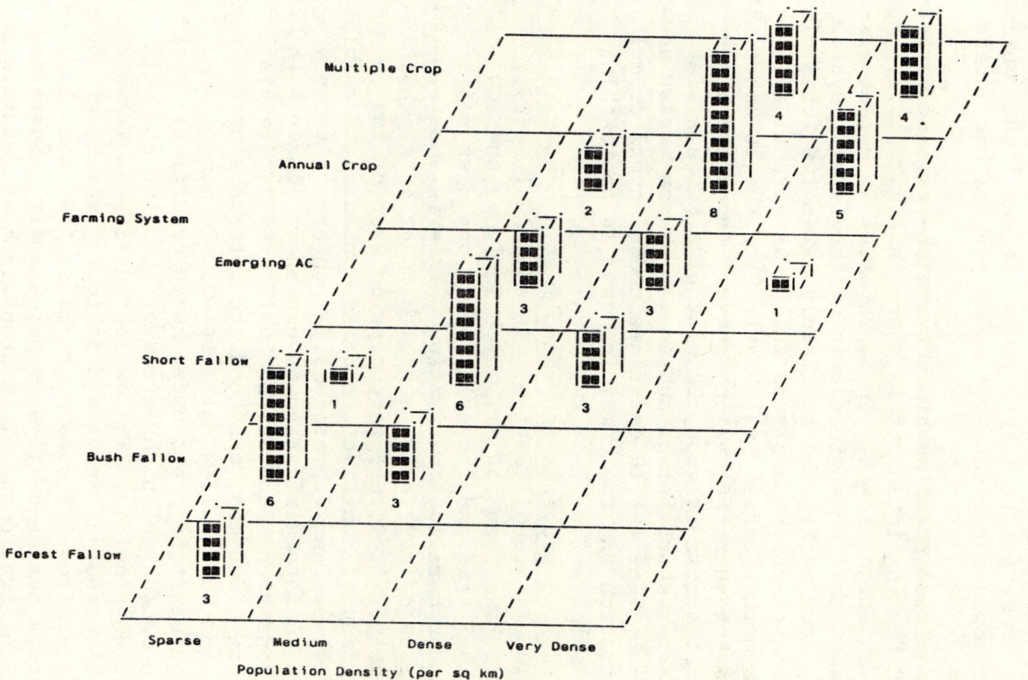

Figure 10.1. Population density and the intensification of farming systems

The classification of our cases shows a very definite trend toward agricultural intensification as population densities increase. Neither in our field visits nor during our literature survey have we come across any cases of sparsely populated areas under annual cultivation or cases of very densely populated areas under forest or bush fallow systems. Presumably, the former could occur under sparse population densities if market access is excellent, while the latter could not occur even under poor market access conditions.

The positive relationship between the ease of market access and farming intensity is shown in figure 10.2. For a given population density, an improvement in market access results in further intensification of the farming system. Under very poor market access, mainly extensive forms of farming such as forest and bush fallow are practiced. Under fair market access, we observe 9 of the 16 cases with intensities of short fallow and above. Among the cases where market access is good, 22 of the 24 cases are at intensities of short fallow and above. Where market access is excellent, 9 of the 10 cases are under annual or multicrop cultivation and the remaining case is approaching permanent cultivation.

Figure 10.3 presents the agroclimatic constraints on the process of intensification. At an annual average rainfall of less than 750 mm, which includes the low rainfall semi-arid and the arid zones, one does not see forest and bush fallow cultivation. This is due to: (1) the slow rate of vegetative regrowth under arid conditions that do not permit forest regeneration even at low densities, and (2) cultivation being concentrated mainly on the lower slopes and depressions, which are relatively more responsive to intensification investments. Under high rainfall conditions such as the humid tropics (rainfall greater than 1,200 mm), one tends to observe a predominance of forest and bush fallow cultivation. Permanent cultivation of field crops under humid conditions is hard to sustain because of high levels of leeching and soil acidification problems. The exceptions in the graph are from the highlands of Kenya and Ethiopia, which, of course, do not suffer from the same problems as the humid lowlands and therefore can be cultivated permanently. Sustained permanent cultivation of field crops is most feasible in the medium rainfall zones (751-1,200 mm), which is consistent with our empirical evidence.

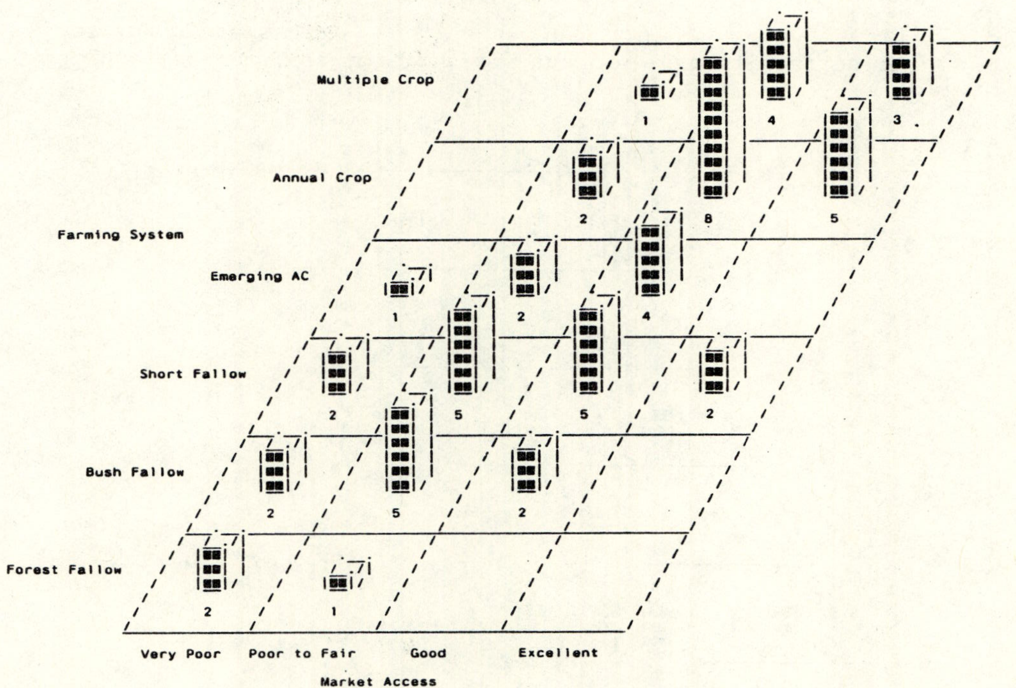

Figure 10.2. Market access and the intensification of farming systems

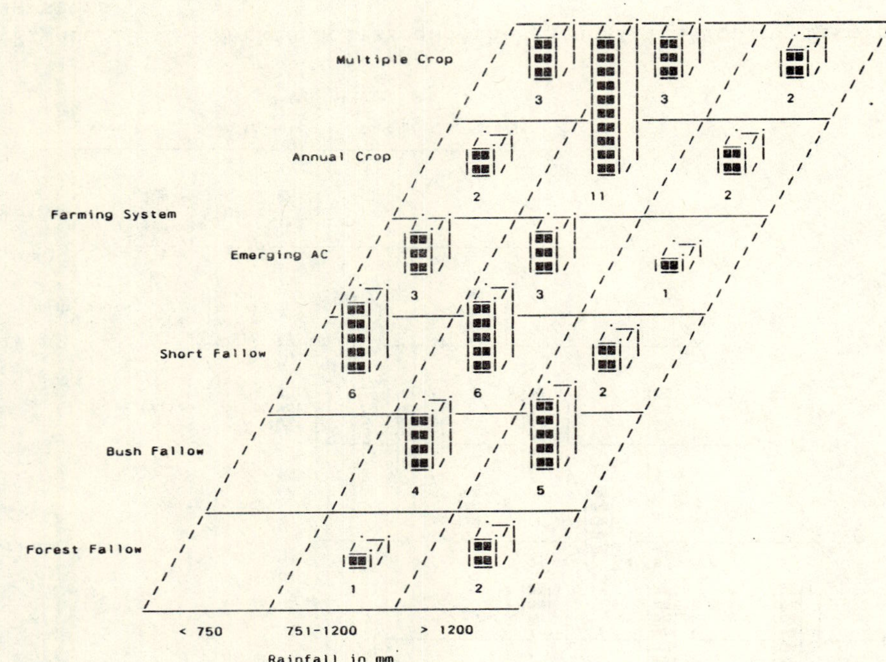

Figure 10.3. Agroclimatic zones and the intensification of farming systems

THE CONSEQUENCES OF AGRICULTURAL INTENSIFICATION

Agricultural Intensification and Soil Preference

The intensification of agricultural systems is constrained by climatic and soil factors. Table 10.1 illustrates the impact of climatic factors on the intensification of the agricultural system. For given agroclimatic conditions, the extent of intensification is conditional on the relative responsiveness of the soils to inputs associated with intensive production such as land improvements, manure, and fertilizers. The responsiveness of intensification is generally higher on soils with higher water- and nutrient-holding capacity. This is primarily because higher water-holding capacity reduces drought risk. Water-holding capacity is higher the deeper the soils and the higher their clay content. It is low on shallow sandy soils.

Figure 10.4 presents a stylized picture of the differences in soil types across a toposequence for given agroclimatic conditions. Soils on the upper slopes are relatively light and easy to work by hand and tillage requirements are minimal. The clay content and hence the heaviness of the soils increases as one goes down the toposequence; consequently, power requirements for land preparation increase. Movement down the slope also reduces yield risks because of increased water retention capacity of the soils. The soils are heaviest in the depressions and marshes at the bottom of the toposequence. These bottomlands or bas fonds are often extremely hard to prepare by hand and are often impossible to cultivate in the absence of investments in water control and drainage. The extremely high labor requirements for capital investments and land preparation make the bottomlands the least preferred for cultivation under low population densities, and they are often found to be under fallow. As population densities increase, however, the bottomlands become intensively cultivated because of the relatively higher returns offered to labor and land investments, especially in rice cultivation. Also, as population densities increase, labor supply increases, making it possible to undertake the labor-intensive investments in irrigation, drainage, etc.

Soil type differences across a toposequence that are characterized here could be micro-variations limited to a few hundred meters or a few kilometers, or they could be

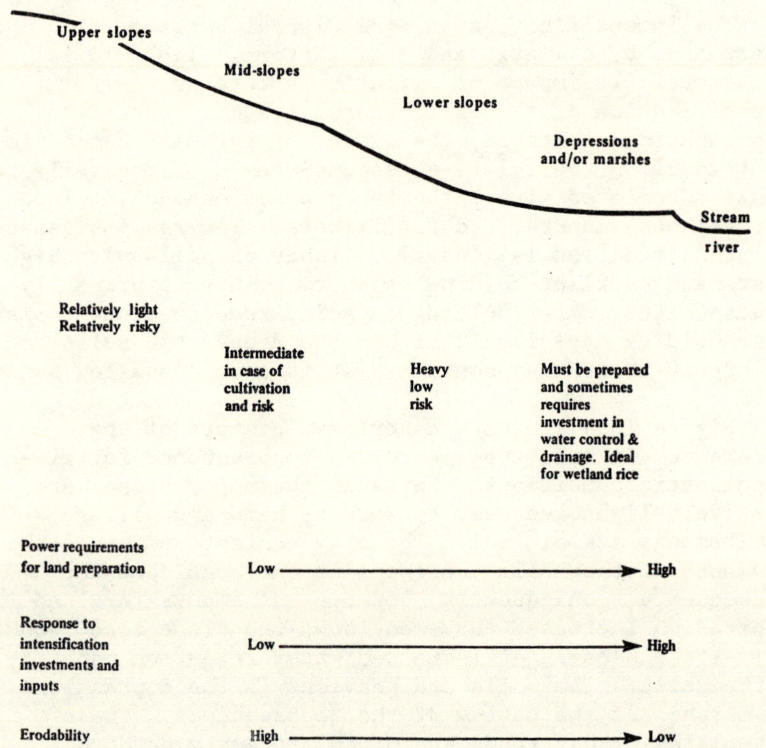

Figure 10.4. Toposequence and soil type

macro-variations where entire regions are part of one toposequence level. For example, large parts of northeastern Thailand can be characterized as upper slopes, although there are, of course, depressions in northeast Thailand within more toposequences. The central plains of Thailand are largely flood plains and have the characteristics of bottomlands and marshes.

Preferences for cultivating different points of the toposequence are also dependent on the agroclimatic conditions. Table 10.3 presents soil preferences by farming intensity and agroclimatic zones. Under arid conditions lower slopes and depressions are the only land

TABLE 10.3
Farming Intensity, Agroclimates, and Soil Preferences

Agro-climates	Farming Intensity: Forest and Bush Fallow	Grass Fallow	Permanent Cultivation
Arid	Lower slopes and depressions only	Lower slopes and depressions only	Lower slopes and depressions only
Semi-arid	Mid-slopes	+ Lower slopes	+ Depressions
Sub-humid	Upper slopes	+ Mid and lower slopes	+ Depressions
Humid	Upper slopes	+ Mid and lower slopes	+ Depressions

that can be cultivated because only here is water retention capacity sufficient to sustain a crop at very low rainfall levels. This is the reason for the intensive cultivation systems of an oasis type that one observes in arid areas even under low population densities. Pockets of arid farming in primarily pastoral areas of Botswana are a good example of this phenomenon.

Under semi-arid conditions the midslopes are the first to be cultivated. As population densities increase, cultivation replaces grazing on the lower slopes and eventually in the depressions. Power sources for tillage are first used in the bottomlands, generally around the time when population pressure makes these lands valuable for cultivation. The reversal of land preferences is quite dramatic. In the semi-arid zones of Africa, where population density is low, the lower slopes and depressions are left for grazing and contribute only minimally to food supply. In the semi-arid zones of India, on the other hand, the depressions are intensively cultivated, usually with rice, using elaborate irrigation systems and animal traction.

Yield risks due to low water availability are not a major problem in the sub-humid and humid tropics, hence one finds cultivation starting at the upper slopes and gradually moving downwards as population pressure increases. At high population densities the swamps and depressions become the most important land sources for food production, often associated with extremely intensive rice production. One observes such labor-intensive rice production in South and Southeast Asia and could expect the same for Africa as population densities increase.

Population pressure leads to a sharp reversal in preference (price) of different types of land in all but the arid zones. As population densities increase, one observes the cultivation of land that requires substantially higher labor input but which at the same time is also more responsive to the extra inputs.

Cultivation Techniques and Labor Use

As discussed by Boserup (1965) and Ruthenberg (1971), the total labor input per hectare on a given crop is positively correlated with the intensity of farming, holding technology constant. Table 10.4 presents examples of labor use with farming intensity in rice cultivation. In West Africa, the movement from forest fallow to annual cultivation using the hoe results in an increase in total labor input per hectare from 770 hours in Liberia to 3,300 hours in Cameroon. The increase in labor input occurs due to an increase in intensity with which certain tasks have to be performed (for example, land preparation and weeding) and due to an increase in the number of operations performed (e.g., manuring, irrigation, etc.). A discussion of labor use across intensities of farming is provided below. Table 10.5 presents the increase in operations performed with the intensification of the farming system.

In the forest and bush fallow systems of cultivation, land clearing, planting, and harvesting are the major tasks performed. Fire is the most prevalent technique for land clearance. This form of land clearance, in addition to regenerating the soil, also reduces weed growth. Land clearance by fire requires very low levels of labor input: 300 to 400 hours per hectare for forest fallow systems in Liberia and Ivory Coast. The ground, being under tree cover, is soft and hence no further land preparation is required prior to sowing with a digging stick or a hand hoe. Such systems of cultivation require almost no weeding or cultivation, and the period between planting and harvesting is virtually task free.

As the fallow period becomes shorter and the land under fallow becomes grassy, fire can no longer be used for land clearance. Fire cannot destroy grass roots, hence grasses persist through the growing season. The intensive use of a hoe for land preparation becomes essential to clear the grass roots. Land preparation and sowing take up

TABLE 10.4
Examples of Labor Use with Farming Intensity for Rice Cultivation

Country	Liberia	Ivory Coast	Ghana	Cameroon	India	Java	Philippines
Region	Gbanga	Man	Begora	Bamunka	Ferozepore	Subang	Laguna
Itensity of Farming	11	24	40	100	121	200	180
Technique	Hoe	Hoe	Hoe	Hoe	Animal plow	Animal plow	Tractor
Time/Operation (hours/hectare)							
Land clearing	418.4	300.8	665	--	--	--	--
Land preparation	--	--	--	714	86.4	494.4	73.6
Sowing/planting	107.2	142.4	207	536.8	129.6	146	80.0
Fertilizing and Manuring	--	--	--	--	12.8		
Weeding	36.8	292	276.8	113	57.6	218	213
Plant Protection	44	222	--	1,393			96
Harvesting	164	218.4	280	264	128.8	324.4	222.4
Threshing		84		280	76.8		
Other	--	--	--	--	136[a]	70	--
Total	770	1,259.2	1,432	3,300	627.2	1,252	685

[a]Irrigation.

TABLE 10.5
Comparison of Operations and Technology Across Farming Systems

Operations	SYSTEM				
	FF	BF	SF	AC	MC
Land clearance	Fire	Fire	None	None	None
Land preparation	No land preparation, digging sticks used to plant roots and sow seeds	Land is loosened using hoes and digging sticks	Use of plow for preparing land	Animal-drawn plows and tractors	Animal-drawn plows and tractors
Manure use	- Ash - Household refuse for garden plots	Ash, burnt or unburnt leaves, other vegetable matter and turf brought from surrounding bushland	- Animal and human waste - Green manuring - Composts - Silt from canals	- Animal and human waste - Composting - Cultivation of green manure crops - Chemical fertilizer	- Animal and human waste - Composting - Cultivation of green manure - Chemical fertilizer
Weeding	Minimal	Required as the length of fallow decreases	Weeding required during the growing season	Intensive weeding required during the growing season	Intensive weeding required during the growing season
Use of animals in farming	None	As length of fallow decreases animal-drawn plows begin to appear	- Plowing - Transport - Interculture	- Plowing - Transport - Interculture - Post-harvest tasks - Irrigation	plowing, transport, interculture, post-harvest tasks, irrigation
Seasonality of labor demand	None	None	Land preparation, weeding and harvesting	Acute seasonal labor demand concentrated around the rainy season and harvest	Acute seasonal labor demand conctrated around land preparation, weeding, harvest, and post-harvest tasks
Fodder supply	None	Emergence of grazing land	Abundant free grazing land	Free grazing during fallow period, crop residùes	Intensive fodder management and fodder crop production

almost 40 percent of the total labor input for the annual cultivation of rice in Cameroon. Under short fallow systems of cultivation, the need for early season weeding and plant protection becomes pronounced. Also, manure use is required to complement fallow periods for maintaining soil fertility.

Permanent cultivation of land requires labor investments for irrigation, drainage, and leveling or terracing. It also requires the development of more evolved manuring techniques to restore soil fertility. Land preparation and intercropping and weeding become much more important tasks.

Intensification, therefore, leads to both an increase in agricultural employment and an increase in yields per hectare. However, intensification of farming, in the absence of a change in tools used, would probably lead to a decline in yield per man-hour. This can be deduced from the observation that the greater proportion of the additional labor input is used for maintaining soil fertility, weeding, and plant protection. In other words, labor input per hectare may increase at a faster rate than yield per hectare in the movement to more intensive systems of farming. Pingali and Binswanger (1984) found a significant positive increase in labor use with farming intensity using data from fifty-two specific locations in Africa, Asia, and Latin America. We also show a significant downward shift in labor use when hand hoes are replaced by animal-drawn plows and a further shift when animal draft is replaced by tractors.

The Evolution of Tool Systems

The transition from digging sticks and hand hoes to the plow is closely correlated with the evolution of the farming system and cannot be understood by using choice of techniques analysis familiar to economists. The emergence of mechanical tillage is generally observed at late bush fallow and early grass fallow stages and not before. The switch from one set of tools to the next would occur when the resulting labor-saving benefits exceed the costs of switching to new tools.

The simplest form of agricultural tool, the digging stick, is most useful in the extensive forest and bush fallow systems where no land preparation is required. As the bush cover begins to recede, the ground needs to be loosened before sowing and at this stage hand hoes replace digging sticks. Hand hoes are used for land preparation and weeding in the latter stages of bush fallow, grass fallow, and even some instances of annual cultivation. Land preparation using the hoe becomes extremely labor intensive and tedious by the grass fallow stage because of the persistence of grass weeds. The use of a plow for land preparation becomes almost indispensable. A switch to the plow during grass fallow results in a substantial reduction in the amount of labor input required for land preparation. The net benefits of switching from the hoe to the plow are conditional on soil types and topography. The benefits are lower for sandy soils and for hilly terrain.

The above discussion on the evolution from hand hoes to animal-drawn plows is formalized in figure 10.5. This graph compares the labor costs under hand and animal-powered cultivation systems and shows the point at which animal traction is the dominant technology.

The overhead labor costs in the transition from hand to animal power are the cost of training animals, the cost of destumping and leveling the fields, and the cost of feeding and maintaining the animals on a year-round basis. The cost of training the animals is independent of the intensity of farming. The cost of destumping is extremely high under forest and early bush fallow systems. As the length of fallow decreases, the costs of destumping decline because of reduced tree and root density. Destumping requirements are minimal by the grass fallow stage. The costs of feeding and caretaking of draft animals are also very high during forest and early bush fallow, primarily due to the lack of grazing land and to the prevalence of diseases such as trypanosomiasis. As the fallow becomes grassy, grazing land becomes prevalent and so does animal ownership; hence the costs of maintaining draft animals decline. By the annual cultivation stage, however, grazing land becomes a limiting factor, necessitating the production of fodder crops, which in turn lead to an increase in the cost of feeding and maintaining draft animals.

The total cost of using draft animals for land preparation, early season weeding, and manuring is given by the curve T_p. The curve T_h shows how total labor cost using hand hoes increases for land preparation and weeding, while T_h' adds in the cost of maintaining soil fertility. The shape of the T_h' curve depends on: (1) the ease of producing compost; (2) the rate of decay of organic matter;

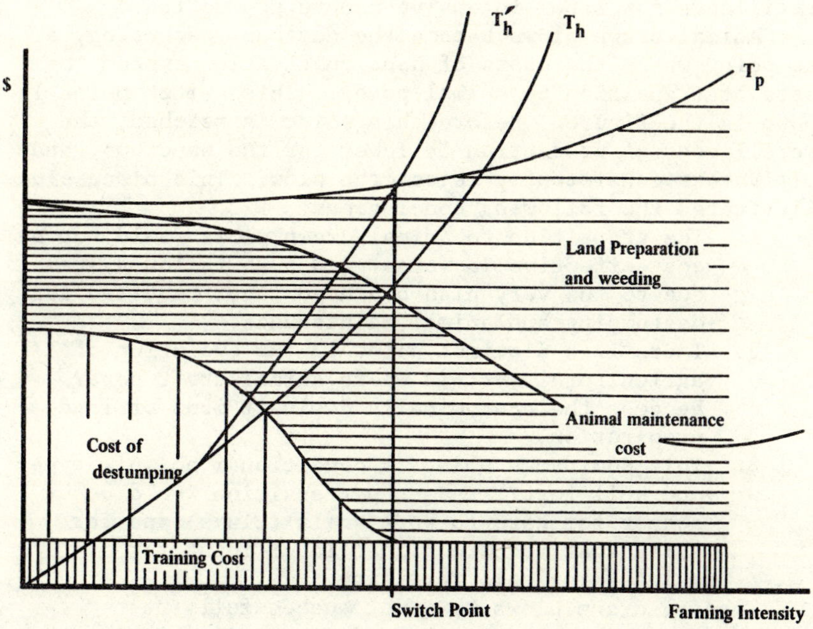

T_p = Labor costs for land preparation, early season weeding, and manuring using animal traction
T_h = Labor costs for land preparation and early season weeding using hand hoes
T_h' = T_h plus labor costs for maintaining soil fertility without manure from draft animals
Switch point = Farming intensity at which animal traction is the dominant technology

Figure 10.5. A comparison of labor costs under hand and animal-powered cultivation

humid areas it is easier to produce compost and manure relative to semi-arid and arid areas because of an abundance of natural vegetation; hence the labor costs involved in the production of manure are lower and the T_h' curve is flatter. In hot tropical areas, high temperatures cause the organic matter to decay at a faster rate relative to the more temperate highlands, and hence additional compost and manure inputs are required, making the T_h' curve steeper. The T_h' curve becomes flatter the cheaper chemical fertilizers are, due to the substitution of fertilizers for labor-intensive manure production.

Animal-drawn plows become the dominant technology at the point where the costs of hand cultivation exceed the costs of transition to animal power. This switch point is shown in the figure. Before this point is reached, the overall cost of production is lower for the hand hoe, and cultivators consistently reject the plow. This discussion illustrates the following conclusions:

1. The transition to animal-drawn plows would not be cost effective in forest and bush fallow systems due to the very high overhead labor required for destumping and animal maintenance.
2. There is a distinct point in the evolution of agricultural systems where animal draft power becomes the economically dominant mode of land preparation.
3. This dominance point is conditional on soil types and soil fertility: the transition would occur sooner for hard-to-work soils (clays) and for soils that require higher labor input for maintaining soil fertility.

Tractor-drawn plows have not successfully been introduced into any systems prior to the grass fallow stage. Indeed, as we show in Pingali, Bigot, and Binswanger (1987), it is almost impossible to bypass the animal traction stage and move directly to tractors. This is because the quality and hence the cost of destumping is much higher for tractor operations than for animal-drawn plows. Moreover, the societies concerned are often characterized by extreme capital scarcity and cannot usually afford the substantially higher capital costs associated with tractors.

Once the stage of annual cultivation has been reached, however, tractors and animals become almost perfect substitutes for plowing. The choice of techniques analysis now finally becomes relevant. The factors involved are the relative costs of land, labor and capital, the seasonality of agricultural production, and the cost of tractors.

Using the data set from Sub-Saharan Africa, we provide a frequency chart of tool use with the evolution in farming systems in figure 10.6. As discussed above, hand cultivation is predominant in the forest and bush fallow systems--none of the forest fallow cases use any other form of tillage, while 6 of the 9 bush fallow cases use hand tools. The majority of cases under short fallow use animal draft power (7 out of 14 cases), three locations use a

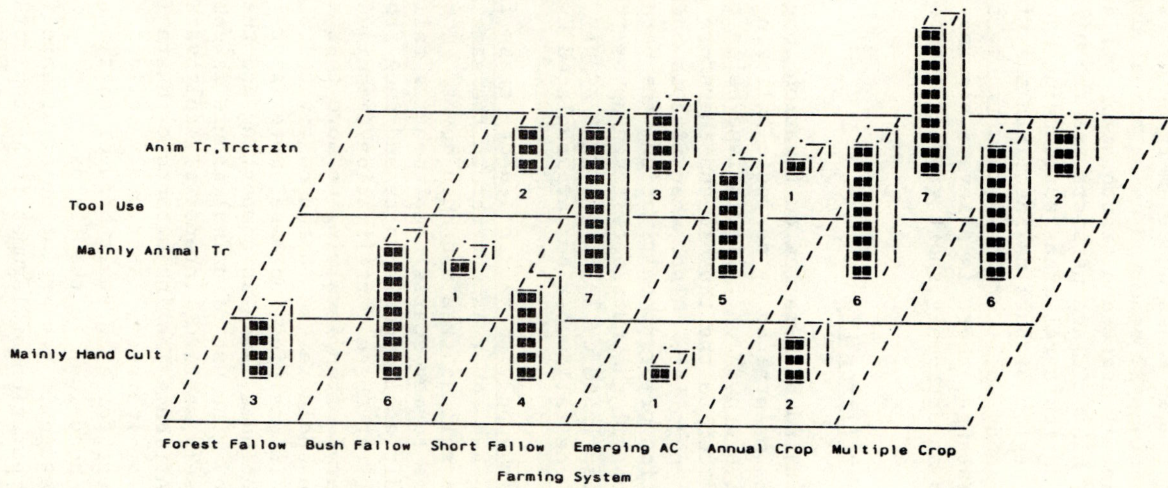

Figure 10.6. Evolution of farming systems and tools used

combination of animal and tractor power, while the remaining four continue to use hand hoes. The dominance of mechanical tillage (both animal draft and tractors) becomes more prominent as short fallow is replaced by permanent cultivation. Of the 30 cases under emerging or established permanent cultivation, 27 reported using animal draft or a combination of animal draft and tractors for tillage. The exceptions where hand tillage persists are the hill slopes of Sukumualand, agricultural areas surrounding Kano city, and the Yatenga region of Burkina Faso. In the first 2 cases the light sandy soils are easy to work by hand and in the last case, Yatenga, there is a severe plowing-sowing trade-off due to an extremely short growing season.

Development of Fertilizer Use

Under forest and bush fallow cultivation long-term soil fertility is maintained by periodic fallowing of land. Renewed vegetative growth on fallowed land helps to return fresh organic matter to the topsoil and therefore recharges it with nutrient supplies. Also, when fire is used for clearing vegetation prior to cultivation, the burnt ashes return to the topsoil the nutrients taken up by tree and bush cover. This closed cycle of nutrient supply is disrupted when long fallow periods are replaced by grass fallows.

The nutrient supply to the soil under grass fallow declines since grass cover cannot return the same amount of nutrients to the soil as tree and bush cover. Accordingly, at this stage the farmer starts complementing fallow periods with additional organic waste and dung from livestock. At first these fertilization techniques are fairly rudimentary, often involving no more than a periodic transport of household refuse to the cultivated plots.

As farming intensities increase, more labor-intensive fertilizing techniques, such as composting and then manuring, evolve. The use of animal manure is common in most of the densely settled, intensively cultivated pockets of Sub-Saharan Africa. The inhabitants of Ukara Island in Lake Victoria, for instance, laboriously collect three tons of manure per year for each adult head of cattle and transport it by head load to the fields.

The final stage in the evolution of organic fertilizer use is the incorporation of legumes in a crop rotation cycle as green manuring. Green manuring, along with other fertility-restoring measures, is a common practice in several parts of India and China. The use of cowpeas in the rotation cycle is becoming increasingly common among the permanently cultivated areas of Africa.

At the multi-cropping stage, one tends to observe increased use of chemical fertilizers as a substitute for the labor-intensive manuring techniques. Such general use of chemical fertilizers is still very rare in Sub-Saharan Africa, although the use of fertilizers for select crops, such as cotton and groundnuts, which are produced for the market, is becoming common.

Figure 10.7 shows the evolution in organic fertilizer use with an intensification of the farming system. Under the forest fallow system we observe no organic fertilizer use, and under bush fallow there is a move toward the use of kraal dust and one case of compost use, although the majority of cases still do not use any organic fertilizer. By the grass fallow stage, however, there is a marked switch with the majority of cases (8 out of 14) using kraal dust or other more intensive techniques. Most of the cases under annual and multi-cropping use some form of organic soil-fertility-restoring techniques, with the majority using animal manure. The two exceptions are from the Arusha region of Tanzania, where the very fertile volcanic soils do not yet require fertility restoration.

The patterns of chemical fertilizer use are presented in figure 10.8. The categories of chemical fertilizer use are: no use; use on select crops; and general use for all crops. The general use of chemical fertilizer is still not common, and only 7 of the 56 cases report such use; of these, 5 are under permanent cultivation systems. The use of chemical fertilizers for marketed crops is more generally prevalent, and 34 of the 56 cases reported such use. It is interesting to note that there is no general relationship between farming intensity and chemical fertilizer use for select crops. This is perhaps because even under low intensities of farming, some market crops are grown on small permanent plots of land. The cases where no chemical fertilizer is used follow farming intensities more closely; 10 of the 13 cases of no fertilizer use practice forest, bush, or short fallow cultivation. Of the 3 annual cultivation cases not using chemical fertilizers, 2 are from the volcanic soil locations of Arusha region, Tanzania, and 1 is an exclusively subsistence crop production case in South Nyanza, Kenya.

Figure 10.9 highlights the complementarity between organic and chemical fertilizer use. The striking feature

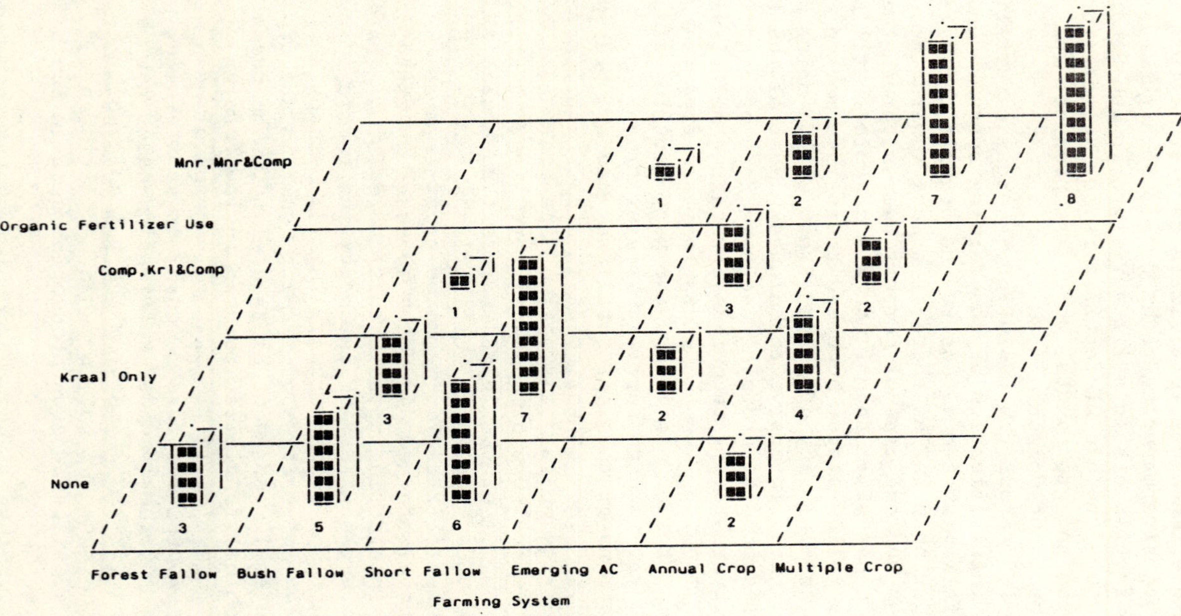

Figure 10.7. The development of organic fertilizer use

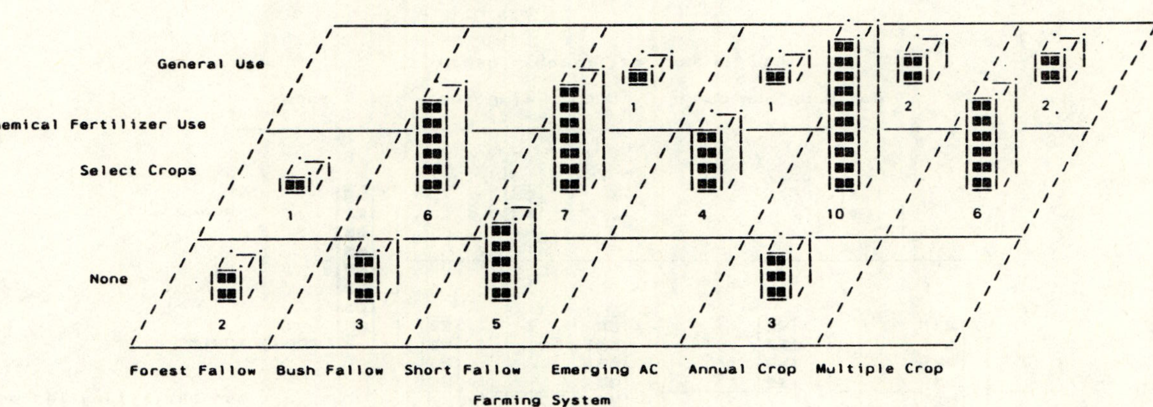

Figure 10.8. Farming intensities and chemical fertilizer use

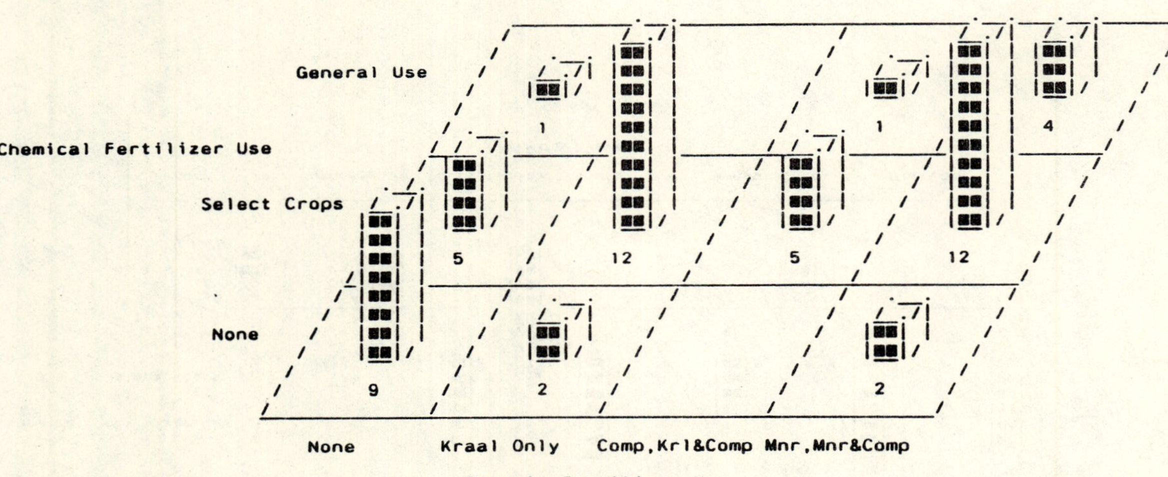

Figure 10.9. Complementarity between organic and chemical fertilizer use

here is the selective use of chemical fertilizers occurring in association with the general use of organic fertilizers. We have no cases of general chemical fertilizer use in the absence of organic fertilizers, although we have two cases of exclusive organic fertilizer use.

Changes in Land Rights

It is well known that the most prevalent organization of land rights under low population density is "communal property" where members of particular lineage groups have general cultivation rights; that is, they are assured the right to cultivate a plot, but when they abandon it again to fallow, the cultivation right to that plot reverts to the lineage. On the other hand, intensive high population density areas typically have "private property" rights; that is, cultivators have a large array of rights to specific plots. The distinction between communal and private property, however, is a rather harmful oversimplification, as it focuses on a simplistic legal codification of land rights rather than on the process of "induced institutional innovation" by which societies move from general land rights to specific land rights by the gradual addition of one specific right after another. While there are enormous variations in how this process proceeds and while political powers have often interfered with it or speeded it up, there are general tendencies as described below.

With general land rights, cultivators typically own only the right to cultivate in a particular region. A lineage head assigns the right to use a specific plot to cultivators who retain this right as long as they actually cultivate the plot; when the current cultivator departs--usually to leave the plot fallow--the use right to the plot reverts to the lineage. In very land-abundant environments, outsiders are often welcome and general cultivation rights may thus be available to lineage groups that are broadly defined to include almost everyone. When population densities increase, the groups or lineages become more narrowly defined and more people are regarded as outsiders who are excluded from the general cultivation right.

With the development of specific land rights, the cultivator can begin to assert certain rights in specific plots, starting with the right to resume cultivation of the specific plot after a period of fallow. At a later stage he asserts, and will receive, the right to assign the plot to an heir or temporarily to another member of the same lineage; the use right in the plot no longer reverts to the lineage head. Such changes often occur at different times in different regions or subregions of a country. Often a system is still described as a "communal property rights system" when all land already is passed from generation to generation within individual families. With increasing population density, the rights assignable by the individual cultivator become more extensive and may include the right to rent out the land to other lineage members, or even to outsiders. Rights to graze cattle on fallow plots or stubble shift from all members of the community to the individual. Eventually individuals acquire the right to sell land to other lineage members and even to outsiders.

The transition to specific land rights improves incentives to undertake investments in specific plots, investments that are required for the intensification of production and preservation of fertility.

Population growth is not the only process that moves societies toward more specific land rights. Any of the other factors that cause intensification have the same effect. We therefore often see large variations in the number of specific rights allocated to individuals within different parts of the same country, depending on the degree of land fertility and market access.

During our field visits in Sub-Saharan Africa, we asked a series of questions that allowed us to determine if and how outsiders could acquire land in a particular society. For this paper we categorized land acquisition as follows: (1) easy access to land, where outsiders can obtain land merely by asking; (2) delayed access to land, where outsiders have to work for a year or more as farm labor before acquiring the privilege to cultivate their own plots; (3) no access to land, where outsiders cannot acquire any land in the village but there are as yet no direct sales of land; (4) private property, where direct sales of land are possible and prevalent.

The relationship between access to land and population density is presented in figure 10.10. As expected, it is relatively easier to acquire land in areas with sparse populations than in areas with very dense populations. Among the very densely populated locations, direct sale of land is prevalent in 9 of the 10 cases; the remaining case has no access to land at all. Of the 14 cases under dense population, one observes the transition to codified private property: six locations have no access to land and seven

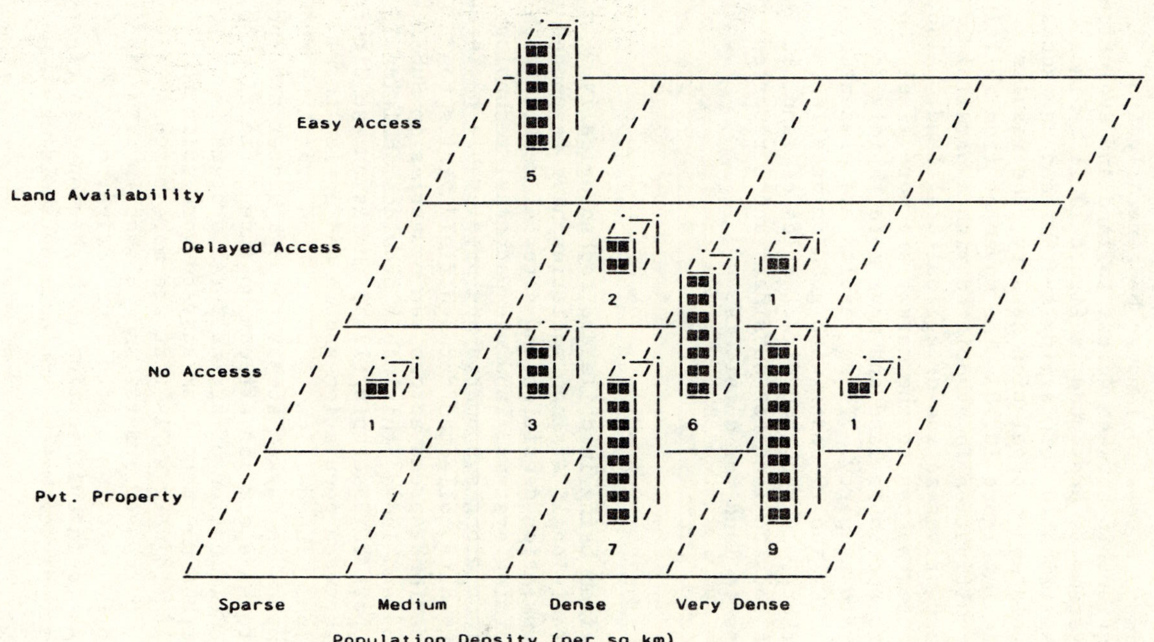

Figure 10.10. Population density and access to land

have switched to private ownership of land. Moving to medium population locations, we find an absence of private ownership and a transition from delayed access to no access to land. It is only in the sparsely populated locations that we find easy access to land to be generally true.

The relationship between access to land and evolution of farming systems is presented in figure 10.11. Since both the intensification of farming systems and the access to land are related to population density, the results match the ones in figure 10.10. The movement from shifting to permanent cultivation is associated with a parallel movement toward privatization of agricultural land. Also, areas with better access to the market are more likely to deny outsiders access to land and move more rapidly toward codified private property. These results are presented in figure 10.12. Of the 29 cases with good or excellent access to markets, 14 reported direct sales of land, 12 reported no access to land for outsiders and 2 reported delayed access to land for outsiders.

CONCLUSIONS

1. Far from being immobile and technologically stagnant, "traditional" African societies have responded to changes in population densities and external markets with changes in farming systems, land use patterns, technology, and institutions along systematic and predictable patterns.

2. Using data collected through field visits to fifty-six locations spread across ten countries in Sub-Saharan Africa and India, this paper provides additional empirical support to Boserup's (1965) thesis on the direct relationship between population growth and agricultural intensification. In addition to population growth we show that improvements in market access through better transport infrastructure have similar effects on intensification.

3. Intensification of agricultural systems is associated with a movement from easy-to-work soils to relatively hard-to-work soils that are more responsive to labor inputs, purchased inputs, and to such investments in land as drainage, erosion control, or irrigation. The responsiveness to intensification is higher on deep clayey soils that have higher water- and nutrient-holding capacity.

4. Intensification is generally associated with increased labor requirements per hectare of cultivated area. Increased labor is required not only for cultivation

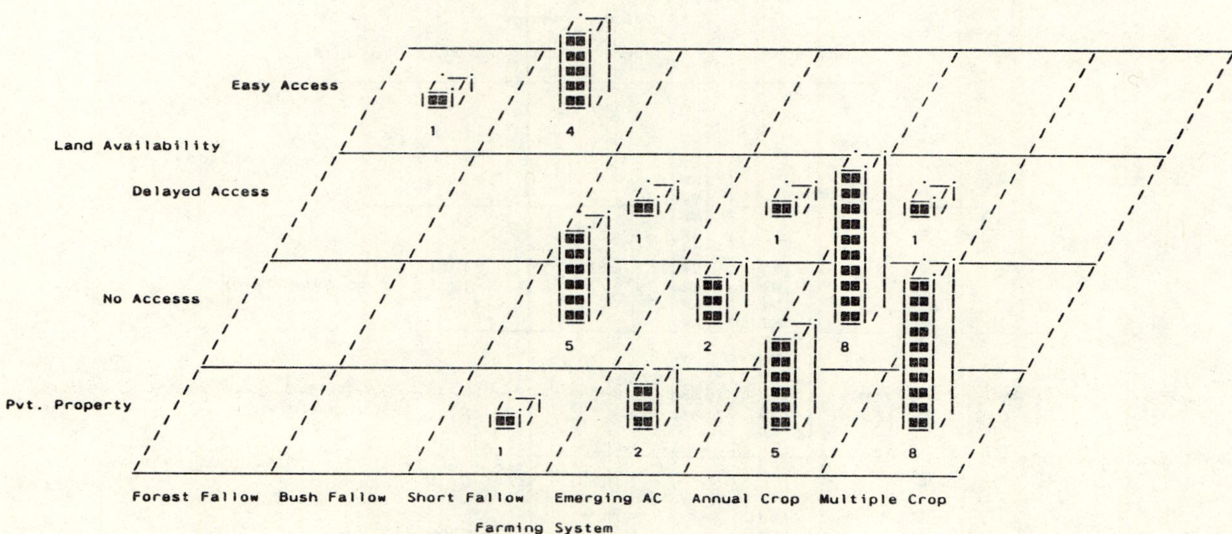

Figure 10.11. Farming intensity and access to land

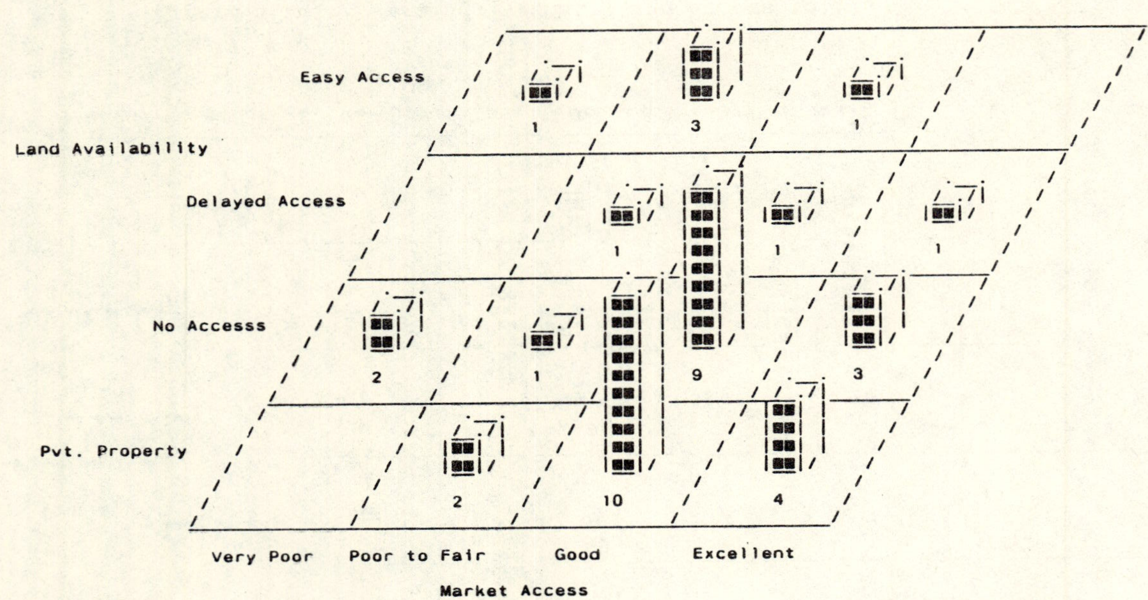

Figure 10.12. Market access and access to land

but also for land investments such as terracing and irrigation, for maintaining soil fertility through intensive manuring techniques, and where feasible, for the maintenance of draft animals.

5. The switch from the hand hoe to animal-drawn plows and later to tractors is closely associated with the evolution of farming systems. The switch to animal-drawn plows occurs around the short fallow stage and not before, because it is only at this stage that the overhead labor costs for destumping and leveling fields and for training, feeding, and maintaining animals are the lowest. At the stage of annual cultivation, tractors and animals become almost perfect substitutes, as it is here and not before that the choice of techniques analysis becomes relevant.

6. The substitution of fallowing first with simple and then with more evolved manuring techniques is likewise related to the evolution in farming systems. The general use of chemical fertilizers, although associated with intensification, is not yet common in Sub-Saharan Africa. The select use of fertilizers for specific market-oriented crops, however, is on the increase almost irrespective of farming intensity.

7. Finally, we show that the institutional arrangements for the acquisition of land by individuals within the group and by "outsiders" are not rigid but do change as increasing population densities or improved market access make land scarce. Land acquisition that is extremely easy under shifting cultivation becomes more and more difficult as intensification leads to more narrowly defined groups or lineages and therefore results in the exclusion of large numbers of people from acquiring the rights to cultivate. The ultimate institutional change, and one that commonly occurs under high population densities, is one of clearly defined private property rights with the ability to buy and sell land.

NOTES

*This paper is a shortened version of Hans P. Binswanger and Prabhu L. Pingali, "The Evolution of Farming Systems and Agricultural Technology in Sub-Saharan Africa." It was presented at the Alexander von Humboldt Award Colloquium at the University of Minnesota in 1984. See also, Hans P. Binswanger, "Evaluating Research System Performance and Targeting Research in Land-abundant Areas of Sub-Saharan Africa."

REFERENCES

Binswanger, Hans P. "Evaluating Research System Performance and Targeting Research in Land-abundant Areas of Sub-Saharan Africa." World Development 14(4)(1986): 460-75.

Binswanger, Hans P. and Prabhu L. Pingali. "The Evolution of Farming Systems and Agricultural Technology in Sub-Saharan Africa." In Technology, Human Capital, and the World Food Problem, 29-52, edited by G. Edward Schuh. University of Minnesota Publication 37, Department of Agricultural and Applied Economics, 1986.

Binswanger, Hans P., and Vernon W. Ruttan. Induced Innovation: Technology, Institutions and Development. Baltimore: Johns Hopkins University Press, 1978.

Boserup, Ester. The Conditions of Agricultural Growth. London: Allen and Unwin, 1965.

________. Population and Technological Change: A Study of Long-Term Trends. Chicago: University of Chicago Press, 1981.

Hayami, Yujiro, and Vernon W. Ruttan. Agricultural Development: An International Perspective. Baltimore: Johns Hopkins University Press, 1971 and 1985.

Pingali, Prabhu L., and Hans P. Binswanger. Population Density and Agricultural Intensification: A Study of the Evolution of Technologies in Tropical Agriculture. Agriculture and Rural Development Department, Research Unit, Report No. ARU 22. Washington, D.C.: World Bank, 1984.

Pingali, Prabhu L., Yves Bigot, and Hans P. Binswanger. Agricultural Mechanization and the Evolution of Farming Systems in Sub-Saharan Africa. Baltimore: Johns Hopkins Press, 1987.

Ruthenberg, Hans. Farming Systems in the Tropics, 3rd ed. Oxford: Clarendon Press, 1980.

Ruttan, Vernon W. Agricultural Research Policy. Minneapolis: University of Minnesota Press, 1982.

11
Low Input Technologies for Sustainable Agricultural Systems

Richard R. Harwood

The terms low input and sustainable represent characteristics of farming systems that tend to become all-inclusive as they gain popularity. This chapter will attempt to restrict their meaning, illustrate their usefulness and importance in specific environments, and show their interrelationships and how they translate into specific technologies.

SUSTAINABILITY DEFINED

The concept of sustainability has the following dimensions:

The Time Dimension. Some aspects of farming systems sustainability have a five- to ten-year horizon, while others extend to centuries. Farmland preservation and soil conservation are examples of the distant horizons, while farm ownership transfer from parents to children and groundwater contamination are aspects of near-term sustainability.

Social Sustainability. The most commonly discussed social element is the ability of a farm family to maintain farm ownership on an indefinite basis through changes in the overall farm economy. Critics of this idea hold that in an industrial society stability of ownership is unnecessary or even undesirable. It is felt that turnover is inevitable or even desirable in response to the changing cost and demand for capital. Proponents of farm family sustainability hold that not only should ownership remain with families for extended periods, but that the technologies used should allow for a "reasonable" amount of human participation. Both of these aspects are seen as related to the viability of rural communities.

It would appear that in developing countries farm family stability would receive high priority for the major reason that alternative employment and living conditions are probably not as attractive to displaced farm families. Those technologies or economic management practices that may place landholding at risk must be viewed with caution if this social element is important. In addition, agriculture must provide suitable employment opportunities. The productive use of labor is highly important in most Third World countries. Finally, agriculture must be safe for those participating in it as well as for those using agricultural products.

Economic Sustainability. Economic sustainability refers to the long-term economic viability of the farm unit.

A first requirement, then, is that farm investment (and the associated risk) should be modest, leading to production levels that may be less than maximum for the available technology. Fluctuation (and hence risk) in the agricultural economic environment has been consistently underestimated by the scientific community.

Second, the farm unit should have several production options and levels open to it.

A third dimension of economic sustainability is that of the entire agricultural sector or a portion of that sector. Individual commodities such as sugar, livestock, or feedgrains may come under significant stress. Sustainability of entire sectors or subsectors depends to an extent on national policy as well as on the international marketplace. Within each sector, individual farmers can only increase efficiency or diversify into other sectors.

Maintenance of Soil and Genetic Resources Bases. The importance of maintaining soil and genetic resources bases is in little dispute. Both resources are important to the developed and developing world.

Minimization of Environment Pollution. As population pressure on the land increases, there is increasing demand for clean water and the absence of biocides in the environment. There is increasing demand that agricultural production materials be kept within the farm and field boundaries and in the upper layers of the soil profile.

Lowered Use of Consumable Industrial Inputs. Most consider a lowered or more efficient use of consumable industrial inputs to be an essential element of sustainability for the intermediate and long term, as it impacts both the environment and the fossil fuel resource base. This must be done while maintaining a high production potential. It can only be achieved through use of "low input" technologies.

The bottom line is that a sustainable agriculture must make optimal use of the resources available to it to produce an adequate supply of goods at reasonable cost; it must meet certain social expectations, and it must not overly expend irreplaceable production resources.

LOW INPUT CHARACTERISTICS

"Low input" systems, as they are most commonly defined, minimize the use of consumable industrial resources. I have injected the word consumable, which is not otherwise commonly used, to include inputs such as fertilizer, pesticides, or similar materials that are expended on a given crop or livestock enterprise over a short time span. The low input "movement" in U.S. agriculture originated in the organic philosophy and has been spurred by pressures for reduction of adverse environmental impact and by the high cost of inputs relative to product price. The approaches being taken fall into two categories: the use of input substitution technologies and systems integration efficiencies. The goal is to achieve a high level of productivity at competitive prices. A brief outline of those technologies follows.

Input-Substitution Technologies

Input-substitution technologies are technologies that are utilized, to an extent, by all farmers, but some of which, such as biological nitrogen fixation, tend to be used more by farmers whose systems are structured for low inputs.

Biological Nitrogen. In U.S. agriculture (extensive land use, capital intensive, and power based), biological nitrogen fixation (BNF) comes nearly exclusively from cash crops when nitrogen prices are low. With the disassociation of livestock and feed production, the rotation of legume-hay crops with cash grain is seen less frequently. As nitrogen price rises with respect to product price, rotations with hay and with leguminous cover crops are again used. In Delaware, hairy vetch came into use as a winter cover crop in corn-growing areas when nitrogen reached seventeen cents to twenty cents per pound. The use of vetch and clover in feedgrain rotations began in Pennsylvania when nitrogen went above twenty cents per pound. In a power-based system where tillage poses little problem or where zero-till planting can be used, the use of herbaceous perennials in rotation is governed by price of nitrogen and length of growing period. In colder zones, nitrogen fixation is less, with a subsequently higher cost per unit of nitrogen fixed.

In tropical Third World countries, the crop pattern does not change rapidly in response to nitrogen price. The lack of short season legumes for rotation is one factor. Seed production difficulties seem to preclude the use of small-seeded legumes. Second, the increased tillage required for conventional cover cropping is difficult where power is limited. Third, the intensity of cropping and the moisture limitation on the cropping season reduce potential. Under lowland conditions the large-seeded sesbania seems to offer promise. Under such conditions leguminous trees seem to offer significant potential. In general, there has been considerably less development of technology in the area of leguminous cover crops for the tropics than for temperate zones, but the potential has not been carefully explored. A workshop at IRRI in May of 1987, entitled "Sustainable Agriculture: The Role of Green Manure Crops in Rice Farming Systems," constituted a state-of-the-art summary of available technology and research.

Host Plant Resistance. The incorporation of both disease and insect resistance into crop plants is an obvious input-substituting approach with significant environment-protecting implications.

Increase of Crop Plant Nutrient Uptake Efficiency. The breeding or selection of cultivars tolerant to low soil-available nutrient levels is a significant step toward lower input levels. The work in breeding rice for tolerance to low levels of zinc and the selection of forage species in Colombia for low-phosphorus soils are prime examples. Genetic engineering studies to select more efficient strains of root-rhizosphere-inhabiting bacteria could provide technologies for the future.

The Use of Biocontrol Agents. While some inputs must be used in biocontrol, they tend to be small and of a nature that usually has less potential for environmental disruption. Current bioengineering research has significant potential for creating technologies in this area. The direction of technology movement in both the biocontrol and host-plant resistance area is to internalize biological control in the crop system, reducing the need for inputs. Such crop systems become more information intensive; they have more genetic information, as well as management information, embodied in them.

Botanical Insecticides. The use of locally grown botanicals has been largely bypassed in recent years with the advent of low-cost synthetics. With insecticide development costs escalating and research budgets being diverted to other areas, the future could well see a resurgence of interest in the little-exploited region of locally produced and locally processed botanicals. A wide range of insecticide-producing plants is known, but they have received little research attention (Stoll 1986).

Integration Efficiencies

Integration efficiencies of systems are both complex and sensitive to the socioeconomic environment. This area is so complex that only a brief summary will be given here. A more complete discussion is given in Francis, Harwood, and Parr (1986). These effects are the most important in affecting efficiency of resource use and hence productivity!

Crop Rotation/Intercrop Efficiencies. There has been a resurgence of research on rotation effects during the past few years in the United States in response to the need for cost reduction, particularly with respect to feed grain production costs. The studies tend to be very interaction specific. Several studies show corn yields to be increased by about 10 percent following soybeans, regardless of the fertilizer level. Several other studies have looked at corn production costs following alfalfa hay.

The most striking and all-inclusive studies, however, have been the comparisons of organic and conventional production. Organic field crop systems of field crop production always include complex four-to-six-year rotations, sometimes including hay crops. Chemical inputs are not used in the organic systems. The complex rotations are characterized by a combination of input substitution technologies and rotation effects that substitute for at least most of the need for industrial inputs. Overall production costs range from 10 percent to 30 percent lower. There is usually somewhat more labor and machinery cost, and production ranges up to 10 percent less. These methods are known and well developed among practicing farmers in temperate zones where mechanization and a legume-seed industry are available. It is felt that they also depend on a warm season/cold season annual shift. Some of the integration efficiencies of the rotation appear to be:

- increased use of BNF crops (hay, cover crops, etc.)
- significant cost reduction in weed control, insect control, and soil-borne diseases
- improved nutrient cycling within the field (soil/crop/organic matter/soil flora and fauna/crop)
- improved soil tilth, water-holding capacity, and reduced erosion

In developing countries, many of these integration efficiencies are built into traditional systems. In the northern zone of the tropics--for instance, in northern Pakistan, Bangladesh, and in Nepal--common rotations are maize-wheat or rice-wheat. Here one has both warm season/cool season and wet/dry effects on weeds. Under such conditions weed control is usually of minimal concern if draft animal use and animal-drawn implements are at all adequate. Here the rotation effects dominate if crop management is even modest. In the continuously wet upland areas such as portions of Indonesia and the Philippines, rotational effects on weed control are far less effective and herbicide input is much more relevant.

Intercrop Effects. There have been sporadic attempts to study various intercrop systems over the past several years. Such systems tend to be very specific in their adaptation to particular climatic and socioeconomic conditions. The most important advantage of intercrops of annuals seems to be for extension of the growing season, especially where power for tillage is limiting. Such systems tend to have modest, if any, advantage in terms of crop growth efficiency. Intercrops of perennials, through use of the multi-story effect, can often have significant advantage.

Crop/Livestock Integration. Crop/livestock integration is a very commonly found integration effect in much of Asia. Integration efficiencies increase markedly when the transition is made to increased stall feeding and the collection and careful handling of manure. The livestock-carrying capacity of such systems often depends very much on off-season feed availability.

Efficiencies of such integration include an increase in the value of such crop by-products as feed, and in the feed value of weeds and local grasses. In animal systems a range of new crops assumes greater economic value, increasing the crop options available to farmers.

Multipurpose fodder/fuelwood trees, as well as peanuts and various grain legumes, all have higher value in the cropping system when used as animal feed supplements during the off-season. Livestock, on the other hand, increase the options for nutrient concentration in manure and for field-to-field nutrient movement.

Nutrient Cycling. There are two basic types of cycling efficiencies: those operating between fields and those operating within a field. Between-field flow is seldom practiced to a major extent unless livestock are involved, with feed coming from the fields and the accompanying removal of large amounts of nutrients. The manure and bedding then is returned to selected fields. In field-crop systems with farm size over one hectare and with at least some access to fertilizer, farmers often will not collect and use manure from their own stall-fed animals. They begin to make use of manure first on high-value crops. Philippine farmers fall in this category. With farm sizes well under one hectare and fertilizer either expensive or not available, farmers will use manure extensively on field crops. Nepal is a case in point. In Bangladesh, where bullock carts are available for hauling, there seems to be a willingness of farmers with somewhat larger holdings to move manure. It is extremely rare for farmers to move material between fields if it is not in the form of either animal feed or manure. The cycling of compost in China is one example, but then the source of nutrients is always a concentrated one.

Within-field nutrient use efficiency depends heavily on crop rotation, use of cover crops, crop residue handling, and tillage methods. Some traditional systems are characterized by a ridge-and-furrow or a mound type of bed formation with localized or concentrated placement of crop residues and animal manure. This is especially effective where the soil is extremely poor, farms are small, or organic matter and nutrients are scarce, as in high-elevation Himalayan areas.

Composting is sometimes used in tropical systems as part of a nutrient-cycling process. Where it is most common, it serves to reduce possible health problems in the field application of human waste, as in China. Its most common functions are to make nutrients from plant residues more readily available to nutrient-demanding crops such as vegetables, or to tie up concentrated nutrients from animal manure by composting with higher carbon-content material such as straw.

Fuel/Crop Nutrient Links. In rural areas of high population density, there is often an increasing shortage of cooking fuel. When this occurs crop residues and often dried animal manure are burned. The loss of nitrogen (and organic matter) from the farm system increases the nutrient drain. It becomes imperative to replace each loss from outside sources, yet these areas typically have less access to commercial sources of nutrients. The fuel (firewood) problem must be solved before a longer-term solution is possible for crop fertility. This condition seems to become particularly acute in areas of moderate-to-low rainfall where rural population pressure is high.

Conditions under Which Low-Input Technologies Apply

Input substitution technologies, with the possible exception of biological nitrogen fixation, are applicable across the board. Crop genetic improvement by incorporation of insect or disease resistance fits nearly everywhere. Research costs are typically not borne by farmers, so such technologies come essentially as a free good to the farmer (assuming that the "improved" cultivar is equally valuable in all other aspects).

Biological nitrogen often depends heavily on system integration if it is to make significant contributions. Integration, in turn, depends on the balance of factors shown in figure 11.1. The combination of lower labor and management costs, along with high input costs relative to product price, favors system integration. Small farm sizes with a demand for maximum productivity per unit of land, if mechanization costs are not a factor, also favor integration.

A summary of the relationship between input costs and farm size, with respect to low-input technology application, is seen in figure 11.2. It is assumed that internalization of the cost of environmental contamination acts the same as an increase in input price.

A summary of the economic environments relative to low-input technologies is as follows:

A. Third world areas of moderate to high production potential. Rural infrastructure is well developed and inputs are available at moderate prices. Examples are Central Luzon in the Philippines, irrigated rice areas

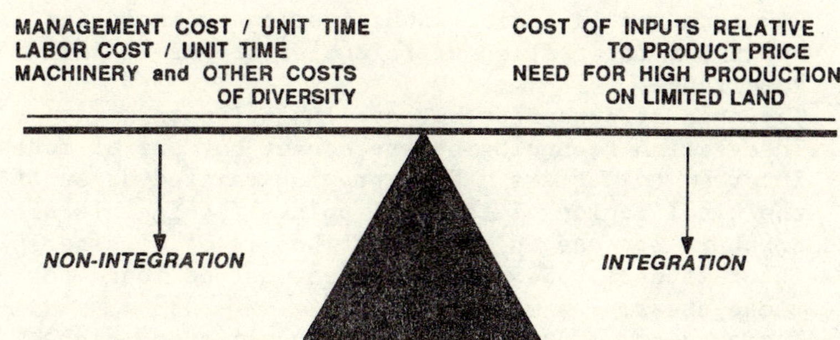

Figure 11.1. The balance of factors that influence systems integration

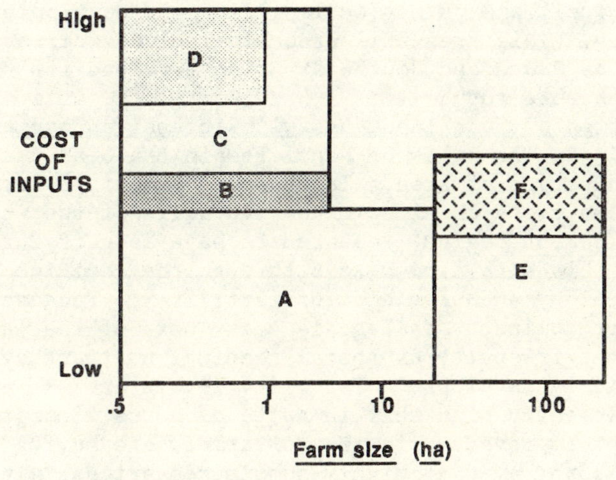

Figure 11.2. Input cost/farm size relationships to low input technology application

in the upper central plain of Thailand, the Punjab in Pakistan and in India. Substitution technologies apply but integration efficiencies are of minimal importance.

B. <u>The area of expensive but available inputs</u>. Integration technologies are sought but are of modest interest to farmers. Rice-growing rainfed areas of the Bicol region of the Philippines fit in this area. Sesbania for use in rice rotations is of considerable interest here. New technologies must be found for these areas.

C. <u>The area of undependable input supplies and high prices</u>. Farmers depend on intensive crop rotation and livestock integration. The Terai and inner valleys of Nepal, much of the Philippines, Indonesia, Bangladesh, and other countries fall in this zone. This is an area of great potential for low-input technologies. Its extent depends both on world prices for fertilizer (and fuel) and on national pricing and development policy. This area will probably include well over half of South and Southeast Asian farmland for the foreseeable future.

D. <u>The area of zero input use with intensive nutrient cycling</u>. The hills of Nepal remain the best examples of this type of area. In Phumbdi Bhumbdi, a farming systems research site in the mid hills of the western development region, researchers have been trying to move this site from D to B through the addition of technology and provision of fertilizer. Because of the predominant small-scale dairy base of the farms and the intensity of animal feeding, nutrient cycling is highly developed. Fertility levels are sufficiently high that in spite of quite widespread use of improved varieties no farmers are buying fertilizer even at highly subsidized prices. In this case the nearby milk market and small farm sizes have encouraged a high level of crop/livestock integration, which results in minimal returns on cash inputs of any kind.

E. <u>The area of high capital, industrial agriculture in developed countries</u>. This area has minimal integration and few biological interactions.

F. <u>The area of integration and mechanization</u>. This area is the direction that a portion of U.S. field crop agriculture is taking as rotations and other integration practices are introduced. Organic agriculture is on the upper boundary of zone F, placed there because farmers, for a variety of reasons, attach high cost to the purchase and use of consumable industrial inputs. A part of that "cost" may be concern with health or environmental impact. They have, in effect, internalized those costs and have minimized use of the products.

Farming Systems of the Future

It is felt by many that future systems will minimize the need for biocide inputs through "internalization" of control of various biological processes. Many of our low-input technologies lead in this direction (Harwood 1984). Secondly, they will have a high degree of closure of nutrient flow loops (Edens et al. 1982).

Such systems will reduce input requirements, will have higher efficiency in terms of output per unit of input (especially for nonlabor inputs), and will have less adverse environmental impact.

The Inclusion of Low-Input Technologies in Development Projects

A first step in the inclusion of low-input technologies in development projects is to determine the relevance of specific technology or groups of technology. Are supply and price situations demanding of such technologies and likely to remain so? If this is the case, the conceptual and specific technology goals should be stated and included in terms of reference for a given project. National development institutions should then be provided the incentives and (or) resources to produce and extend the relevant technologies. Where technical assistance is involved, the teams should be constituted to focus on relevant integration aspects. The technologies should be discrete and of measurable impact. The nitrogen-specific technologies coupled with those that reduce soil loss should usually have first emphasis. This will often be done in the context of crop-livestock integration. The potential and demand for such integration seems enormous in most of the Third World.

In summary, our capabilities in farming systems research, coupled with a growing conceptual understanding of farming systems integration processes, are opening new doors for development in the face of continuing resource shortages. We are employing new tools in an age-old way, it is hoped, with ever-increasing precision. While integration efficiencies are complex, the principles appear to be universal. Lessons learned from systems integration will have different expression in each environment, but can have extremely wide application. They signify the coming of the next revolution in scientific agriculture.

REFERENCES

Edens, T. C., and D. C. Haynes. "Closed System Agriculture: Resource Constraints, Management Options and Design Alternatives." Annual Review of Phytopathology 20(1982): 363-85.

Francis, C. A., R. R. Harwood, and J. F. Parr. "The Potential for Regenerative Agriculture in the Developing World." American Journal of Alternative Agriculture 1(2)(1986): 65-74.

Harwood, R. R. "The Integration Efficiencies of Cropping Systems." In Sustainable Agriculture and Integrated Farming Systems. East Lansing: Michigan State University Press, 1984.

Stoll, G. Natural Crop Protection Based on Local Farm Resources in the Tropics and Subtropics. Gaimersheim, F. R. Germany: Verlag Joseph Margraf, Tropical Scientific Books, 1986.

12
Agricultural Research in Nigeria: Organization and Policy

Francis S. Idachaba

The recent performance of Nigerian agriculture has not been impressive. Agricultural GDP in real terms declined at an annual rate of 0.4 percent during 1960-1970 (1960 base year) and at an annual rate of 1.1 percent during 1970-1980 (1970 base year).

Domestic production has been unable to match recent increases in population and income. The result has been steep increases in food and fiber prices. Food imports have risen dramatically in attempts to augment supplies and moderate domestic price inflation. Nigeria moved from being a net exporter of 300 tonnes of maize per annum during 1952-1955 to becoming an importer of 153,346 tonnes per annum during 1978-1982; from being an exporter of an annual average of 188,234 tonnes of palm oil during 1958-1960 and 894,485 tonnes of groundnuts during 1964-1966 to becoming a large importer of vegetable oils in the 1970s. Imports of food and live animals have risen from an annual average of 8.22 percent of total imports during 1967-1970 to 15.91 percent during 1981-1983.

The decline in food production could be attributed to falling productivity, a succession of adverse macroeconomic policies, and the failure of the national agricultural research system to transform modes of production, among other causes. Why, after almost a century of agricultural research, has the Nigerian agricultural research system not succeeded in significantly raising total or partial productivity?

This paper (1) reviews the historical evolution of agricultural research organization in Nigeria, (2) examines the meaning of agricultural research policy within a developmental context, (3) analyses agricultural research resource allocation priorities, and (4) identifies the key issues in agricultural research policy in a developing economy.

The Nigerian experience has relevance for other African countries for at least three reasons: first, it is a large country spanning different ecological zones needing different agricultural research outputs; second, development phases of the agricultural research system are closely related to the country's agricultural development experience (a replica of agricultural development systems on the African continent); and finally, Nigeria devotes relatively large amounts of resources to agricultural research. The historical evolution of Nigeria's agricultural research system is presented in the first section, while the key issues in agricultural research policy in a developing country are presented in the second. The third section contains the Summary and Conclusions.

HISTORICAL EVOLUTION OF NIGERIA'S AGRICULTURAL RESEARCH SYSTEM

The evolution of Nigeria's agricultural research system conforms with Ruttan's classification of the origins of research institutes (Ruttan 1982). From modest beginnings in 1893, when a botanical station was established in Lagos, the national system has grown to eighteen separate agricultural research institutes (table 12.1).

Salient Features of the Development of the Nigerian National Research System

The first research institutes in Nigeria's national system were created to cater largely to the needs of export crops. The National Cereals Research Institute (NCRI) owes it origin to the erroneous belief of the British Empire Cotton Growing Corporation in 1905 that Ibadan, in a rain forest zone, was a suitable location for cotton research. After a few years of unsuccessful trials, cotton research moved north, eventually settling in Samaru, Zaria, in the Savannah zone. The Institute for Agricultural Research (IAR) dates back to 1922, when a regional research station was established at Samaru as headquarters of the Department of Agriculture of the northern provinces. Over the years,

TABLE 12.1
The Distribution of Agricultural Research Institutes, Nigeria, 1986

	Institute	Research Emphasis	Headquarters Location (State)
1.	Institute for Agricultural Research (IAR)	Sorghum, millet, wheat, barley	Samaru (Kaduna)
2.	National Cereals Research Institute (NCRI)	Rice, maize, grain, legume, surgane	Ibadan (Oyo)
3.	National Root Crops Research Institute (NRCRI)	Yams, cassava, cocoyams	Umudike (Imo)
4.	Institute of Agricultural Research and Training (IAR & T)	Cereals, legumes	Ibadan (Oyo)
5.	National Institute for Horticultural Research (NIHORT)	Fruits, vegetables	Idi-Ishin-Ibadan (Oyo)
6.	Cocoa Research Institute of Nigeria (CRIN)	Cocoa, kola, coffee, cashews	Gambari-Ibadan (Oyo)
7.	Rubber Research Institute of Nigeria (RRIN)	Rubber	Iyanomo (Bendel)
8.	Forestry Research Institute of Nigeria (FRIN)	Forests, wild flora, fauna	Ibadan (Oyo)
9.	Nigeria Institute for Oil Palm Research (NIFOR)	Oil palm, vaphia, coconut, dates	Near Benin Cith (Bendel)
10.	Kainji Lake Research Institute (KLRI)	Fish, irrigated crops	New Russa (Kwara)
11.	Lake Chad Research Institute (LCRI)	Fish	Maiduguri (Borno)
12.	National Institute for Oceanography and Marine Research (NIOMR)	Fish	Lagos
13.	National Animal Production Research Institute (NAPRI)	Livestock	Shika-Samaru
14.	Nigerian Institute for Trypanosomiasis Research(NITR)	Trypanosomiasis, onchocasiasis	Kaduna (Kaduna)
15.	National Veterinary Research Institute (NVRI)	Cattle	Vom (Plateau)
16.	Leather Research Institute Nigeria (LRIN)	Leather	Zaria (Plateau)
17.	Agricultural Extension Research Liaison Service (AERLS)	Extension services	Samaru-Zaria
18.	Nigerian Stored Products Research Institute (NSPRI)	Storage of food and export crops	Lagos

it has done most of the research on cotton and groundnuts (traditional export crops), in addition to food crops. The Nigerian Institute for Oil Palm Research (NIFOR) had its origins in the establishment of the Oil Palm Research Station in 1939. This was replaced by the West African Institute for Oil Palm Research (WAIFOR) in 1951 and by NIFOR in 1962. The Rubber Research Institute of Nigeria (RIN) was established for the purpose of research on rubber, a major export. The purpose of the Nigerian Stored Products Research Institute (NSPRI), the successor to the West African Stored Products Research Institute, was to do research on the storage needs of export crops, while the Cocoa Research Institute of Nigeria (CRIN) was established to cater to the needs of cocoa, a leading export crop.

Second were the pan-territorial research institutes, meant for research on export commodities of broad agro-ecological regions in Anglophone West Africa. The West African Research Organization, the umbrella organization for these institutes, was dismantled in 1962 in the wake of the nationalist euphoria that accompanied independence. This gave rise to the establishment of such national institutes as the Nigerian Institute for Trypanosomiasis Research (NITR), successor to the West African Institute for Trypanosomiasis Research; and NSPRI, CRIN, and NIFOR, already discussed. Consequently, Anglophone West African countries today have only the international agricultural research centers to fall back on for cross-country, broad agro-ecological research. An inevitable duplication of research efforts among countries has resulted.

Third were the research institutions that grew out of the research arms of ministries/departments of agriculture. IAR belongs to this category. So does the National Animal Production Research Institute (NAPRI), which owes its origins to the establishment of the Shika Stock Farm in 1927. Other examples include the Forestry Research Institute of Nigeria (FRIN), successor to the old Federal Department of Forestry Research; the National Veterinary Research Institute (NVRI), successor to the Federal Department of Veterinary Research; the National Cereals Research Institute (NCRI) and the National Root Crops Research Institute (NRCRI) both successors to the Federal Department of Agricultural Research.

Finally, there were research institutes that grew out of external assistance--especially from the joint Food and Agriculture Organization/United Nations Development Program (FAO/UNDP). These include the Kainji Lake Research Institute (KLRI), the Leather Research Institute of Nigeria (LRIN), and the National Institute for Horticultural Research (NIHORT).

The adoption of a federal constitution (1954) marked the beginning of the development of a cooperative federal-state agricultural research system with the creation of research institutes from the federal departments of agriculture and the regional ministries of agriculture. The constitution placed "scientific and industrial research" on the concurrent list, implying that both federal and state governments could engage in agricultural research. Yet there was an anomaly in that agriculture itself was left as a regional responsibility, with the federal government playing no role. There was no real coordination of agricultural research within the country. The Technical Advisory Committee, inaugurated in 1955, could act only in an advisory role, and even this appeared to have been a limited exercise within the federal agricultural research establishment.[1] The regions saw their role in agricultural research in the applied and experimental application of the results of basic research, which was seen as the proper domain of the federal government.

Laws and Decrees Establishing Research Institutes

Upon the dissolution of the West African Research Organization in 1962, the Nigerian Research Institutes Act, 1964, established four research institutes: CRIN, NIFOR, RRIN, and NITR.[2]

The Agricultural Research Council of Nigeria Decree, 1971, vested responsibility for coordinating all agricultural research in the council. The Agricultural Research Institutes Decree, 1973, empowered the Federal Commissioner for Agriculture and Rural Development to establish institutes to conduct research and training in any field of agriculture, veterinary sciences, fisheries, forestry, agro-meteorology, and water resources, *and to take over any existing state research station* (emphasis the author's). This was a major development because the 1963 constitution had placed "scientific and *technological* research" on the concurrent list, meaning that regional/state as well as the federal government could engage in agricultural research.

What appeared to be the *modus operandi* of agricultural research after the 1954 constitutional change was that federal departments would handle basic research while the regional departments would conduct applied research. The International Bank for Reconstruction and Development (now the World Bank) recommended in its influential 1955 report on Nigeria that all basic research on livestock, crops, fisheries, and forests should be done by the federal government, while the regions should concentrate on applied research or the experimental application of research findings (World Bank 1955). In one fell swoop, the 1973 decree destroyed the foundation being laid for a cooperative federal-state agricultural research system. One notable consequence was that no state government has, since 1973, established any agricultural research facility, either for fear of arbitrary seizure by federal authorities or because the decree provided a ready excuse for states to abdicate their responsibility, especially since state governments no longer had the power to tax farm produce.

By the Research Institutes (Establishment) Order, 1975, fourteen research institutes were created: three under food crops (NCRI, NRCRI, NIHORT); four in tree crops (CRIN, RRIN, NIFOR, FRIN); four in veterinary and livestock (NVRI, NAPRI, NITR, LRIN); and three in fisheries (LCRI, KLRI, NIOMR). Of the fourteen, four were created by the Research Institutes Act, 1964 (CRIN, NIFOR, NITR, RRIN); four were converted from federal research departments to full-fledged research institutes (NCRI, NRCRI, FRIN, NVRI); three were converted from research units of federal departments into full-fledged research institutes (LRIN, LCRI, NIOMR); and in one instance, a research arm of a university was converted to a research institute (NAPRI).

The National Science and Technology Development Agency Decree, 1977, repealed the 1973 decree and set up an executive agency to coordinate all research in Nigeria, agricultural and nonagricultural. The National Science and Technology Development Agency (NSTDA) was to advise the federal government on national science policies and priorities, prepare plans for the development of science and technology, prepare annual budgets for scientific research, and receive grants for allocation to research institutions. It was also to advise the government on the creation of new research institutes and centers and the reorganization of existing ones, as well as allocate special research projects to the universities. The agency was to facilitate the application of research results and advise on scientific and technical manpower requirements. The decree also empowered the NSTDA commissioner to take control of any existing federal or state research establishment, thus confirming the provisions of the 1973 decree that assigned a lame-duck role to the states in the establishment of research institutes. By the decree, all research institutes established under the 1975 decree were brought under the aegis of the NSTDA. The Research Institutes (Establishment) Order, 1977, established the Nigerian Stored Products Research Institute to conduct research on bulk storage of export products as well as locally consumed foodstuffs.

The 1979 constitution (now suspended by the military) placed "industrial and agricultural research" on the concurrent list, meaning that agricultural research could be conducted jointly by both federal and state governments. The ousted civilian administration scrapped the NSTDA and replaced it with a full Ministry of Science and Technology. The Buhari military administration scrapped the Ministry of Science and Technology and merged it with the Ministry of Education, but the Babangida military administration has once again delinked the two.

Unlike the development of the decentralized cooperative system in the United States, the research institutions did not develop in response to local, state, and national movements and pressure groups. The decentralized system in Nigeria developed largely to cater to the regional export economies, prompted by the needs of imperial transnationals and nurtured by a colonial civil service. The achievements of research institutions have rarely become political issues in Nigeria in the same way that they became campaign issues in state elections in the United States in the 1880s and 1890s (see Peterson and Fitzharris 1977). Thus, research institutes in Nigeria have not developed a tradition for quick, flexible responses to farmers' needs.

The Nigerian agricultural system has been a beneficiary as well as a victim of the oil boom. The creation of new institutes and the dramatic increase in resource allocations in the last decade were made possible by the petroleum revenue windfall. However, it also created a "fiscal superman" syndrome at the federal level and resulted in the federal takeover of all state-owned agricultural research institutes in 1975. This destruction of a decentralized, cooperative joint federal-state agricultural research system remains a major handicap of the national agricultural research system today.

The evolution of the national agricultural research system has mirrored developments in the agricultural sector. First was the precolonial low-level equilibrium era when aggregate supplies and demands for farm produce equilibrated at low levels, with farming systems evolving out of intergenerational experimentation. Second was the export crop era during which research focused mainly on export crops such as cocoa, groundnuts, cotton, and oil palm. This was followed by the postcolonial era marked by massive rural-urban migration, declining per capita agricultural GDP, soaring food prices, and mounting food imports. Internal socio-political pressures were generated to tailor the research system to the needs of the national food economy.

KEY ISSUES IN NIGERIA'S AGRICULTURAL RESEARCH POLICY

The inability of the national agricultural research system to spearhead the structural transformation of the Nigerian agricultural economy suggests a need to identify the critical issues involved in developing an effective system.

Nigeria did not have an explicit or coherent national agricultural policy for most of her history. Any policy as such could only be implied, and it consisted of ad hoc creations of research facilities that were largely the product of history and the perceived needs of the moment. What could be the main elements of a national agricultural research policy within a developmental context?

Objectives of Agricultural Research Policy. A clear statement of agricultural research policy objectives is required to rationalize research resource allocations and to ensure consistency with the objectives of the agricultural sector, the macroeconomy, and society's goals, values, and aspirations, as well as the nation's factor endowments. Within the African context, such objectives include (1) creation of new knowledge and technologies in the form of new production and consumption processes, new inputs, and new outputs; (2) liberalizing access to new high pay-off inputs through the introduction of new technologies and inputs that are largely scale-neutral; (3) substituting abundant and cheap resources for scarce and expensive inputs; (4) reducing food and agricultural production costs; (5) raising farmers' incomes; and (6) making the agricultural sector more responsive to price and other policy incentives by increasing its dependence on productive inputs that are more supply price elastic. Some or all of these objectives could be translated into quantitative targets in a national plan for agricultural research.

Agricultural Research Policy Instruments. Agricultural research policy instruments include the creation and location of research institutes and institutions, together with statutory allocations of commodity/input research responsibilities to manage the national agricultural research system. Other policy instruments include relative research resource allocations, wages and service conditions of research staff, training facilities to build up the institutional and intellectual capital of research institutes, and subsidies on inputs created from new knowledge and technology from the research system. Other complementary policy instruments include tariff policy on research materials and other imports, farm credit, support for output and input markets, and resource allocations to rural infrastructures. Farm input subsidies are also an important policy instrument, though they also serve some other macro policy objectives. The role of farm input subsidies is discussed more fully in Idachaba (1974, 1981b).

Policy must then be implemented, monitored, and evaluated, the evaluation requiring prior specification of performance indicators and measures.

Agricultural Research Policy Impact. The allocative consequences of agricultural research policy are relatively easy to handle--especially the effects of new technologies on relative factor scarcities and prices. The distributional consequences are harder to measure, though they are no less important.

Two observations are pertinent. First, agricultural research policy objectives are derived from agricultural sector objectives, which are derived from overall macroeconomic objectives. Macroeconomic objectives are derived from overall societal goals, objectives, and aspirations, which are conditioned by the domestic and international environments. Viewing policy objectives in this way--through a set of transformation functions--ensures consistency between "lower level" objectives (e.g. agricultural research) and "higher level" objectives (e.g. agricultural sector). It also ensures that changes in higher level objectives are sequentially transmitted to lower level objectives. For example, a new policy objective to become self-sufficient in main food staples could be translated into new research objectives to breed and disseminate high-yielding varieties of those food commodities of which the country is a net importer.

Second, agricultural research policy has concentrated on the supply side (top-bottom) to the neglect of the demand side (bottom-up), thereby causing several defects, some fatal, in the agricultural policy process in much of Africa. Policy objectives have been articulated largely by bureaucrats, with farmers and farmers' organizations playing no role. Policy instruments were identified and utilized by bureaucrats with little or no consultation with farmers and farmers' organizations. Consequently, the end users of research output have been unable to monitor research work and to articulate their needs. Thus there is a long time lag between the emergence of a problem and the widespread adoption of a solution.

Unless the national agricultural system operates within a general policy frame, programs and projects tend to be characterized by institutional and other policy "ad hockery" with frequent policy revisions, modifications, and, quite often, complete policy reversals. The absence of an explicitly articulated agricultural research policy statement largely explains the erratic mode in which agricultural research has evolved in Nigeria and other African countries over the years.

Resource Allocation Priorities

The following criteria, stated as working rules, can be identified for allocating resources to agricultural research, among commodities and inputs, and over time.[3]

Foreign Exchange Contribution of a Commodity. The working rule is, for a given target for foreign exchange savings/earnings, allocate research resources to a commodity in direct proportion to its relative importance as a foreign exchange earner/saver. But this leaves unanswered the question of the relative resource allocations to domestically traded goods as a class and internationally traded goods as a class. The colonial phase witnessed a very skewed allocation process in favor of export crops, almost to the neglect of food crops, both in aggregate allocations as well as in micro-resource allocations at the research institute level (tables 12.2 and 12.3). Relevant weights for allocation to export crop research under this working rule are the share of agricultural exports in total export earnings and the share of agriculture in GDP.

Fiscal Role of Crop. The real world in many African countries is filled with policy-induced distortions in the form of marketing board taxes that have become institutional realities. To this extent, these countries face a second-best problem of constrained optimization so as not to kill the goose that lays the golden egg. A working rule is, allocate research resources to crops in direct proportion to their relative importance in government revenues. The objective is to expand the production and revenue base of the crop.[4]

TABLE 12.2
Federal Government Allocations to Food Crop, Export Crop, and Industrial Crop Research, Nigeria, Second National Development Plan, 1970-1974

	Allocations ₦ mill.	Share of Total Expenditures on Crop Research %	Share of Total Federal Government Expenditures on Agriculture %	All Sectors %
Research on food crops	2.286	33.02	3.71	0.21
Research on export crops	4.376	63.20	7.10	0.39
Research on industrial crops	0.262	3.78	0.42	0.02
Total expenditure on all crop research	6.924	100.00	11.23	0.62
Total expenditure on agriculture (federal)	61.670	--	100.00	5.55
Total expenditure on all sectors (federal)	1110.188	--	--	100.00

Source: Underlying data from Second National Development Plan 1970-74 (Lagos: Federal Government Printer, 1970).

Value of Production. The working rule here is, allocate resources to commodities in proportion to their share of agricultural GDP.

Value of Urban Consumption. Foods with higher per capita urban consumption (relative to rural areas and probably reflecting a higher income demand elasticity) tend to command higher research priority. This is because of the high social costs of urban riots and political unrest. Nigeria has launched special programs to boost rice and dry season wheat production. Yet both rice and wheat are far below sorghum and millet in value of production.

Regional Development. Large sections of the country should not be left behind in the development process. The implied working rule here is that research resource allocations to crops should be proportional to their relative shares of total cropped area, to achieve balanced regional development.

Employment Generation Potential. Those African countries with abundant and cheap sources of farm labor supply should allocate research resources to crops that will fully utilize the abundant resource. In Nigeria,

TABLE 12.3
Export Crop Research Bias in Fertilizer Trials in Food and Export Crops Compared with Their Relative Importance in Hectarage, Northern Nigeria, 1952-1961

Crop	Fertilizer Trials, 1952-61		Crop Area		Crop Area Per Trial
	no.	%	ha	%	ha
Sorghum	87	13.59	3,189,375	36.63	36,659
Millet	8	1.25	3,249,000	37.22	405,000
Groundnuts	186	29.06	894,321	10.27	4,808
Cotton	23	3.59	477,090	5.48	20,743
Yams	90	14.06	370,980	4.26	4,122
Maize	28	4.38	352,755	4.05	12,598
Rice	59	9.22	138,105	1.59	2,341
Soyabeans	46	7.19	47,385	0.54	1,030
All Export Crops	300	46.88	1,418,796	16.30	4,729
Food Crops	340	53.12	7,287,215	83.70	21,433
All Crops	640	100.00	8,706,011	100.00	13,603

Source: F. S. Idachaba, Agricultural Research Policy in Nigeria, Research Report No. 17 (Washington, D.C.: International Food Policy Research Institute, 1980).

labor is expensive, and research resources should be allocated to enhance the utilization of the relatively abundant resource, land.

Politically Visible Crops. Social responses in urban areas to shortfalls in supply vary from food item to food item. Among the politically most visible are the convenience foods (bread and garri (processed cassava)), sugar, milk, and rice. The more strategic a crop is in political visibility, the higher the research priority that tends to be accorded it. This may be an act of self-preservation on the part of the ruling political class but it may also mark the beginning of endogenous technical change.

Nutritional Significance of Crop. The higher the nutritional significance of a food item or class of foodstuffs as a source of calories or protein, the higher the research priority that should be accorded it, using a given base period.

Value Added and Import Substitution. Research resources should be allocated in direct proportion to the share of the crop in domestic value added and also in proportion to the contribution in raw materials for import substitution.

Narrowing Wealth and Income Inequality. Research resource allocation should accord relatively high priority to those commodities whose production narrows existing wealth and income inequalities. This suggests research priority for new inputs that are within the economic reach of poor farmers; it also suggests research into the wealth and income distribution consequences of new technologies.

Foods that Poor People Eat. Research priority should be accorded those commodities that poor people eat, especially foods that are relatively income elastic at low-income levels. This will uplift the welfare of the poor.

Self-Reliance Strategies. Research resources should be allocated in direct proportion to food import dependency, meaning that research priority be accorded those commodities for which a country has high food import dependency ratios. In Nigeria, wheat, sugar, maize, and rice are obvious candidates for priority attention.

How to resolve conflicts in allocative criteria remains an unresolved issue--especially when conflicts arise from limited supplies of scarce resources. One possible solution is to streamline the transformation of relations between "lower level" objectives/priorities and "higher level" objectives/priorities so that the latter would implicitly assign weights to the former.

The Integration of Research, Extension, and Training

The historical circumstances of the evolution of Nigeria's agricultural research system have almost guaranteed the lack of integration of research, extension, and training. Initially, there were the departments of agriculture for the northern and southern provinces. The federal constitution (1954) and the regionalization of agriculture gave birth to three regional departments and their relevant research arms, in addition to the research institutes created by the Research Institutes Act, 1964. Since agriculture became a regional responsibility--and by implication extension--the federal government was left with an impressive list of research institutes but was constitutionally "barred" from any agricultural extension activities. By 1975, the federal government had taken over all agricultural research institutes but no provisions were made for liaison with state extension services. The system was left with the worst of both worlds: it had neither the unique features of a Rothamsted Experiment Station nor the integrated system of the United States land grant college. While the Morrill Land Grant College Act (1862) and the Hatch Experiment Station Act (1887) provided the basis for America's decentralized cooperative federal-state agricultural research system (Peterson and Fitzharris 1977), Nigeria's Agricultural Research Institutes Decree (1973) and the National Science and Technology Development Agency Decree (1977) guaranteed that Nigeria would have one monolithic (federal) agricultural research system that remains unviable and unworkable in a heterogeneous federal setup.

With one or two exceptions, Nigeria's agricultural research institutions remain in total isolation from the universities--a great contrast to the United States system in which the state college of agriculture formed the base while the agricultural experiment station, attached to the college, conducted research (Peterson and Fitzharris 1977).

Institutional Responsibility for Agricultural Research, Extension, and Training

It is presumed that the political leadership in a state within a federal setup does not allocate scarce resources to agricultural research without concern as to whether residents in the state are in a position to reap the full benefits therefrom. That is, state decision makers are indeed worried about the "free rider" problem, about the benefits of research that "spill over" to residents of other states at no cost to them. Yet, transactions costs of exclusion of free riders appear prohibitive. The result is that there would be underinvestment in basic research--whose benefits are generally freely appropriable--if it were left solely to state governments (see figure 12.1 and Idachaba 1980). This suggests that the federal government should handle mainly basic research, while state governments should concentrate on applied and adaptive research, and local governments on local field trials to suit particular environmental niches.

National Research Institutions and International Research Centers. The International Agricultural Research Centers (IARCs) see their role primarily in the applied research area because of constraints imposed on them by their sponsors to show practical research results of regional and worldwide relevance. Research institutions in the developed countries feel they have the capability for basic research that ought not be duplicated by the IARCs or the national institutions. But this leaves two unanswered questions: who does basic research on problems that are specific to the tropical environment, and, what is the research portfolio of a mature IARC? The complementarities between national systems and IARCs also center around the sequencing of research activities: how do they share the sequencing between production, maintenance, and processing research?

Extension services are best rendered through a decentralized system because of their location specificity--implying that this is the primary responsibility of state and local governments. However, extension knowledge and technology that have general applicability across ecological zones and social systems should attract federal resources. More or less the same arguments apply to training.

Public and Private Sector Involvement in Agricultural Research. Active private sector involvement in agricultural research by colonial multinationals ended in the 1960s and local agencies of multinationals have not picked up this responsibility. No new grounds have been broken with respect to joint public-private sector participation in agricultural research since the exit of the British Empire Cotton Growing Corporation, the cessation of cotton exports, and the establishment of the Nigerian Cotton Board (a public parastatal). Are there

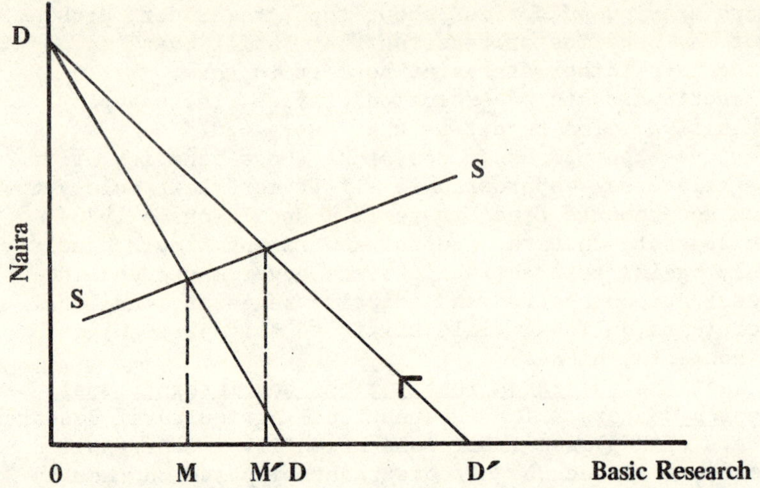

Note: Demand curve DD is the demand for basic agricultural research by the residents of one state; curve DD' is the demand for basic agricultural research by all Nigerians; curve SS is the supply of basic agricultural research; OM is the optimum amount of basic agricultural research for one state; OM' is the optimum amount of basic agricultural research for Nigeria.

Figure 12.1. Demand and supply of basic research

incentives for domestic private sector firms to engage in concerted, meaningful agricultural research, beyond quality control and formulation activities? If not, what incentives would they require? Initially, the small and fragmented farm input markets and the "free rider" problem act as disincentives for private sector participation, and government encouragement is required. There is also a structural disincentive. In the colonial system, when overall economic policy (including tariffs and subsidies) was in colonial hands, it was easier and considered "safer" by multinationals to engage in substantial agricultural research. With independence and the change of guards at the helm of economic affairs, multinationals and their local agencies have been unwilling to invest in desired research activities. They have relinquished all research responsibilities to nationalist governments, which have not, however, been forthcoming in the required research support. How much are transnationals prepared to invest in a developing country's future?

Funding of Research, Extension, and Training

Funding of agricultural research in Nigeria has moved from the cooperative, joint federal and state government-private sector funding of the colonial system to the present monolithic system in which the federal government funds all research. As mentioned above, the present system, dating back to 1975, is unviable and is insensitive and unresponsive to the pressing research problems of local economies within the country.

Recent trends in the funding of agricultural research in Nigeria are shown in table 12.4. Two observations are pertinent. First, funds actually released were almost always less than budgeted sums. Second, relative allocations in recent years reflect little about the research purpose--rather, budgeted sums were in most cases the amounts required just to pay overheads such as salaries, wages, etc. Thus the older research institutes with larger overheads tended, on the average, to have larger budget allocations but not necessarily any more research activity.

Funding of research has not only been inadequate, it has been unstable. Uncontrollable and undesirable fluctuations in research funding result in half-hearted research programming, uncompleted projects, and abandoned projects. It is difficult for researchers to have a long time horizon under these circumstances. Annual reports of research institutes contain long lists of projects that are "rolled on" from one year to the next for lack of funds or for unstable funding. Closely related to this is the untimeliness of the release of funds, which introduces much uncertainty into research programming and reduces its overall effectiveness. During 1975-1980, funds released to the research institutes as a percentage of budgeted sums ranged from 25-50 percent (Report of Research Institutes Review Panel, 1980/81, vol. 2, p. 20). On the average, funds have been provided along disciplinary lines rather than programmatic lines. Actual expenditures are not related to commodity priorities. The program performance budgeting system is just being introduced in some institutes.

State governments and the private sector must take decisive steps to provide funds for applied and adaptive

TABLE 12.4
Federal Government Resource Allocations to Research Institutes, Nigeria, 1975/76-1983

	1975/76	1976/77	1977/78	1978/79	1979/80	1980	1981	1982	1983	1976/77 to 1979/80	Percentage of All Allocations	1980-83	Percentage of all Allocations
	₦ Million										%	₦	%
Food Crops													
IAR	5.452	6.331	7.746	3.600	5.483	--	5.000	3.000	4.000	23.160	10.383	12.000	5.686
NCRI	--	4.300	11.300	9.900	5.800	2.200	5.000	4.000	5.000	31.300	14.031	16.200	7.677
NRCRI	1.851	2.808	3.028	4.757	4.424	4.583	6.000	6.100	4.500	15.017	6.732	21.183	10.038
NIHORT	1.737	2.750	3.017	1.785	2.233	3.300	4.000	4.000	2.750	9.785	4.387	14.050	6.658
IAR&T	--	2.085	5.648	2.158	2.887	--	4.000	3.700	2.600	12.778	5.728	10.300	4.881
Subtotal	9.040	18.274	30.739	22.200	20.827	10.083	24.000	20.800	18.850				
Subtotal %	27.096	33.025	48.605	40.384	42.062	47.389	23.566	33.465	33.784				
Tree Crops													
CRIN	3.593	2.601	4.350	3.718	3.246	0.667	3.000	3.400	3.000	13.915	6.237	9.400	4.454
NIFOR	4.030	10.433	4.624	3.255	2.949	--	5.000	4.800	3.800	21.261	9.531	13.600	6.444
RRIN	1.500	3.600	1.100	2.450	2.250	1.600	3.000	3.258	3.500	9.400	4.214	11.358	5.382
PRIN	3.800	5.300	3.600	4.700	5.000	--	5.000	3.100	3.150	18.600	8.338	11.250	5.331
Subtotal	12.923	21.934	13.674	14.123	13.445	2.267	16.000	14.558	13.540				
Subtotal %	38.735	39.639	21.621	25.691	27.153	10.655	22.378	23.422	24.241				
Livestock													
NVRI	6.574	6.176	6.010	4.468	2.962	3.669	5.000	5.550	3.000	19.616	8.794	17.219	8.159
NAPRI	--	2.900	1.270	1.485	2.058	--	5.000	0.521	4.400	7.713	3.458	9.921	4.701
NITR	2.600	2.080	3.215	2.319	1.194	--	4.000	3.500	3.000	8.808	3.949	10.500	4.976
LRIN							3.000	2.150	1.500	--	--	6.650	3.151
Subtotal	9.174	11.156	10.495	8.272	6.214	3.669	17.000	11.721	11.900				
Subtotal %	27.498	20.161	16.595	15.047	12.550	17.244	23.776	18.858	21.305				
Fisheries													
LCRI	--	0.796	1.988	2.166	2.026	1.000	4.000	4.530	2.765	6.976	3.127	12.295	5.826
KLRI	0.824	1.136	1.693	3.092	1.432	2.175				7.353	3.296	2.175	1.031
NIOMR	0.787	1.028	2.763	2.980	2.150	0.966	4.000	4.800	3.800	8.920	3.999	13.566	6.428
Subtotal	1.611	2.960	6.444	8.238	5.608	4.141	8.000	9.330	6.565				
Subtotal %	4.829	5.349	10.189	14.980	11.326	19.462	11.189	15.011	11.754				
General Services													
AERLS	0.536	0.568	1.189	1.286	1.617	1.117	2.000	2.200	2.000	4.660	2.089	7.317	3.467
NSPRI	0.079	0.442	0.702	0.854	1.804	--	4.500	4.500	3.000	3.802	1.704	12.050	5.710
Subtotal	0.615	1.010	1.891	2.140	3.421	1.117	6.500	5.750	5.000				
Subtotal %	1.843	1.825	2.990	3.893	6.909	5.250	9.091	9.251	8.952				
TOTAL	33.363	55.334	63.243	54.973	49.515	21.277	71.500	62.155	55.855	223.064	100.000	211.034	100.000

Source: Recurrent and Capital Estimates of the Federal Republic of Nigeria, various issues.
Note: Acronyms are identified in table 12.1.

agricultural research on farm production and input problems. New legislation is urgently required to provide and protect state government and private sector investments in agricultural research.

Finally, there is the issue of the extent to which actual expenditures at the institute level reflect program priorities at the institute or national level. This borders on research management, to which we now turn.

Research Management

Research management probably constitutes the most important constraint on Nigeria's national agricultural research system. The management problem is at two levels: macro and micro. The macro level refers to the management of all the agricultural research institutes by national institutions (ministries, parastatals, office of the president, etc.). The institutional modalities for managing the nation's agricultural research system have been in a state of flux, beginning with the Nigerian Research Institutes Act (1964), but more so since the Nigerian Council for Science and Technology Decree (1970) and the Agricultural Research Council of Nigeria Decree (1971). Frequent changes in institutional arrangements have left the national agricultural research system with no consistent and clear sense of direction (table 12.5). In the process, no durable overall plan for agricultural research has emerged.[5] Many institutes have had directors appointed on grounds other than professional merit and demonstrated potential for research leadership. This has had disastrous consequences for morale within the institute research community.

Closely related to the macromanagement problem is the problem of inadequate appreciation of the intrinsic value of research by policymakers outside the agricultural establishment and by the society at large--especially the discipline and commitment to fund research of no immediate value. This has given rise to grossly inadequate and unstable funding and excessive delays in the release of funds.

Research institutes have not been subject to regular research management audit that evaluates the institute's capabilities to achieve research objectives and targets.

TABLE 12.5
Frequent Changes in Institutional Arrangements for Managing Agricultural Research in Nigeria, 1964-1985

1985 Law/Decree	Year	Provision	Remarks
1. Nigerian Research Institutes Act	1964	Established Cocoa Research Institute of Nigeria, Nigerian Institute for Oil Palm Research, Rubber Institute of Nigeria, and Nigerian Institute for Trypanosomiasis Research	Following dissolution of West African Research Organization in 1962
2. Nigerian Council for Science and Technology Decree	1970	Umbrella organization to coordinate research grouped into physical sciences, agriculture, medicine	--
3. Agricultural Research Council of Nigeria Decree (ARCN Decree)	1971	Established ARCH to coordinate all agricultural research	--
4. Agricultural Research Institute Decree	1973	Vested power to establish institutes to conduct research and training in any field of agriculture, veterinary sciences, fisheries, forestry, agro-meteorology, and water resources in Federal Commissioner for Agriculture; also power to take over any existing state research station	Watershed in state, federal funding of agricultural research; destroyed all initiatives for states to fund agricultural research
5. Research Institutes (Establishment) Order	1975	Established 14 research institutes: NCRI, NRCRI, NIHORT, CRIN, RRIN, FRIN, NVRI, NAPRI, NITR, NIFCR, LRIN, LCRI, KLRI, NIOMR	Clear Commodity mandate for each institute

6. National Sciences and Technology Development Agency Decree (NSTDA)	1977	Set up an executive agency to coordinate all research in Nigeria, agricultural and nonagricultural; all research institutes established by the 1975 decree brought under the aegis of the NSTDA	Repealed the 1973 decree but still vested powers to take control of any existing federal or state research establishment in NSTDA Commissioner;
7. Constitution of the Federal Republic of Nigeria	1979	--	placed "industrial and agricultural research" on its concurrent list
8. Federal Ministry of Science and Technology Act	1979	Scrapped the NSTDA, created the Federal Ministry of Science and Technology	
9. Federal Ministry of Education, Science and Technology Decree	1984	Scrapped the Federal Ministry of Science and Technology and merged it with Federal Ministry of Education	Military suspended constitution
10. Federal Ministry of Science and Technology Decree	1985	Created separate Federal Ministry of Science and Technology	

In particular, no system exists for forecasting future research priorities or for regularly ensuring that existing national priorities are closely adhered to in research programming at the institute level. Closely related is the problem of priority distortions during periods of budget squeeze. Rarely do research directors, in response to a severe budget squeeze, close down whole programs, units, or sections in the face of new realities and revealed national priorities. The normal approach of a private concern whose existence is threatened by falling profits or mounting debts is: mass reorganization is effected, heads roll, and new objectives and targets are set. How much of this "survival strategy" can research institutes, or any public institution, for that matter, adopt?

Management at the micro (institute) level poses equally formidable problems. Some directors quickly abdicate their responsibility to provide required research leadership and to motivate young researchers. Others have allowed the proliferation of sections and units, especially those developed to satisfy the aspirations of particular individuals.

The national research leadership has failed to develop a reward system that is tied to some measure of researchers' productivity. Consequently, professional advancement has been linearly tied to length of service, as in the regular civil service. This has created a sort of identity crisis for the civil servant-scientist.

Unless institute research leadership can manage available physical and human resources to effectively produce new knowledge and technology, the substantial resources being allocated to the national agricultural research system could be wasted.

The problem of institutional obsolescence at the macro level affects research institute performance, as the research leadership receives conflicting signals from successive managers at the headquarters level. Frequent changes in policy directives not only result in half-completed and poorly executed research projects but also disrupt long-term research programming.

Organizationally, all the evidence is not in on the best model, whether along disciplinary lines or program lines. Some institutes have overemphasized the disciplinary setup almost to the neglect of meaningful programmatic content while others have emphasized programs with little or no disciplinary core to back them up.

<u>Locations of Research Facilities</u>

The historical error in locating the precursor to the National Cereals Research Institute at Ibadan, on the mistaken belief that Ibadan would be suitable for cotton production, graphically illustrates the need for more work on modeling optimal locations of agricultural research institutes. In an earlier effort, six out of fourteen agricultural research institutes were found to be unsuitably located for the commodities on which they were doing research (table 12.6).

On the supply side, an institute should be located where there is a good research environment that includes adequate administrative support in order to exploit complementarities in research personnel, materials, good library facilities, and equipment. Thus expensive laboratory equipment can be shared and consultations held on research methodology within a large scientific community. Such complementarities are most significant in the early stages of growth of a research institute, when it is primarily engaged in maintenance and production research to select and breed varieties suited to the ecological region. The existence of a university and other research institutes serves to attract other research institutes. This is particularly the case for "discipline-based" as opposed to "commodity-based" research institutes.

There must be adequate supplies of physical, social, and institutional infrastructures conducive to innovative and inventive activity (Ruttan 1982). There must be adequate and reliable supplies of electricity, water, and postal and telecommunication services, as well as laboratory software (laboratory technicians and technologists, instrumentation engineers, etc.). The location should have reasonably low maintenance and operational costs per scientific man year.

The ease with which land can be acquired for the research needs of an institute is another important determinant of location--areas with low capital costs per scientific man year in land compensation fees would tend to attract research institutes, other things being equal.

For a private sector research facility, the availability of government (state or federal) subsidies on infrastructural development could encourage particular choices of locations.

The institute must be relevant to the agroeconomy of the area and there must be a production environment against

TABLE 12.6
Suitability of Agricultural Research Institute Headquarters by Crop, 1968/69 to 1974/75

Institute	Nearest Three States to Headquarters	Crop	Index Location of Suitability (percent)
IAR	Kaduna,	Groundnuts	65.60
	Kano,	Sorghum	55.14
	Sokoto	Millet	50.29
		Melons	7.35
		Benniseed	15.65
NCRI	Oyo,	Maize	40.75
	Ondo,	Rice	35.10
	Ogun	Beans	3.95
		Soyabeans	0.00
		Sugarcane	--
NCRI	Imo,	Yams	22.80
	Anambra	Cassava	29.49
	Cross River	Cocoyams	65.98
NIFOR	Bendel Ondo, Imo	Palm kernels (1969-71)	56.40
RRIN	Bendel, Ondo, Imo	Rubber	--
NIHORT	Oyo, Ogun, Ondo	n.a.	--
CRIN	Oyo, Ogun, Ondo	Cocoa (1968) (69-70/71)	90.05
NAPRI[a]	Kaduna, Kano, Sokoto	Cattle	n.a.
NVRI[a]	Plateau, Kaduna, Bauchi	Cattle	n.a.
NITR[a]	Kaduna, Niger, Plateau	Cattle	n.a.

Source: F. S. Idachaba, Agricultural Research Policy in Nigeria, Research Report No. 17 (Washington, D.C.: IFPRI, 1980).
Note: Full names of the institutes are: Institute for Agricultural Research (IAR), National Cereals Research Institute (NCRI), National Root Crops Research Institute (NCRI), Nigerian Institute for Oil Palm Research (NIFOR), Rubber Research Institute of Nigeria (RRIN), National Institute for Horticultural Research (NIHORT), Cocoa Research Institute for Nigeria (CRIN), National Animal Production Research Institute (NAPRI), National Veterinary Research Institute (NVRI), and National Institute for Trypanosomiasis Research (NITR).

[a]Though indexes of suitability are not available, indications are that the locations of livestock research institutes in Kaduna State are suitable.

which to test new varieties against various stresses (soils, climate, etc.).

On the demand side, an institute should be located in the core producing area that constitutes the demand for new knowledge and new technology. Through intergenerational, unorganized experimentation, the core producing area's selected varieties adjust to the environment and form the point of departure for the new research institute's varietal selection trials in the subecological zones in the core producing area. The core producing area (where the crop is no longer new) is also where farmers have attained allocative equilibrium in the Schultzian sense (Schultz 1964) such that substantial incremental production gains can only come from massive infusion of new inputs and management practices generated by the research institute.

A strand of the above argument is that a research institute should be located in a region where there is maximum development potential and a huge potential demand for the crop. The experience with maize in Nigeria illustrates this point. Maize was traditionally grown in the savannah belt of Southern Nigeria. Beginning in 1975/76, the World Bank-assisted Agricultural Development Projects (ADPs) introduced new varieties of fertilizer-responsive maize into the farming system of the northern projects. The production effects were dramatic--in fact, the northern ADPs now constitute the major source of surplus maize for the poultry industry concentrated mainly in the south.

Closely related to the core producing area on the demand side is the need to minimize the costs of technology dissemination/diffusion per producer/beneficiary. The farther away the core producing area is from the research institute, the higher the dissemination costs.

The ultimate test of any new technology is its adoption by farmers on the basis of proven profitability. The location of a research institute in the core producing area minimizes the costs of adaptive, operational, and on-farm research.

<u>Conflicts in Research Institute Location Criteria</u>. An area may qualify for institute location on demand-side considerations because of the concentration of producers, production, and demand for new technologies but may fail woefully as a candidate for institute location because, located in some remote part of the country, it has no scientific community, electricity, water, roads, etc. (Idachaba 1980; <u>Report of Research Institutes Review Panel</u> 1981). Under such circumstances there is a conflict, a

trade-off between suitability indicators viewed from the supply and demand sides (figure 12.2). For example, Badeggi in Niger State qualifies as a rice research station on demand-side considerations but fails woefully on supply-side considerations. The result is that such remote research stations are manned only by relatively junior research staff whose potential for inventive activity in scientific research is rather limited. What is needed is a set of policies that will shift the trade-off curve downward and to the left. Such policy measures include provision of necessary infrastructures, special allowances, and pay incentives for research staff working in such remote areas.

Historical Errors of Location: Guidelines for Constrained Spatial Optimization. Given that historical errors of location of some research institutes have become institutional realities, what are the guidelines for a second-best optimization? Scrapping such institutes and completely relocating them has never attracted the required political will and courage. The second-best solution is to place a limit on the physical and human resources expansion of the headquarters location of the institute and to decentralize its research activities to the core demand areas of actual and potential production.

In research environments with grossly inadequate infrastructures, economies of scale from initial increasing returns are rapidly exhausted, and the internal institute production process quickly encounters diseconomies of scale from diminishing returns as bottlenecks occur in the supply of technologists and technicians, electricity, and other utility supplies. Under such circumstances, and especially for crops spanning different ecological zones, a network of small to medium-sized research institute facilities is recommended over large institute facilities.

SUMMARY AND CONCLUSION

We have been concerned in this paper with the role of a national agricultural research system in the structural transformation of an agriculture that can no longer expect major output increases from extensive agriculture utilizing present resources or from mere reallocation of traditional factors of production.

In trying to understand why, after almost a century of organized agricultural research, there are still no breakthroughs in food production and the food situation

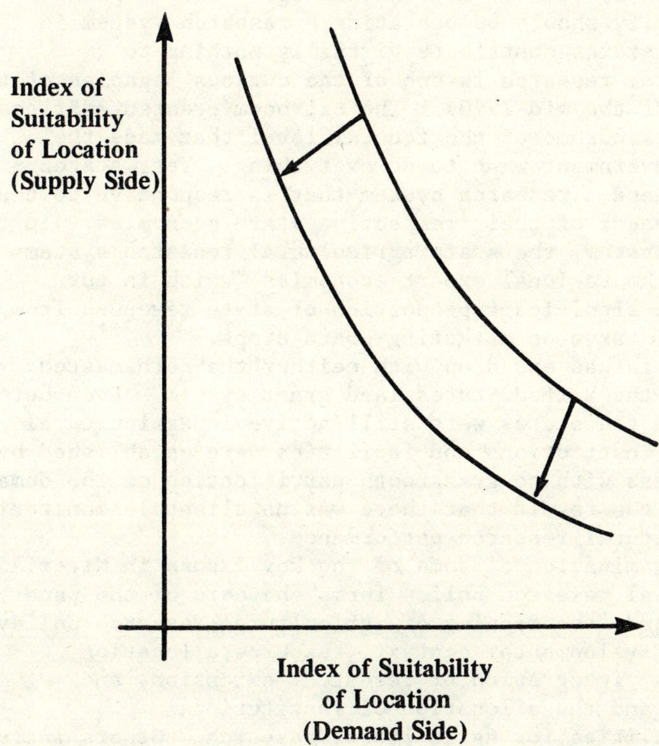

Figure 12.2. Trade-offs between supply side and demand side criteria for locating research institutes

appears to be worsening, we examined the origins of the policy environment of the agricultural research system. Though some impressive gains in research were made in cocoa, cotton, groundnut, and oil palm research, certain institutional developments almost guaranteed failure. These include, among others: provisions for constitutional responsibility for agriculture between state and federal governments; frequent and confusing changes in the institutional arrangements for managing the nation's agricultural research systems; and the Agricultural Research Institutes Decree, 1973, which, in one fell swoop, destroyed what was left of a cooperative joint federal-state agricultural research system and established a monolithic federal agricultural research system that remains financially unviable, ineffective, and unworkable. That a country with ecological and agroclimatological heterogeneity should be operating a research system in which the states contribute virtually nothing to agricultural research is one of the curious legacies of the oil boom of the mid-1970s. The oil boom created a "fiscal superman" syndrome at the federal level that made the federal government want to do everything. Yet, states urgently need a research system that is responsive to the peculiar needs of their respective state economies. In the colonial system, the state agricultural research systems serviced the regional export economies, which in turn provided a significant proportion of state revenues from a variety of taxes on marketing board crops.

Nigeria has ended up with neither the Rothamasted model nor the United States land grant system. Even before 1973, when the states were still active in agricultural research, institutions and facilities were established by policymakers with no grassroots participation on the demand side with the result that there was no clientele monitoring of agricultural research performance.

An examination of some of the key issues in Nigerian agricultural research policy forms the core of the paper. These include the meaning of agricultural research policy within a developmental context, resource allocation priorities, integration of research, extension, and training, and the allocation of institutional responsibilities for agricultural research. Others include funding, research management, and the location of research institutes.

NOTES

1. The Federal Department of Agricultural Research was located in the Federal Ministry of Economic Development, the sole voice for agriculture at the federal level during 1954-1968. The Federal Ministry of Natural Resources and Research was established in 1965 to coordinate federal programs at the federal level--"agriculture" was deliberately omitted from the title of the new ministry in order not to hurt the political sensibilities of the regions.

2. See table 12.1 for full names of acronyms.

3. Note that allocative criteria are essentially country specific, reflecting a country's peculiar constraints and opportunities, and are therefore not necessarily applicable to other countries. Also, allocative criteria are essentially dynamic, reflecting changes over time in a society's goals, opportunities, and constraints.

The literature on allocative criteria for agricultural research is extensive. For a small sample, see Fishel (1971); Arndt, Dalrymple, and Ruttan (1977); Daniels and Nestel (1981); and Ruttan (1982). For specific Nigerian case studies, see Idachaba (1980, 1981a).

4. In Nigeria, 7.5 percent of operating revenue of produce marketing boards was statutorily to be expended on research. Grants by the regional marketing boards during 1955-1961 were: Western Nigeria--₦10 million, mainly for cocoa research and extension; Northern Region--₦5.6 million, mainly to Samaru Research Station with ₦3.2 million for general research and ₦2.4 million specifically for cotton development. The World Bank had earlier recommended that the recurrent costs of the West African Institute for Oil Palm Research (WAIFOR) be met from an endowment to which the Nigerian Oil Palm Produce Marketing Board contributed 82 percent. The Cocoa Marketing Board also provided funds for WAIFOR: in 1953/54 the Cocoa Marketing Board provided ₦474,000 for cocoa and soil survey. In Northern Nigeria, the marketing board provided ₦226,000 for cotton development in 1953/54.

5. Almost a century after the launching of scientific research, the nation is still being promised a comprehensive policy on science and technology. Among the policies that would be pursued during the Fourth Plan period is the development of a comprehensive policy on science and technology. "Such a policy will facilitate the transformation of the Nigerian Society into one in which science and technology will form a fundamental part of the thinking apparatus of the average citizen..." (Fourth National Development Plan 1981-85, vol. 1, p. 208).

REFERENCES

Arndt, T. M., D. G. Dalrymple, and V. W. Ruttan, eds. Resource Allocation and Productivity in National and International Agricultural Research. Minneapolis: University of Minnesota Press, 1977.

Daniels, D., and B. Nestel, eds. Resource Allocation to Agricultural Research, IDRC 182e. Ottawa: International Development Centre, 1981.

Fishel, W. L., ed. Resource Allocation in Agricultural Research. Minneapolis: University of Minnesota Press, 1971.

Fourth National Development Plan, 1981-85. Vol. 1. Lagos: Federal Ministry of National Planning, 1981.

Idachaba, F. S. "Policy Distortions, Subsidies and African Rural Employment Creation: A Second-best Approach." Indian Journal of Agricultural Economics 29(2)(1974): 20-32.

________. Agricultural Research Policy in Nigeria, Research Report No. 17. Washington, D.C.: International Food Policy Research Institute, 1980.

________. "Agricultural Research Resource Allocation Priorities: The Nigerian Experience." In Resource Allocation to Agricultural Research, edited by D. Daniels and J. Nestel. Ottawa: International Development Research Centre, 1981a.

________. "Farm Input Subsidies in Nigeria's Green Revolution Progamme: Lessons from Experience." Food Policy Research Paper 2, Department of Agricultural Economics, University of Ibadan, 1981b.

Peterson, W. L., and J. C. Fitzharris. "Organization and Productivity of the Federal-State Research System in the United States." In Resource Allocation and Productivity in National and International Agricultural Research, edited by T. M. Arndt, D. G. Dalrymple, and V. W. Ruttan. Minneapolis: University of Minnesota Press, 1977.

Report of Research Institutes Review Panel. Vols. 1 and 2. Ibadan: Green Revolution National Committee, 1981.

Ruttan, V. W. Agricultural Research Policy. Minneapolis: University of Minnesota Press, 1982.

Schultz, T. W. Transforming Traditional Agriculture. New Haven: Yale University Press, 1964.

World Bank. The Economic Development of Nigeria. Baltimore: Johns Hopkins University Press, 1955.

13
Mobilizing Political Support for the Brazilian Agricultural Research System

Eliseu Alves

The question addressed in this paper is, from society's point of view, should agricultural research be considered a priority for the public sector? In other words, should a major objective be to increase total funds allocated to agricultural research and, if so, how should support from the public authorities be obtained?

The analysis is based on the idea that it is stress (or crisis), be it economic or social, that induces government to make decisions--clearly, not all decisions, but most of them, when heavy investments are needed (De Janvry 1973).

The question is, then, how do we detect signals of stress and make them known to society and government? Also, how does stress relate to lack of investment in research, and can society be motivated to support research?

This paper addresses these questions. Brazilian agriculture supplies the background for the study, but the conclusions are intended to be applicable to other situations.

PROCESSES FOR DEVELOPING AGRICULTURAL TECHNOLOGY

For the purposes of this paper two processes for developing agricultural technology are relevant. What separates them is the presence or not of organized research (public or private).

The first process is illustrated by model A, figure 13.1. Organized research is not present. The growth in the stock of knowledge and technology is due to the accumulated experience of the farmers and to their ability

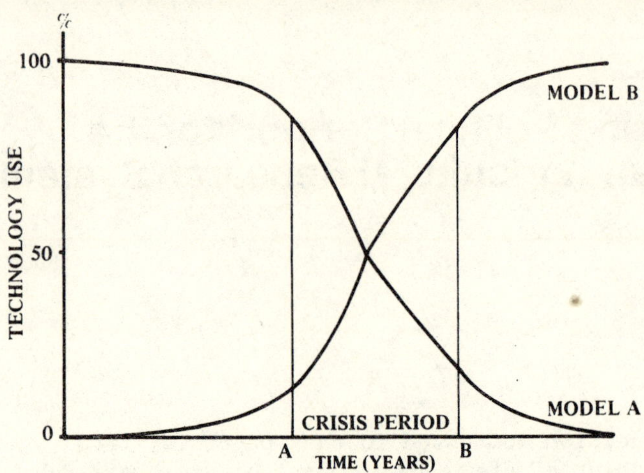

Figure 13.1. Models of technology in use by farmers

to import and adapt technology to the conditions of the country. The power to increase productivity (of labor or land) is small. The objective is to maintain the productivity level and to create conditions for expanding the cultivated area whenever possible. The process is limited, however, if the new environment is very different. For instance, the technology brought by the European and Asian immigrants was successful in developing agriculture in the northeastern and southern regions of Brazil but failed when it was applied to the Cerrados (Brazilian savannas--25 percent of Brazilian territory, most of it located in the central part of the country) and to the Amazon region. Those regions present problems peculiar to them that are very difficult to solve without the help of science.

Until the beginning of the century, this unsystematic way of generating knowledge (by trial and error) was responsible for the majority of technologies developed in Brazil. Here and in other developing countries its influence is still present, sometimes to a large degree. But it began to lose importance in the beginning of the 1960s.

The second process is illustrated by model B, figure 13.1. Knowledge and technology are generated by organized research (public and private). Technology is then science based. The transfer of knowledge and technology is through the research system, which adapts them to the needs of the country. The purpose of the technological development is to increase the productivity of land or labor, and to make possible the growth of the cultivated area when there is land to incorporate. Depending on the situation, one goal prevails over the other.

Figure 13.1 illustrates the agricultural technology development process. The vertical axis measures the amount of technology (models A and B) used by farmers. It varies from 0 to 100 (percent of farmers that use the technology generated by models A or B). The horizontal axis is time.

At the beginning, all technologies used by farmers are generated by model A. They begin to lose importance, slowly at first, then at growing speed when point A is reached. Modernization of agriculture speeds up, and after point B most of the technologies in use by farmers are generated by model B. Eventually a point may be reached when model B is the only source of the generation of technology. After point B, agriculture is science based.

The interval AB (fig. 13.1) indicates the period when stress (or crisis) is mounting. It coincides with a great acceleration of the agricultural modernization process. The crisis may be the consequence of a food deficit, the loss of the country's ability to export, or the disturbance of rural life when some farmers progress at a faster rate, improving their social status, while the great majority lag behind. In other words, income is being concentrated by the modernization process and land is in fewer hands. The rate of migration speeds up, and if the industrial and service sectors do not develop, unemployment and underemployment, with all of their consequences, emerge in the cities.

The stress is a consequence of society's delay in investing in research or in importing technology. The crisis is also a consequence of the lack of response of agriculture to a growing demand or to the increase in competition in international markets. How long the stress will last depends on the ability of society to remove the chief block to the modernization of agriculture, the lack of investment in research.

There is need for a great increase in the speed of agriculture modernization (at point A and after it occurs) because investments were not made in the right amount in the previous period. Hence the crisis can be avoided if the government learns when model A loses the capacity to bring about supply growth according to the growth of demand and makes the necessary investments.

TYPES OF CRISIS

The crises are classified as the following types, which may be present at the same time:

Food Deficits. Demand for food is growing at a faster rate than supply and the country is not able to import food because of balance of payments problems brought about by a large accumulated deficit in foreign currency. This is the most serious type of crisis, in the sense of causing damages to economic and social systems.

The most important factors behind demand growth are population and per capita income increases. Urbanization shifts demand to such products as animal protein and fruits and vegetables. Edible beans, manioc, and other domestic products become less and less important. The need to export is another important factor. Among the factors that explain the lack of response of supply are investment in research and extension, price, credit, and export policies.

By helping to solve food crises, research institutions can show government and society the importance of increasing the investment in them.

Loss of Capacity to Export. Agricultural exports are an important sector of most developing countries. After the Second World War the developed countries experienced a fast growth in exports of agricultural products as the result of agricultural modernization and huge amounts of government subsidies. Since then, competition has mounted and prices have fallen. Since 1975, prices of the most important export commodities, soybeans, corn, wheat, sugar, rice, and cotton, have fallen to levels that are now the lowest in modern times.

The developing countries have not been able to meet this competition and, hence, are losing their position in international markets. Some of them are becoming net importers of food or are importing food at increasing rates.

The loss of important markets and the import of food trigger the mechanism of a deep crisis in rural areas that reflects over all society. Income in rural areas decreases and the rate of migration increases. The balance of payments worsens. Society sees in the import of food a loss of prestige and an increased vulnerability to fluctuation in international trade.

LAND TENURE

The growth in rural population, the lack of employment opportunities in other sectors of the economy, and the concentration of land in the hands of a few are responsible for another type of crisis that is common in the developing countries: the pressure by peasants for land. When there is a food deficit and the migration rate is very high, society may see land distribution as the main cause of the crisis. Then comes the quest for agrarian reform as a way to modernize agriculture and to reduce income disparities. In modern times, research and extension are seen as important tools for the success of the program.

When there is a frontier where agriculture can expand, one way out of the crisis is to put more land into cultivation. This happens at the expense of investments to increase productivity. When research institutions face such a situation, two courses of action are advisable. First, demonstrate how much society loses because productivity does not increase; second, show that research is an important tool in solving problems posed by a new environment in which agriculture is expanding.

Factor Price Change. Factor price changes, or increasing costs of production, also cause stress. Farmers react by stimulating the generation of technologies that save the factor that has its price increased the most. Also, the research institutions are induced to shape their research programs according to the factor price changes. It is not only a question of reallocation of funds. The government is also induced to increase its investment in research. The literature on this subject is vast; one of the best references is Hayami and Ruttan (1985).

GENESIS OF THE CRISES

The factors that explain the crises vary according to type of crisis. For instance, the policy to expand acreage, coupled with high rates of migration to the cities and the restrictions of the immigration laws, causes labor to become relatively scarce to land and, consequently, shifts the factor prices.

The discrimination against agriculture that is so common among developing countries, to foster industrialization, restricts the ability of agriculture to respond to demand increase and causes a growing food deficit, in conjunction with a loss of exporting capacity.

The policy to subsidize modern inputs through credit discriminates against small farmers and favors the concentration of land in fewer hands.

If the increase of productivity is the best avenue to increase supply and the government does not support research to the extent that it is needed, agriculture loses the capacity to grow and to compete in international markets.

From the standpoint of research institutions that want to see their funds increased, it is important to know the nature of the crises that are present and, moreover, to be able to foresee when they start, if they are not already present.

SOURCES OF SUPPLY GROWTH

Increases in agricultural production can be achieved by increasing output per unit area (yield), by increases in land area per worker, and by growth of the labor force (table 13.1).

Thus there are two paths for output growth: yield and area increase. They need not be mutually exclusive. However, from the standpoint of costs to society, one may be preferable to the other. If yield is preferable to area increase, and if the investments are made only to increase acreage, then p will be smaller for a given set of prices and technologies.

From table 13.2, which shows the change in the rate of output growth of Brazilian agriculture for four decades, it can be seen that acreage increase is the main source of growth, except for the last decade, when yield becomes the most important source.

Yield increase started at a very low level and recently speeded up to become the most important source. The opposite is true for acreage, which decreased in the last two decades.

About 66 percent of Brazil's agricultural land is not in agricultural use. The slowdown in acreage increase is a consequence of the difficult environment in which the new frontier, the Amazon region, is located. It is expensive to bring this land under production, and the operating

TABLE 13.1
Sources of Growth in Agricultural Production

The following identities are useful in identifying the sources of growth. The symbols are annual rates (geometric) of growth.

$p = (a+r) + [(a)(r)]$

p = production; r = yield; a = acreage

$a = (t+k) + [(t)(k)]$

t = area that each worker can cultivate;

k = labor force in agriculture.

The two identities may be combined to give:

$$p = (r+t+k)+[(r)(t) + (r)(k) + (t)(k) + (r)(t)(k)] \quad (1)$$

The bracketed terms measure the interaction effects, which tend to be small in relation to additive ones.

To increase acreage it is necessary to have $(t+k) + [(t)(k)] > 0$. If $k < 0$ (labor force in agriculture is decreasing), then it is necessary to have $t > |k+(t)(k)|$. This means that mechanization must be stimulated.

If $r = 0$, which was common among the developing countries until the 1960s, production will grow only by acreage increase. If $t=0$, which was true also in some developing countries, production growth is equivalent to labor force growth.

If there is a technological barrier to acreage expansion or if the frontier is already exhausted, then $a = 0$, and $t > 0$ implies $k < 0$. This means that labor is displaced, and the size of displacement varies with t.

costs of production there are higher than in any place else in Brazil. To these negative factors, one must add the high transport costs to the important markets of the country, which are far away. Nonetheless, acreage increase is one of the important sources of production expansion in

TABLE 13.2
Decomposition of the Geometric Rates of Growth of Production between Acreage and Yield, Brazil, 1940-1980

Sources of Growth	Decades			
	1940-50 %	1950-60 %	1960-70 %	1970-80 %
Yield (r)	.53	1.58	1.89	3.49
Acreage (t+k)	2.58	4.16	3.46	2.96
Production (p)	3.11	5.74	5.35	6.45

Source: Adapted from Alves (1985).
Note: The interaction effects were small and were included in r and t+k.

Brazil, and research institutions need to show society how they can contribute to solving the problems associated with the new frontier.

ACREAGE OR YIELD INCREASE

Figure 13.2 clarifies the conflict between acreage and yield growth as ways to increase production. The vertical axis measures acreage cost and the horizontal axis yield cost. The path MR indicates the social costs to obtain one additional unit of production through yield and acreage. Each point over the path reflects a different date. Time increases from 0 on. Line OC is the locus of points where the two costs are equal.

Included in acreage are the usual costs of investments to bring land into production, such as roads, infrastructure, forest clearing, etc. Included in yield are the costs of research and extension, credit subsidies to stimulate the diffusion of innovations, and the usual costs of production.

In the beginning, acreage costs grow slowly because the land first put into production is more fertile and easier to cultivate. Costs then increase at a faster rate because the new land is far from markets, is difficult to clear, and is not as fertile as the older areas. Such technologies as transport, land clearing, and drainage,

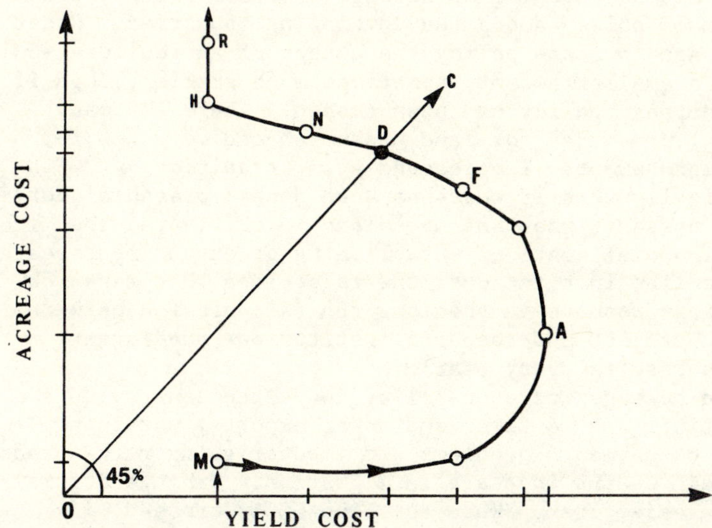

Figure 13.2. Social cost to obtain one additional unit of production by yield and acreage increase

etc., can reduce such costs by delaying the increase in acreage costs to produce one additional unit of output.

Yield cost grows fast in the beginning. Traditional technologies, model A, have very limited power to increase yield. The possibilities for importing technologies are also very restricted. When the country shifts to model B and improves its capacity to generate knowledge and technologies, yield cost decreases and eventually stops. At point D, yield cost and acreage cost are equal. At point A, yield cost begins to decrease and the rate of decrease speeds up after that point.

Eventually the frontier is exhausted and there is no peaceful way to incorporate additional areas. This happens at H, when acreage cost grows without limit. At this point yield cost reaches a minimum and the path becomes vertical.

The convex part of the path MR faces the line OC. If a country is at a point below line OC, let us say point A, one additional unit of output costs more if obtained through yield than through acreage increase; above line OC, the opposite occurs.

Long before the frontier is exhausted, it may be cheaper to increase output by yield increase, although the country may be pursuing an acreage increase policy, the most common policy among the developing countries. Under the acreage increase policy the output of agriculture would grow by a smaller amount, sometimes much smaller, than if a yield increase policy had been chosen. There are many political pressures for land expansion and very few for yield improvement. The reasons are: tradition (historically this is what has been done); pressures for land by peasants who want more land to cultivate; the high cost (financial cost) of establishing or expanding research and extension institutions; the false idea that research only brings results in the long run (a confusion between research project and research institution--the latter produces results every year).

The acreage increase policy leads the country to a food deficit and to fewer and fewer exports, worsening balance of payments problems and impoverishing rural areas, with a consequent increase in migration. Urban and rural crises are set free. When the line OC is crossed, if investments in yield increase have not previously been made, the country will be moving toward crisis. It is also important to note that it is difficult to convince the authorities to invest in research when the point reached is below line OC. The research institutions established in this period may not survive because of lack of support.

To take advantage of the crisis to improve the support to research institutions, it is important to know whether or not the country is urban. Different strategies are used, depending on whether the political power is in the hands of the urban or the rural people.

Figure 13.3 contains two paths. Path L_1 is for a country that created employment opportunities in the cities. It was able to continuously attract the rural population. Initially, a majority of the population was located in rural areas. Migration, sometimes at growing rates, shifted the population to the cities. At line OC, urban population equals rural population, and then becomes larger. Countries that followed path L_1 also pursued a policy of draft industrialization. The idea of the policy was to apply the savings (internal and external) to the industry energy and road construction sectors and let agriculture grow by acreage expansion, with a very small amount of investment in modernization, to avoid competition with the urban sectors for the saving. Points above line OC (refer to figure 13.2) were reached before the agriculture policy shifted to yield increase.

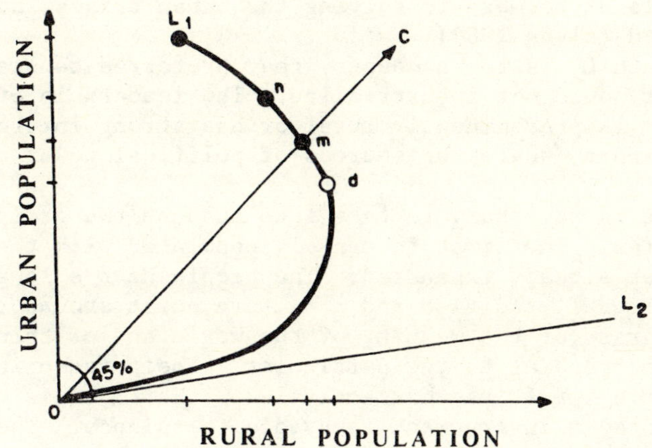

Figure 13.3. The growth of urban and rural population: Number of inhabitants

When line OC is reached, political power also shifts to the cities--to the leaders of the industrial and financial sectors and to organized labor, which are political groups with the same interests. These groups are important targets for obtaining research support. It is important to give them information that will lead to their understanding that the urban crisis will worsen if the agricultural policy does not shift to yield increase and that without investments in research this shift cannot be achieved.

The themes for this understanding are the following:

1. Linkages between violence and food deficit
2. Reduction of worker productivity and ability to learn as a consequence of nutritional problems
3. Loss of the capacity to export
4. Increase in migration as a consequence of income problems in rural areas
5. The need to import food and the consequent devastating effects on the balance of payments and the morale of the country
6. The worsening of equity problems as a consequence of the increase in food prices

Competent professionals, hired from the press, should be used to reach the target groups. The media should include television, newspapers of wide circulation, and radio. The message should interest the urban population, but not to the neglect of rural people. On the contrary, the role of farmers in solving the urban crises should be stressed (Alves 1984).

Path L_2 is for a country that preferred to stay rural or that could not industrialize. The leadership of the country is predominantly rural or has strong interests in rural areas, where the sources of political power are located.

It is necessary to take into account two groups of countries. The first is densely populated with the frontier already exhausted. The people have a long agricultural tradition and a culture built around religion. Their grasp of the meaning of the world and of things is from the general to the particular. Their reasoning is much more intuitive than analytical. Here, it is not easy to create an environment favorable to science. Their methods, sometimes, are not democratic; they may or may not work, and most of the time they do not work.

They are poor countries; the level of savings is low. The investment in research institutions needs the support of the donor community. Hence to show the relevance of research it is important not to overlook the donor community. On the other hand, the possibilities for internal financing should not be neglected.

One needs to be careful not to present science as opposed to religion and to avoid comparisons, whether implicit or explicit, between scientific and religious reasoning.

The themes to be discussed are linked to poverty problems, famine, well-being, and the role that modern agriculture can play in improving the income in rural areas. In those societies, the question of equity is a major issue and, hence, should be addressed carefully. The press, churches, and meetings with important leaders are the main communication channels with society.

The second group of countries has not yet exhausted the frontier through which agriculture can expand. However, the frontier presents problems to agricultural production and may require huge investments to become productive. This is a theme for discussion, as is the question of how science can be helpful in developing the frontier land. One needs to be careful in adopting scientific reasoning to the problems since the farm people tend to be suspicious of the paradigms of science.

SIGNS OF CRISES

Four types of crises have been analyzed: food deficit, loss of export capacity, land tenure, and factor price changes. Simple statistics may be used to substantiate them. The growth of output (p) can be computed, also yield and acreage. The identity given below is true for each period:

$p = r + [(a)(r)]$

a = annual rate of growth of area (geometric).

If there are statistics of the labor force, identity (1) can be applied. Yields can be computed on a crop by crop basis. With this approach the effects of crop substitution on yield are lost. The evidence shows whether or not yields are stagnated, what the contribution of acreage to output growth is, and, finally, how the contribution of acreage splits into labor productivity and labor force increase.

It is much more complicated to evaluate whether line OC in figure 13.2 has been crossed. In other words, where are the returns to investment higher: through acreage or yield increases? Statistics on the rate of return to research and extension provide some hints. But the best approach is through a direct evaluation of the return of investment in acreage or yield growth. The literature is lacking in this respect.

CONCLUSION

It has been shown here how crises can be helpful in obtaining society's support for research. The ideas discussed are very general so that they can be broadly applied. A given research institution, however, should find and use the best strategy for its own situation. The point to be stressed is that support for research institutions can be increased if the correct strategies are chosen, and this involves the discussion of broad economic and social issues.

REFERENCES

Alves, E.R.A. "Making Agricultural Research a Priority for Public Investment." In Brazilian Agriculture and Agricultural Research, edited by Levon Yeganiantz, 175-89. Brasilia: EMBRAPA, 1984.

________. "Produtividade Animal e Novas Tecnologias." In Pesquisa Agropecuaria - Novos Rumos, edited by Levon Yeganiantz, 93-157. Brasilia: EMBRAPA, 1985.

De Janvry, A. "A Socioeconomic Model of Induced Innovations for Argentine Agricultural Development." Quarterly Journal of Economics 87(3)(1973): 410-35.

Hayami, Y., and V. W. Ruttan. Agricultural Development: An International Perspective, rev. ed. Baltimore: Johns Hopkins University Press, 1985.

14
The Use of Events Analysis in Evaluating National Research Systems

Howard Elliott

This paper describes a methodology for systematically recording and analyzing information about significant events in the development of an agricultural research system. It is an especially useful documentation method for analysts who may not be particularly cognizant of the detailed events--both historical and current--that have molded the institutional development of a given agricultural research system.

The methodology uses a relational database management program to explore the relationships between technical, institutional, and political factors associated with individual events. The database management approach allows us to array information about a large number of events in different ways, emphasizing in turn the chronological, institutional, political, and technological aspects of the system's development. By cross-referencing different types of information we can not only identify patterns of interaction, but at any moment can also document the supporting evidence that will have been drawn from a wide variety of sources. The method is somewhat detailed and time-consuming for a brief review mission, but a national research institute could carry out an events analysis as a one-time basic exercise and update it regularly as a tool for researchers, policymakers, and the frequent consultants seeking information.

In this analysis, an event is essentially defined by those who originally cite it. Any occurrence (e.g., introduction of new techniques, development of germplasm, or change in policy or institutional structure) that is mentioned, either in the literature or by informants in interviews, is stored in the database in standardized fashion. The fact that the event was cited means that in the mind of its author it was important in explaining some phenomenon. By standardizing the information about the event, we are able to use the event to analyze relationships that may not be apparent when the event is considered in isolation or only in its original context. Clearly "trivial" events can be excluded by whoever constructs the database.

The following information was obtained for each event:

1. Description of the event

2. Nature of the event (agronomic, biological, chemical, mechanical, economic, institutional)

3. Crop to which the event relates (e.g., rice, maize, dairy cattle)

4. Date of event (or period during which event occurred)

5. Sector in which event originated (public or private; external or domestic)

6. Organization principally responsible for the event (name)

7. Subsector of the agricultural technology management system to which originator of the event belongs (policy-making sector, donor sector, technology-generating sector, technology transfer sector, service sector, post-harvest users, or marketing sector)

8. Cooperating organizations (names)

9. Subsector to which the cooperating organizations belong

With such information about hundreds of events, drawn from the literature or interviews, it is possible to relate technological changes to policy changes, institutional changes, and exogenous events occurring in the non-agricultural sector that have an impact on the system.

Each of these pieces of information becomes a field in the database record for an individual event. The record describes the event and identifies the associated institutional and technical characteristics needed for studying the interrelationship between policies, institutions, and technologies. After providing a description of the event, the researcher classifies it by its nature: biological, chemical, agronomic, economic, or institutional. The analyst is also required to record supplementary information of an institutional nature: the component of the Agricultural Technology Management System (ATMS) in which it originates; whether this entity is public or private, foreign or domestic; the organization principally responsible for the event; the sector of the

ATMS to which it belongs; and the names of cooperating institutions. Dating of the event is important. One of the hypotheses, commonly advanced in Latin America, is that technological change tends to occur in waves of innovation and that institutional forms and policies are affected by the nature of the technologies being introduced. The methodology allows the researcher to retrieve all documented events that bear on such hypotheses.

In particular, the following analyses are possible:

1. A chronology of technological events by commodity, their nature, and the nature of participating institutions
2. An analysis through time of the interaction between public and private sector institutions (either in the system as a whole or for particular commodities)
3. A chronology of major institutional changes in the system or principal policy changes

AN APPLICATION OF THE METHODOLOGY: THE CASE OF PANAMA

The International Service for National Agricultural Research (ISNAR) and Rutgers University collaborated in a review of the Agricultural Technology Management System of Panama (ISNAR/Rutgers 1985). Experience from Latin America led the review team to record information about individual events that would contribute to an analysis of public and private sector interactions, the evolution of technology through time, and the role of external forces in the technology adopted.

Chronology by Commodity

By way of example, table 14.1 lists a subset of the events occurring in Panama in an important food commodity, rice. The database on rice contains ninety entries spanning the period 1932-1985. In table 14.1 we show only the years 1960-1985, as an example of the way information can be displayed.

The chronology of events during 1932-1985 demonstrated that technological change before 1950 was primarily mechanical. There were also several economic and institutional events in the late 1940s and early 1950s related to the efforts of growers to gain price supports and other benefits.

In the 1950s there was increasing reliance on imported chemical technology to overcome the difficulties of upland

TABLE 14.1
Events Occurring in Rice Research and Development Programs, Panama, 1960-1985

Year	Case	Nature	Description of Event
1960	rice	inst	Central Agricola organized "Centrales" to supply inputs and credit
1960	rice	inst, econ	Central Agricola organized Sociedades de Siembra for seed and commercial production
1960+	rice	biol	*Pyricularia* (blast) attack leads to effort on rice
1961	rice	chem	Introduction of 2,4,5T and mixtures with 2,4D
1963	rice	biol	Introduction of Surinam varieties tolerant to *Pyricularia*
1963	rice	chem	Introduction of herbicide Propanil
1963	rice	agron	Wide use of certified seed
1964	rice	inst	Seed program transferred to Ministry of Agriculture, increased storage installations
1965	rice	biol	Danger to rice production due to attack of blast
1965	rice	biol	Massive change from American to Surinam varieties resistant to blast
1965	rice	chem	Mass use of Propanil by rice growers to combat weeds, 2,4D ineffective
1965	rice	biol	Faculty of Agronomy (FAUP) begins first breeding work
1966-1969	rice	mech	Irrigation project in Rio del Villa
1967	rice	biol	Introduction of first HYV rice IR-8
1968	rice	biol	Beginning of rice breeding program at INA-Divisa
1968	rice	chem	Introduction of first fungicides by Japanese experts at INA
1969	rice, other	inst	Research at INA-Divisa given national coverage, team strengthened

(continued)

TABLE 14.1 (cont.)

Year	Case	Nature	Description of Event
1969-1970	rice	mech	Study of possible irrigation project by English mission in Cocle
1970	rice	chem	Introduction of carbamate insecticides, incompatible with Propanil, less dangerous to humans
1970	rice	chem	Introduction of systematic insecticides
1970-1975	rice	econ	Role of asentamientos campesinos in rice production rises to 15%
1972	rice	biol	Widespread change for tall to dwarf varieties (CICA 4)
1972	rice	chem	Massive use of fungicides for Pyricularia due to change to susceptible varieties
1973	rice	econ	Panama self-sufficient in rice seed
1974	rice	econ	Creation of first "empresa de segundo grado": rice mill in Chiriqui
1975	rice	agron	IDIAP begins production of basic and registered seed
1975	rice	inst	Creation of IDIAP, transfer of leading rice breeders to IDIAP
1977	rice	biol	First modern rice varieties developed nationally (Damaris, Anayansi)
1977	rice	biol	IDIAP releases first blast-tolerant varieties (Damaris, Anayansi)
1980	rice	chem	Introduction of pyrethroid insecticides (compatible with Propanil)
1980	rice	agron	IDIAP introduces on-farm methodology
1982	rice	biol	CIAT program projected with great force to Panama
1982	rice	biol	CIAT/IDIAP collaboration intensifies on upland rice
1985	rice	agron	CATIE-IDIAP working on low-input systems
1985	rice	econ	MIDA/IMA lowers support price for rice

rice cultivation. Insecticides were followed by herbicides and later by fungicides, with the private sector playing an important role in their distribution. The mid-1950s also saw the dissemination of American varieties of rice, which required chemical protection. In the early 1960s fertilizer use became more widespread, but the success of introduced varieties was limited by an attack of rice blast.

The disastrous effect of the susceptibility of American varieties to blast led to increased attention to breeding and selection of introduced varieties in the mid-1960s, and to improved seed production. The Faculty of Agronomy at the University of Panama (created in 1959) began its rice breeding program in 1965 in response to the blast crisis. The first introduction of a dwarf variety (IR-8) in 1967 was followed rapidly by others. At the same time, the research program at Instituo Nacional de Agricultura (INA) was strengthened and given a national mandate. By 1972, dwarf varieties were in general use, but they required increased use of fungicides because they remained susceptible to blast.

This continuing problem led Panamanian scientists to maintain a national breeding program, not only for its results in developing resistant varieties, but also for its didactic benefits. The Instituto de Investigaciones Agropecuarios de Panama (IDIAP), created in 1975 with staff from the Ministry of Agriculture and the Faculty of Agronomy, was able to release the first blast-resistant national varieties in 1977. Since that time, the Centre International de Agricultura Tropical (CIAT) program has collaborated with Panama on problems of breeding upland rice. The Centro Agronomico Tropical de Investigacion y Ensenanza (CATIE), meanwhile, has provided assistance in research on rice farming systems for small farmers.

Similar analysis may be performed for other commodities. Comparisons of technological events across commodities highlight the forces that produce different levels of "successful" research and adoption from one commodity to another.

For example, an analysis of events in the development of pastures in Panama shows that there were three separate approaches to improving pastures. From 1968 to 1972, FAO and the Ministry of Agriculture attempted to develop forages species, but these required a high input of fertilizer and heavy establishment costs. This approach foundered with rising oil prices. From 1972 to 1975, CATIE used the production systems approach, which analyzed management practices to discover the optimum use of existing pasture and financial resources. Finally, beginning in 1979, the Rutgers/CIAT approach sought germplasm for low-input use on acid soils and gave attention to seed multiplication. Clearly, the approach was changed according to both the source of technical assistance and the changing prices of factors of production.

Public and Private Sector Roles

There is a tendency to overlook the importance of the private sector in promoting technological change while concentrating on breeding and other activities of the public sector. An events analysis is useful in correcting this imbalance. Table 14.2 lists the documented ATMS events that had their origin in the private sector (whether external or domestic). Even though the private sector had a predominant role in the introduction of mechanical and chemical technology, there were several instances where private firms cooperated with public entities in the introduction of improved seed technology (e.g., introduction of IR-8 rice in 1967, introduction of "1-12" tomato in 1975, and of *Andropogon gayanus* pasture grass in 1979).

Conversely, public agencies--both domestic and international--have historically played a major role in the promotion of new mechanical and chemical technologies, although both technologies are usually associated with private sector efforts. The methodology helps trace the contributions of particular institutions. For example, an array of activities in which the Servicios Interamericanos de Cooperaction Agricola en Panama (SICAP), an American technical assistance program, was either the principal organization (Field 6) or cooperating organization (Field 8) provided a history of its contribution. A subset of these activities demonstrated that its collaboration with the private sector was concentrated in well-identified time periods and associated with certain types of technology.

Chronology of Major Institutional or Policy Changes

Finally, events may be categorized to highlight institutional changes in the agricultural technology management system. Any event of an "institutional" nature

TABLE 14.2
Events in Rice Research and Development Initiated by the Private Sector, Panama, 1932-1980

Year	Case	Nature	Description of Event
1932	rice	mech	First mechanized and irrigated production of rice in Alanje
1936	rice	mech	Alfonso Tejeira carries out first mechanized production on Cocle Plains
1945	rice	mech	First mechanized rice fields in Chiriqui with IFA help
1946	rice	mech	Second experience in Chiriqui with irrigation (Sr. Jurado)
1947	rice	mech	First huller in David bought with credit from Central Agricola (Empresas Romero)
1948	rice	chem	Introduction of DDT (chlorinated insecticides)
1948-1949	rice	inst, mech	Consolidation of group of mechanized rice farmers renting machines from IFA
1949-1953	rice	econ	Seed imported from USA covers national needs
1950	rice	econ	Growth in use of American varieties
1952	rice	econ, inst	Formation of first Rice Growers Association to get price support
1952	rice	mech, econ	Mechanization in Chiriqui aided by government financing of group
1952	rice	chem	Generalized use of chlorinated insecticides
1955	rice	chem	Initial use of chemical fertilizer by rice growers
1955	rice	biol	Generalized use of American varieties
1957	rice	chem	Introduction of phosphorous insecticides (Folidol)
1960	rice	chem	Generalized use of phosphorous insecticides
1960	rice	chem, econ	Wide use of fertilizer based on specific formulas

(continued)

TABLE 14.2 (cont.)

Year	Case	Nature	Description of Event
1960	rice	inst	Central Agricola organized "Centrales" to supply inputs and credit
1960	rice	inst, econ	Central Agricola organized Sociedades de Siembra for seed and commercial production
1961	rice	chem	Introduction of 2,4,5T and mixtures with 2,4D
1965	rice	biol	Danger to rice production due to attack of blast
1965	rice	chem	Mass use of Propanil by rice growers to combat weed, 2,4D ineffective
1965-1975	dual-purpose cattle	econ	Rising imports of milk products due to price policy
1967	rice	biol	Introduction of first HYV rice IR-8
1970	rice	chem	Introduction of carbamate insecticides, incompatible with Propanil, less dangerous to humans
1970	rice	chem	Introduction of systematic insecticides
1972	rice	chem	Massive use of fungicides for _Pyricularia_ due to change to susceptible varieties
1979	pastures	agron	Introduction of new pasture species (_Andropogon gayanus_) with BNP, FAUP, CIAT
1980	rice	chem	Introduction of pyrethroid insecticides (compatible with Propanil)
1980+	pastures	econ	BNP, Nestlé, BDA make credit available for improved pastures

that has been cited as having had an impact on the system should appear in this array of events. The event can be associated with the principal and collaborating organizations involved in it. In the case study of Panama, the chronology of institutional events highlighted changes in policy toward extension that occurred in several organizations at the same time, leading the study team to look for reasons why these changes occurred at that particular time. A similar history of political events provides a record of major changes in price or subsidy policies that are important influences on the nature of technology demanded by farmers. The initiators of such policies may be identified and their interests analyzed.

CONCLUSION

The creation of a database of events occurring within an agricultural technology management system is a powerful way of relating political, institutional, and technological forces to the results they produce. A database approach allows the researcher to integrate information collected from a wide range of sources and to identify patterns in events that may not have been apparent to the authors citing such events.

More important, the presentation of the information in ways that highlight relationships of particular importance (e.g., public and private sector collaboration, role of international centers, etc.) provides at a glance the supporting evidence for hypotheses that are put forward. The establishment of such a database is an appropriate exercise for a national research team, and the existence of such a database could save national researchers large amounts of time currently devoted to briefing consultants about the system.

REFERENCES

Cuellar, M., H. Lopez, R. Lass, and F. Estrada. "Impacto de las politicas sectoriales en el cambio technologico en el sector agropecuario: Arroz." IDIAP, Panama, April 1983. Unpublished.

International Service for National Agricultural Research (ISNAR) and Rutgers University. "Identifying Opportunities to Improve Agricultural Technology Management Systems in Latin America: A Methodology and Test Case." Study prepared under AID Contract LAC-0000-C-00-4081-00, December 1985.

PART 4

THE EMERGING ROLE OF THE PRIVATE SECTOR IN AGRICULTURAL RESEARCH

INTRODUCTION

Carl E. Pray

Private sector agricultural research is very important and growing in most developed market economies. It constitutes about two-thirds of the food and agricultural research in the United States. In most developing market economies it plays a less important role but is important in some specific commodities and regions. The private sector is also a very important means of transferring technology between countries.

Our knowledge of the role of private sector agricultural research in developed and developing countries is quite limited because of the lack of quantitative data on private sector research inputs and outputs. Therefore, there are a number of important policy issues for which there are very few studies to provide policymakers with guidance: First, what are the proper roles of the private and public sectors? Second, how can linkages between public and private sector research and extension be improved so that farmers and consumers will receive the benefits of new technology more rapidly? Third, how can governments provide incentives to induce a higher level of private sector research?

Because many companies treat research expenditures and other inputs as confidential information, data are limited and are not readily available. Two of the papers presented here attempt to rectify that situation. The paper by Crosby presents the results of the latest survey of private sector agricultural research by the Agricultural Research Institute. It indicates that in the United States there is about twice as much agricultural research in the private sector as in the public sector. The total expenditure by the private sector was about $2.1 billion. The private sector employs about 2,800 scientists trained at the Ph.D. level.

The paper by Pray presents some preliminary estimates of private sector agricultural research in seven Asian countries. In contrast with the United States, private research is much smaller than public sector research in all of the Asian countries. He reports that it is growing, however, and has been important in producing new technologies in plantation crops, new hybrid seed varieties, some new machinery, and new pest control techniques. He also deals briefly with the first and second issues mentioned above. Given the low level of private research in Asia, he suggests that private research should play a larger role in the future. He also suggests that public research could play an important role in encouraging more private research by improving communication and cooperation with the private sector.

Private companies can also be very important in the transfer of technology between the developed and developing countries. Goodman's paper on the seed industry in Mexico indicates that multinationals and the local private sector can be important suppliers of new technology. This paper concentrates on the policy options facing the government of Mexico in building an effective seed industry. The options are to allow open competition between foreign and domestic firms, keep multinationals out, or create a mixed system with direction from the government. He suggests that the last option is the most viable in the Mexican context.

The Wolek study focuses on the linkages between USDA research and the private sector. The study shows that the USDA projects planned jointly by the public and private sectors had a far greater chance of commercialization than those that originated solely with the USDA scientists. Wolek concludes that the earlier the private sector becomes involved with USDA projects the more likely they are to be commercialized.

The Evenson, Evenson, and Putnam paper deals with the issue of incentives to research. The authors have made a thorough review of patent laws as applied to various types of agricultural inputs and processing technology throughout the world. Their data indicate that there is much less patent protection for agricultural inputs in developing countries than in developed countries. This suggests that there may be opportunities for developing countries to induce more private sector research through stronger patent laws. Unfortunately, there are as yet no statistical studies relating productivity growth to these patent policies. The examples of Taiwan and South Korea suggest that the absence of strong patent policies may not be a constraint to growth. More study is needed on this issue.

These studies are just a beginning. Much more effort is needed to understand the relationship between public and private sector research. In the areas of technology policy, there are many opinions and many policies being tried. Unfortunately, there are still too few studies on the impact of these policies. It is hoped that there will be more research in this area in the future.

15 Private Sector Agricultural Research in the United States*

Edwin A. Crosby

This paper is a revision of the third in a series of surveys conducted by the Agricultural Research Institute to obtain data on the areas of emphasis and magnitude of expenditures by private industry for agricultural research (Crosby et al. 1985). The first of these surveys provided information for the fiscal year 1965 (Wilcke and Sprague 1967). The second covered the period 1976 and 1977 (Wilcke and Williamson 1977). Thus, to the extent that the data reported here are typical of the expenditures by industry, they should provide an indication of the priorities in the various areas of agricultural research and also in budgetary support, and the changes in expenditures that have occurred over the nineteen-year period since the first survey was conducted.

PROCEDURE

This survey was conducted following an expression of interest from the U.S. Department of Agriculture for more up-to-date information on both the nature and extent of industry research relating to food and fiber production, utilization, and marketing systems. A mailing list was compiled by an outside consultant who contacted agricultural trade associations that were national, state, and local in scope. In addition, listings of agricultural companies in two publications were consulted. Such a list obviously does not include every agricultural business but it does include those that are most likely to be conducting and financing research.

The objectives of this third survey were:

1. To bring up to date the data on private sector research to supplement data already available on agricultural research being conducted by public research agencies.

2. To identify for the USDA, the Office of Management and Budget, and members of Congress the major areas of research emphasis by private industry so that public funding of research can be appropriately complementary.

3. To provide valid data to correct misconceptions on the magnitude of support and areas of emphasis in private sector research programs in relation to public sector research.

4. To help guide our efforts to assure that public and private research are directed toward the most pressing agricultural problems.

5. To assist decision makers in avoiding unnecessary duplication in research, thus helping to assure that public funds are directed more efficiently and effectively toward the solution of the most pressing agricultural problems.

A less detailed questionnaire than was used in the first two surveys was developed in the hope of getting a much better response than was obtained in the second survey. This questionnaire was patterned to a considerable extent after the one used by Kalton and Richardson (1983), who made a survey of seed industry research expenditures and scientist years[1] devoted to research in the plant breeding area in 1982 and 1983. They received a response to their questionnaire of over 80 percent, an exceptionally good return. The estimated annual expenditure for plant breeding research was about $115,000,000.

In the ARI survey, a total of 648 questionnaires were mailed, and 326 responses were obtained, including 119 reporting no research, and 6 reporting research programs but not providing adequate data.

Two hundred and one responses provided usable data. This was considered a good response. Data from 155 responses to the Kalton-Richardson (1983) survey are included in this report of the ARI survey, with no identification as to the individual companies reporting. These data were included because it was considered vital that information be obtained from this segment of the agricultural industry, but it was not felt desirable or practical to repeat a survey that had been made just one year previously.

All questionnaires were coded by the Agricultural Research Institute office. Only that office had access to the code, and the code was destroyed as soon as the survey was completed. Thus all information was confidential as to the identity of the companies that provided the individual segments.[2]

SCIENTIFIC PERSONNEL (COMBINED ARI AND KALTON-RICHARDSON DATA)

In table 15.1 we have listed thirteen areas of research that we consider to be distinctive and, in addition, a fourteenth classification for others not falling within the thirteen. In this table we have summarized from the questionnaires returned the number of companies reporting research in each category and the number of scientist years spent in research for that particular category in terms of Ph.D., Master of Science, and Bachelor of Science degree personnel. Some companies, of course, are involved in more than one area of research and therefore there is a discrepancy between the total number of responses received and the total number of companies listed in this table.

On the basis of the companies that reported, it is evident that the pesticide industry, in the combined areas of synthesis and screening and the areas of product development and registration, devotes more scientist years in the Ph.D. category than do other industry classifications. This is followed by plant breeding; biotechnology; and human food product development, processing, and nutrition in that order. The number of M.S. scientist years employed follows the same pattern except that human food research surpasses biotechnology as a major participant at this level. At the B.S. level, again the pesticide industry employs the largest number, followed by plant breeding, human food, and farm machinery.

The larger number of scientist years devoted to research in the pesticide segment of the industry is to be expected, given that a large amount of chemical screening is needed to identify prospective compounds. The companies in this business are well established and to a considerable extent, larger companies than those in the other groupings. Biotechnology is a new industry and is made up largely of small companies; it therefore would not have the extensive staffs employed by some of the older, more established businesses. It would be anticipated that the staffs and research budgets of the biotechnology companies will increase as products are developed and the business becomes

TABLE 15.1
Number of Scientist Years Devoted to Agricultural Research by Degree and Area of Research within the Organization (Combined ARI and Kalton-Richardson Data)

Major Areas of Research	Number of Companies[a]	Ph.D.[b]	Number of Companies[a]	M.S.	Number of Companies[a]	B.S.
1. Agricultural Economics	11	10.4	15	24.7	12	18.9
2. Biologics--Animal Health (Drugs, Botanical Products)	15	74.6	10	69.8	14	131.6
3. Biotechnology	41	287.6	26	134.8	24	167.9
4. Pesticides						
a) Synthesis & Screening	23	408.7	19	144.6	25	222.7
b) Product Development & Use Registration	38	851.6	30	455.0	32	618.0
5. Plant Nutrients (Fertilizers minor elements, etc.)	23	53.9	21	37.0	27	47.7
6. Natural Fiber Processing (Cotton, wool, wood, etc.)	5	5.8	4	7.0	3	23.8
7. Animal Nutrition & Feeds	47	95.5	30	58.7	30	111.5
8. Human Food Product Development, Processing & Nutrition (Foods, vegetable oils, etc.)	57	262.5	53	324.3	51	595.0
9. Farm Machinery & Equipment	11	32.5	22	66.1	32	230.0
10. Tobacco Production & Processing	1	0.5	1	1.0	1	0.7
11. Packaging Materials & Containers (Plastic, paper, metal, etc.	8	14.7	12	23.3	21	68.9
12. Energy Research (Solar, wind, biomass, etc.)	12	31.0	7	42.5	11	10.9
13. Plant Breeding	108	518.9	92	348.6	109	610.8
14. Other than above	31	186.4	25	253.0	23	476.0
TOTAL	431	2,834.6	367	1,990.4	415	3,334.4

[a]Number of total companies varies because of working in multiple areas.

[b]Includes D.V.M.'s.

well established and productive. Several of the companies operating in the pesticide area, for example, are also involved in the biotechnology area.

The number of scientist years reported in the survey provides an estimate of the magnitude of the research effort within companies (table 15.2). The minimum number of Ph.D., M.S., or B.S. scientist years for all companies shows very little difference in range. The minimum range is from 0.01 to 0.5 for Ph.D, 0.01 to 1.0 for M.S., and 0.1 to 1.3 for B.S. in the companies that reported. The maximums, however, show decided differences. The greatest number of Ph.D. scientist years employed by any one company in an identified area is 182 for the product development and use registration area of the pesticide business, ranging down to 0.5 in tobacco. The number of scientist years at the M.S. level ranges from 220 in the unspecified group (other) to 1, also in tobacco production and processing. At the B.S. level, the highest number reported was in the unspecified group (other), where 400 were employed by one company of the unspecified group (other) and only 0.7 in the tobacco classification. Thus, the total scientist years devoted to agricultural research by these 356 companies was 8,159, of which 2,835 were at the Ph.D. level, 1,990 at the M.S. level, and 3,334 at the B.S. level.

It may be of even greater interest to note from the data in table 15.1 the total scientist years spent in each of the 13 major areas of research identified. These range from a total of 2,700.6 in the pesticide area to a low of 2.2 in tobacco production and processing, where a very large amount of the work is done in public institutions but is supported by the tobacco industry, and 36.6 in the natural fiber processing area. The figure of 54 in the area of agricultural economics probably represents only those studies devoted to the general economy, whereas much of the economic analysis work on markets or products is included in the specific product areas.

RESEARCH EXPENDITURES (COMBINED ARI AND KALTON-RICHARDSON DATA)

In table 15.3 we have calculated the research expenditures for 356 companies that supplied usable data on this subject by using the midpoint of the range reported for research expenditures multiplied by the number of companies reporting in each category. In the

TABLE 15.2
Minimum and Maximum Number of Scientist Years Reported
(Combined ARI and Kalton-Richardson Data)

	Ph.D.		M.S.		B.S.	
	Minimum	Maximum	Minimum	Maximum	Minimum	Maximum
1. Agricultural Economics	0.1	2.0	0.7	5.0	0.1	6.0
2. Biologics--Animal Health (Drugs, Botanical Products)	0.1	35.0	0.1	55.0	0.2	60.0
3. Biotechnology	0.2	68.0	0.3	36.0	0.1	54.0
4. Pesticides						
a) Synthesis & Screening	0.1	85.0	0.1	35.0	0.1	58.0
b) Product Development & Use Registration	0.1	182.0	0.1	63.0	0.1	95.0
5. Plant Nutrients (Fertilizers, minor elements, etc.)	0.1	18.0	0.2	8.0	0.1	14.0
6. Natural Fiber Processing (Cotton, wool, wood, etc.)	0.3	2.5	0.5	3.0	1.3	20.0
7. Animal Nutrition & Feeds	0.01	17.0	0.1	15.0	0.1	42.0
8. Human Food Product Development, Processing & Nutrition (Foods, vegetable oils, etc.)	0.1	51.0	0.3	46.0	0.3	110.0
9. Farm Machinery & Equipment	0.3	20.0	0.2	20.0	0.1	80.0
10. Tobacco Production & Processing	0.5	0.5	1.0	1.0	0.7	0.7
11. Packaging Materials & Containers (Plastic, paper, metal, etc.)	0.1	4.0	0.1	8.0	0.1	15.0
12. Energy Research (Solar, wind, biomass, etc.)	0.1	20.0	0.01	40.0	0.1	3.0
13. Plant Breeding	0.2	61.0	0.2	54.0	0.1	95.0
14. Others than above	0.3	90.0	0.2	220.0	0.1	400.0

TABLE 15.3
Total Research Expenditures (In-House and Outside) (Combined ARI and Kalton-Richardson Data)

Dollars	No. Companies Supporting In-House Research	Dollars	No. Companies Supporting Outside Research	Dollars
0			215	
Under 100,000	104	5,200,000	81	4,716,500
100,000 - 500,000	110	33,000,000	30	3,400,000
500,000 - 1,000,000	47	35,250,000	14	9,300,000
1,000,000 - 5,000,000	51	153,000,000	14	34,059,170
5,000,000 - 10,000,000	16	120,000,000	2	15,850,000
10,000,000 - 20,000,000	13	195,000,000		
20,000,000 - 30,000,000	4	100,000,000		
30,000,000 - 40,000,000	3	105,000,000		
40,000,000 - 50,000,000	3	135,000,000		
50,000,000 - 100,000,000	3	225,000,000		
Over 100,000,000	2	200,000,000		
Total	356	1,303,450,000	356	67,325,670
Total Research Expenditure	$1,370,775,670			
Mean for 356 Companies	$3,850,493			

questionnaire, respondents were asked to provide only a range of expenditures and not an exact figure. It was felt that this would be more acceptable to those companies reluctant to provide exact figures on research expenditures. Thus, for the 356 companies, a total of $1,370,775,670 was reported as having been spent per year by industry in agricultural research in the survey period.

EXTRAPOLATION FROM QUESTIONNAIRE DATA TO ENTIRE INDUSTRY (COMBINED ARI AND KALTON-RICHARDSON DATA)

We made three sets of assumptions to provide estimates of total agricultural research expenditures by private industry:

1. The first assumption was that since the incentive of receiving a copy of the final report was offered to induce the return of questionnaires, all of those doing no research responded by checking one box and returning the questionnaire.

Under this assumption, the 322 who did not reply would all be conducting research, and it would be assumed at about the same rate per company as those who did respond. On the basis of $1,303,450,000 having been spent on in-house research by the 356 companies that reported usable data and $67,325,670 in support of outside research, the total would be $1,370,775,670 and the mean expenditure per company would be $3,850,493. The additional 322 companies that did not report would also be operating on the same basis as the 356 that reported, and would have spent $1,239,858,746. By adding this to the 356 total of $1,370,775,670, we arrive at a figure of $2,610,634,416 for the total of all companies doing research. This obviously is not a precise figure since ranges were used in reporting and extrapolations were used to arrive at this figure.

2. Under our second assumption, the 322 companies that did not report included the same percentage of companies not doing research as the original 648 companies that were asked to fill out questionnaires in the ARI survey. On this basis, an additional 197 satisfactory questionnaires would have been received. Since the original 356 companies spent an average of $3,850,493 per company, the additional 197 companies would have spent in total $758,547,121, which, added to the original $1,370,775,670, would provide a grand total of $2,129,322,791 for the 553 companies doing research under this scenario.

3. Under the third assumption, the 322 companies that did not return questionnaires were doing research, but at a lower level of expenditure per company. More recent figures obtained by Kalton and Richardson (1983), reported in the 155 questionnaires included in this survey, indicate an average expenditure of $1,041,613 per company in the plant breeding area. Using this figure as an average for the 322 not reporting, these companies would have invested a total of $335,399,386, which, added to the total of $1,370,775,670 for the 356 companies reporting, gives an industry total of $1,706,175,056.

This is probably a conservative figure since the mailing list used undoubtedly did not include all companies conducting agricultural research. Since ranges of expenditures were used rather than exact figures, it is probable that the second assumption provides the best estimate of private industry annual expenditures in agricultural research--approximately 2.1 billion dollars.

The range of expenditures ($1.7-2.6 billion) indicated by this survey shows a marked increase over the figures from the 1965 industry survey, in which a total of $460 million was estimated from 247 questionnaires. These figures may be compared with the more precise figures supplied by R. Eddleman (pers. com. 1985) for the state and USDA research programs, which increased from $429.9 million in 1967 to $1,703.6 million in 1983. In both periods, industry expenditures appear to be equal to or somewhat in excess of public outlays for agricultural research.[3]

If the total expenditure to support the research of 8,159 scientist years in-house was $1,303,450,000, the average cost per scientist year regardless of academic rank was $159,756. The expenditures for each area of research were calculated by multiplying the total number of scientist years reported for each area by the average cost per scientist year and are presented in table 15.4. It is apparent from these data that the pesticide industry has a larger research budget than any of the other major areas. This is followed by plant breeding, human food, biotechnology, farm machinery and equipment, biologics, and animal nutrition and feeds. If the calculations for those expenditures are made with a differential for the Ph.D., M.S., and B.S. scientist years, the amounts would be changed but there would probably be very little, if any, change in the rankings of the different research areas.

TABLE 15.4
Amount of Money Spent per Area of Research
(Combined ARI and Kalton-Richardson Data)

Major Areas of Research	Total SY	Total Dollars@ $159,756 per SY	Rank
1. Agricultural Economics	54	8,626,824	12
2. Biologics	276	44,092,656	7
3. Biotechnology	590	94,256,040	5
4. Pesticides			
a) Synthesis & Screening	(776)	(123,970,656)	
b) Product Development and Registration	(1,925)	(307,530,300)	
Total Pesticides	2,701	431,500,956	1
5. Plant Nutrients	139	22,206,084	9
6. Natural Fiber Processing	37	5,910,972	13
7. Animal Nutrition & Feeds	266	42,495,096	8
8. Human Food	1,182	188,831,592	3
9. Farm Machinery & Equipment	329	52,559,724	6
10. Tobacco Products & Processing	2	319,512	14
11. Packaging Materials	107	17,093,892	10
12. Energy Research	84	13,419,504	11
13. Plant Breeding	1,478	236,119,368	2
14. Others	914	146,016,984	4
Total	8,159	1,303,449,204	

RESEARCH CATEGORIES (ARI DATA ONLY)

Table 15.5 presents the distribution of research expenditures in the categories of relevant basic research, applied research, and development research as reported in the questionnaires received from the ARI survey. The definitions for relevant basic research, applied research, and developmental research as included in the questionnaire are as follows:

Relevant basic research--Research conducted to determine the basic cause or mechanism of why certain results or reactions are obtained.

Applied research--Research conducted to develop knowledge or information directly relevant to technology, product development, or market possibilities.

Developmental research--Research conducted to develop a new or improved technology or product, to support market

TABLE 15.5
Approximate Percentage of Total Research Budget in Each Category[a]
(ARI Data Only)

Range (%)	Relevant Basic Research		Applied Research		Developmental Research	
	No. Companies	%	No. Companies	%	No. Companies	%
0	82	40.8	15	7.5	29	14.4
1-10	68	33.0	18	9.0	5	2.5
11-20	19	9.4	25	12.4	25	12.4
21-40	16	8.0	49	24.4	41	20.4
41-80	11	5.5	69	34.3	77	38.3
81-99			10	5.0	17	8.5
100	5	2.5	15	7.5	7	3.5
TOTAL	201		201		201	

[a]Plant breeding figures were not included because of lack of usable information on this question.

testing and introduction, to maintain product performance and quality, or to meet regulatory requirements.

The data reveal that 82 of the 201 reporting companies (40.8%) conduct no relevant basic research. Five companies classified their research programs as 100% in this category.

Fifteen companies reported 100% of their programs in the applied research area. There were 15 companies (7.5%) reporting no applied research. This group would depend very heavily on public research institutions to supply the information needed for the development of products and programs.

In the developmental research area, 29 companies (14.4%) reported no expenditures. Seven companies reported 100% of their research budgets being utilized in the developmental research area.

Not all of the companies reporting research expenditures provided complete breakdowns on the percentage of their budgets spent in basic, applied, or developmental research. The usable data obtained from the 201 responses in the ARI survey reveal that the companies reporting spent a total of $1,145,000,000 in agricultural research, allotted as follows: $171,649,930 to relevant basic research, $498,379,030 to applied research, and $474,947,740 to developmental research. This would indicate that private industry is devoting 15.0% of its agricultural research expenditures to relevant basic research, 43.5% to applied research, and 41.5% to developmental research.

If the 119 companies that reported doing no research are added to the 82 doing research but reporting no basic research, we have 62% of the 326 companies that responded in the ARI survey not doing any basic research. This figure becomes important when the complementary roles of public and private research programs are considered.

SUMMARY AND CONCLUSIONS[4]

1. The results of the surveys reported here provide the basis for estimating that the private sector of the agricultural industry spends from approximately $1.7 to $2.6 billion per year on agricultural research, depending upon the assumptions made in extrapolating from the data provided in the completed questionnaires to a total for all companies asked to provide data in the ARI survey. Our

best estimate is that private industry is investing approximately $2.1 billion annually in agricultural research.

2. The 356 companies (201 in the ARI survey and 155 from Kalton, et. al.) that reported their research expenditures provide budgets of approximately $1,303,450,000 per year for in-house research programs.

3. One hundred and forty-one companies that reported supporting research through grants or contacts to universities, foundations, or other private or public organizations reported a total expenditure of approximately $67,325,670. However, since the questionnaire specified that the amount be stated to the nearest $50,000, small grants and fellowships or scholarships would not have been reported. For that reason, this total is probably under-reported.

4. The total research expenditures of the 356 companies reporting was $1,370,775,670 for the support of both in-house and outside research.

5. The 356 companies reporting supported a total of 2,835 scientist years at the Ph.D. level; 1,990 at the M.S. level; and 3,334 at the B.S. level--a total of 8,159 scientist years at all levels.

6. The cost per scientist year for in-house research was $159,756, if the assumption is made that there is no cost differential for the three academic ranks.

7. The greatest expenditure of research funds reported by the companies responding was in the area of pesticides.

8. Eighty-two companies of the 201 providing data reported no relevant basic research while 5 companies reported 100% basic research.

9. Fifteen companies reported no applied research and 15 reported 100% applied research.

NOTES

*This paper is a revision of the report by E. A. Crosby, B. R. Eddleman, R. R. Kalton, V. W. Ruttan, and H. L. Wilcke, A Survey of U.S. Agricultural Research by Private Industry III.

1. One scientist year equals one full-time scientist for one year, including persons with a B.S., M.S., or Ph.D. degree. This is in contrast to the public sector, where scientist year refers to a Ph.D. or equivalent.

2. The term company or companies is used throughout this report as a general term to include the various organizations in the private industry sector that provided data in this survey. Included are companies, corporations, cooperatives, associations, foundations, and other for-profit and not-for-profit entities.

3. Ruttan (1982, 183), in his book *Agricultural Research Policy*, indicated that the private industries were spending in the neighborhood of 1.6-2.0 billion dollars annually for agricultural research in 1979. These estimates were somewhat higher than indicated in the previous ARI surveys for 1965 and 1976-77.

4. Since the Kalton-Richardson survey did not provide enough usable data for a breakdown into relevant basic, applied, or developmental research, observations 8, 9, 10, 11, and 12 are based only on data from 201 ARI responses.

REFERENCES

Crosby, E. A., B. R. Eddleman, R. R. Kalton, V. W. Ruttan, and H. L. Wilcke. A Survey of U.S. Agricultural Research by Private Industry III. Bethesda, Md: Agricultural Research Institute, 1985.

Kalton, Robert K., and Phyllis Richardson. "Private Sector Plant Breeding Programs: A Major Thrust in U.S. Agriculture." Diversity, 1983, 16-18.

Ruttan, Vernon W. Agricultural Research Policy. Minneapolis: University of Minnesota Press, 1982.

Wilcke, H. L., and H. B. Sprague. "Agricultural Research and Development by the Private Sector of the United States." Agricultural Science Review 5(3)(1967): 1-8.

Wilcke, H. L. and J. L. Williamson. A Survey of U.S. Agricultural Research by Private Industry. Washington, D.C.: Agricultural Research Institute, 1977.

16 Private Sector Agricultural Research in Asia*

Carl E. Pray

In 1984 the University of Minnesota initiated a study of research by private agribusiness in the Third World. This study was initiated because it appeared that some countries in Asia were missing out on important opportunities for growth by discouraging the private sector from doing research and transferring agricultural technology. This study confirmed the presence of such opportunities in some countries. In these times of pressure on government budgets and declining donor enthusiasm for government research, private sector research and technology transfer may offer a relatively inexpensive and efficient way to supply new technology to farmers.

This paper is relevant to agricultural scientists and research administrators because (1) the government agricultural research system in most countries can play an important role in inducing more private sector research and making it more productive; (2) the research system of the country as a whole can be made more productive if government research complements rather than duplicates private research; and (3) government scientists usually play an important role in regulating private research, technology imports, and the sale of technology by the private sector.

Before continuing, several definitions are needed. In this study the private sector means for-profit firms and excludes nonprofit organizations and the collective activities of for-profit companies like industry associations. The study was also limited to technology that is applied in production agriculture and excludes technology for post-harvest operations.

The data upon which this paper is based were collected between March 1984 and February 1986. The author visited seven countries in Asia and talked to as many private companies involved in research or technology transfer as possible. About 100 firms were interviewed. In Bangladesh, Pakistan, and Indonesia, where there was little private sector research and technology transfer activity, all that had formal research programs were contacted. In India, Philippines, Thailand, and Malaysia, a sample of firms were contacted. This was not a random sample, rather it was the firms in which we had contacts. The sample may be biased by oversampling the multinationals and undersampling the local firms. We did, however, make special efforts to contact at least one local firm in each industry.

HAS PRIVATE SECTOR RESEARCH AND TECHNOLOGY TRANSFER CONTRIBUTED TO AGRICULTURAL GROWTH?

Private sector research has made major contributions to the growth of some economies in Asia. The largest economic impact has been on the plantation industry. The oilpalm varieties of Southeast Asia that have made that area the world's fastest growing producer of edible oils are largely the result of private research. Many of the clones used in rubber production in Indonesia and Malaysia were developed in Harrison and Crossfield's breeding program. Yields of oilpalm were increased and costs of production reduced by the introduction of the oilpalm pollinating weevil. Plantation research has substantially reduced the cost of pest control and fertilizer use.

After plantation research, it is not clear which type of research has had the most impact. Applied tobacco research in India, Pakistan, and Bangladesh introduced Virginia tobacco, increased yields substantially, and almost eliminated imports of Virginia tobacco. New high-yielding corn hybrids have been developed by the private sector and are being commercially planted in several hundred thousand hectares in Thailand, India, and the Philippines. They are currently spreading to Indonesia and Malaysia. In India, pearl millet hybrids and sorghum hybrids developed by the private sector have spread to about one million hectares and one-half million hectares respectively. These hybrids have raised yield per acre substantially. The Indian machinery industry has developed tractors that run more efficiently and safely under Indian conditions than previous tractors. Research by agricultural chemical companies in Asia has identified chemicals to control Rottboellia exalta, the most serious weed problem in corn in the Philippines; fungicides for seed treatments of corn for downy mildew were developed and are widely used; and new rice herbicides were developed in Thailand. Research in Southeast Asia has developed safer and cheaper methods for applying pesticides.

The private sector has probably had more impact on agriculture by transferring technology than by developing new technology through R&D. Fertilizer and fertilizer production technology were imported from North America, Europe, and Japan. The first generation of tractors and almost all pesticides were developed elsewhere and transferred to Asia. Arbor Acres and Shaver were the pioneers of commercial poultry production in Asia in the early 1960s, and American and European companies continue to be the source of most commercial poultry breeds. The technology for commercial swine production has been transferred from the United States, Europe, and Taiwan to Thailand, the Philippines, and Singapore by the private sector. Private companies have rapidly transferred rubber and oilpalm technology between Malaysia, Indonesia, Thailand, and the Philippines. Banana production for export and pineapple production for canning were both based on introduced technology.

The private sector has also played an important role in the diffusion of public sector technology within Asian countries. In Thailand, Charoen Pokphand and Cargill are selling corn varieties developed by Kasetsart University and Rockefeller. In the early days of the Green Revolution, Esso helped spread wheat and rice HYVs and fertilizer in Pakistan and the Philippines.

Charoen Pokphand's Indonesian subsidiary has conducted three-week practical courses on poultry production for farmers since 1978. It has expanded this course to a second location on Java. Over 1,000 farmers have taken this course. Large integrated poultry operations in Thailand and the Philippines provide technical assistance to their contract farmers. Buyers and technicians in export industries and processing industries like the Bangladesh Tobacco Company provide extension advice on varieties and management practices that will provide high quality cigarette tobacco at a low price.

Two other ways in which the private sector spreads technology are through consulting firms and spinoff companies. The most important agricultural consulting firms are in Malaysia, where they are divisions of the major private plantation companies. They provide advice on planting and management of plantations in Malaysia, Indonesia, Thailand, and elsewhere. The main place that we observed spinoff companies from the public universities was in Thailand. There a number of small seed companies were founded by ex-faculty members and students from Kasetsart University using inbred lines of corn from Kasetsart University.

On the whole it appears that private research and technology transfer activities have had an important positive impact on agricultural production. Further study is needed to quantify that impact. It is also important to note that private research and technology transfer can also have negative effects on some producers and consumers. When a new production technology is widely adopted, almost always someone loses. In general, early adopters and consumers benefit from new technology, while late adopters and nonadopters lose. Increased production of corn, sorghum, and pearl millet took place mainly in the most favored agricultural regions, where these crops are grown, and depressed prices for farmers' products in the poorer areas. Rapid growth of commercial poultry has adversely affected backyard poultry producers. Herbicides and farm machinery for cultivation and threshing have displaced labor in some places. Pesticides have saved lives by eliminating diseases like malaria but have caused death and sickness when used improperly.

WHAT IS PRIVATE SECTOR RESEARCH DOING?

Our best estimates of private sector research expenditures are presented in table 16.1. They are based primarily on our survey but are supplemented by published data where available. They are almost certainly underestimates because it was not possible to interview all companies in most countries. In Malaysia, in particular, we had time to interview only a few firms and most of those were in the plantation sector in peninsular Malaysia.

The amount of private sector research is large relative to previous expectations. Most policymakers and government scientists thought private research and development were almost nonexistent in South and Southeast Asia. The amount of private research is small, however, when compared with the amount of private research in developed economies like the United States, with the size

TABLE 16.1
Private Sector Research Expenditures, Seven Asian Countries, 1985

	India	Philippines	Thailand	Indonesia	Malaysia	Pakistan	Bangladesh	Total
				(US $ 1,000s)				
Seeds	833 (8)[a]	1583 (4)	665 (5)	0	0	182 (3)	Less than 1000 (1)	3264
Pesticides	3500 (20)	1170 (8)	887 (5)	800 (1)	500 (3)	387 (5)	40 (2)	7284
Machinery	6775 (3)	none	none	none	?	none	none	6775
Livestock	2275 (3)	500 (6)	1725 (2)	600 (3)	?	none	none	5100
Processing and Plantations	3324 (25)	1137 (7)	1034 (3)	600 (3)	10000 (9)	234 (2)	50 (1)	16379
Total Private Research	16707	4390	4311	2000	10500	804	90	38802
Government Ag. R&D[b]	248000	7000	73595	6700	44400	56170	8000	
Private as Percent of Govt. Research	7	63	5	3	24	1	1	
Agricultural Value Added ($ billions)	59.7	8.7	5.6	21.1	6.6	6.6	6.7	
Private as Percent of Ag. GDP	.03	.06	.05	.01	.17	.01	.01	

Sources: India, 1983; India, Department of Science and Technology, Research and Development Statistics, 1982-83, New Delhi, 1984. Philippines, 1984; Moises Sardido unpublished statistics collected for UNDP study. Thailand, 1984; Rungruang Isarangkura, "Thailand and the CGIAR Centers: A Study of Their Collaboration in Agricultural Research," Study Paper No. 16, Consultative Group on International Agricultural Research, 1986. Indonesia, 1984; personal communication with Peter Oram, IFPRI. Malaysia, 1980; personal communication with Peter Oram, IFPRI. Pakistan 1984; Pakistan Agricultural Research Council, National Agricultural Research Plan, Islamabad, 1986. Bangladesh, 1985; personal communication with A. Kaul, Winrock International.

[a] Number of firms in parentheses.

[b] These numbers are not consistent in their inclusion or exclusion of capital expenditures. The Philippines does not include capital expenditures but Pakistan does and some of the others are unclear.

of Asian agricultural sectors, or with the amount of public sector research expenditures in Asia. The latest survey of agricultural research in the United States (ARI 1985) suggests that U.S. private firms have spent at least $2.1 billion on agricultural research in recent years. The mean expenditure on R&D of the 356 companies in the ARI survey was $3.8 million, more than is spent by the private sector in Indonesia, Pakistan, and Bangladesh put together. Private research expenditure as a percent of agricultural value added was less than 0.2 percent in all countries sampled (see table 16.1). In comparison, the United States spent 2 percent of agricultural GDP or 3 percent of value added. In the sampled countries, the public sector spends far more on agricultural research than the private sector. Table 16.1 indicates that the Philippines is the only place in which the private sector spends as much as one-third of the total research expenditure of the country.

Across all seven countries the most research was carried out by the plantation and processing industries.[1] These industries include oilpalm millers and planters, rubber planters, cigarette manufacturers, banana growers and exporters, sugarmillers, pineapple canners and producers, and a few others. Only their research on product agriculture, not their research on processing, is included. It should be noted that almost two-thirds of the plantation research is done in Malaysia. Some of the largest plantations are not entirely private. The Malaysian companies are owned by a mixture of government-owned corporations, local private companies, American- and European-based multinationals and Asian-based multinationals. They still seem to be operated as private corporations, however.

The research of processing and plantation firms spans a wide spectrum. Plant breeding and selection are done by oilpalm plantations, rubber plantations, pineapple processors, and cigarette companies. Several companies are using sophisticated tissue culture techniques to clone and multiply oilpalms, and at least two companies are doing research to develop techniques to clone coconuts. Plantation and processing firms also invest much of their research resources in reducing their plant protection costs. Plantations in Malaysia and Indonesia are doing or are financing biological pest control research and integrated pest management research. Research in the Philippines is attempting to reduce the cost of plant protection on banana plantations and to identify safer pesticides and application techniques.

There are major investments in research by the input industries. Pesticide research is dominated by multinational companies. They and their subsidiaries conduct centrally funded research in Asia that is testing new compounds. The local subsidiaries fund research on different formulations and application techniques and test to meet local registration requirements. The only country in which there was some synthesis of new compounds was India, and the two companies there that did any synthesis research have probably stopped in the last few years. Some bioefficacy and registration research is carried on in every country, but there is a tendency for centrally funded research to concentrate in a few countries where regional stations are located. The Philippines has five regional programs for rice and Malaysia five or six for plantation crops.

Regional and local research consists of screening new compounds in field tests after they have passed all of the basic toxicity tests and the initial greenhouse screens back at headquarters. Among the new products that are being tested now are insect growth regulators such as chitin inhibitors. Companies also test compounds that are already being used commercially elsewhere and try them on different pests. In all seven countries covered by this survey, companies are required to prove bioefficacy in local conditions in order to register and sell their chemicals. This requires a large part of the private research money in some countries. There is some research on improved application methods, and a small amount of private research on integrated pest management (IPM). Some of this is done in cooperation with plantations. Other IPM research is done as a result of the buildup of insect pests that are resistant to many pesticides, such as some cotton and vegetable pests in Thailand.

Most of the pesticide research is on insecticides for use in rice and cotton. There has also been considerable research on herbicides for plantation crops. Recently, there has been increased emphasis on rice herbicide and rice fungicide research for Southeast Asia.

Formal R&D on agricultural machinery is done primarily by a few large firms in India that do research on tractors and pumps. The tractor firms concentrate on improving fuel efficiency and increasing their safety for road use because haulage is the major use of tractors in India. Pump manufacturers are trying to increase the efficiency of their pumps, and at least one company is trying to develop solar-powered pumps.

Much informal research on farm machinery is being done in all of these countries. Recent theses on the Philippines (Mikkelsen 1984) and Thailand (Paitoon 1982) substantiate the large amount of innovative activity and the impact of this activity on production. However, because of the short time of our surveys, we could interview only a few small firms to corroborate the findings of other studies. The research in this sector was primarily trying out suggestions by farmers for improved machinery or changes to make production cheaper by substituting cheaper inputs for more expensive ones.

Livestock research includes poultry breeding by one firm in India, swine breeding by a Thai firm, and much work on feed by a number of firms in India, Thailand, and the Philippines. Feed research is done mainly by Asian firms but some is carried out by multinationals. Goals are to improve the quality of the feed and reduce its cost by using inexpensive local materials. Several companies are also doing research on ways of producing shrimp cheaply for export.

The private sector in the Philippines spends more on agricultural research than any other country in our survey. This is largely due to Pioneer's large program in Mindanao, which serves Indonesia and Thailand as well as the Philippines. Multinationals play a very large role in this research in the Philippines, Thailand, and Pakistan but not in India, where several local companies have research programs.

Seed research concentrates on breeding hybrid corn, with some breeding work on hybrid sorghum, sorghum-sudan grass, sunflower, and pearl millet. There is also a small amount of research on hybrid rice. A few companies do some research on plant protection, agronomy, and plant physiology.

It is useful also to remember what the private sector is not doing. There are certain crops that the private sector will not work on. Table 16.2 gives a rough estimate of the distribution of private and public research in the Philippines. The table shows a number of crops on which the private sector spends little research money. These include important subsistence crops like yams and cassava and also white corn and unirrigated rice (not in table). The reason that little private sector research money is directed to these crops is that companies do not sell subsistence farmers many inputs and cannot profit by buying the farmers' products. There is little research on sugarcane because it is unprofitable. The private sector

TABLE 16.2
Philippines Private and Public Research by Commodity (millions of pesos)

	Private (1985)	Public (1984)
Rice	20	15
Corn	30	6
Sugarcane	3	29
Coconut	2	11
Tobacco	2	19
Fruits and vegetables	19	3
Other crops	3	27[a]
Livestock and poultry	5	17

Sources: Private from survey, public from Sardido.

[a]Half of this is root crop research.

also will not do much research in certain disciplines or topics. There is very little research done by the private sector on IPM, farm management, plant nutrition, nonhybrid plant breeding, or social science. Some regions or countries are almost completely neglected by private sector research. Table 16.1 indicates that the least developed economies have little private research. Companies in these economies find it unprofitable to do research because of the low level of modern input use and small marketed surplus.

Finally, it should be noted that private research in Asia is for the most part very applied research. It concentrates on adaptive work that allows technology developed elsewhere, or in the public sector, to fit local agroclimatic and economic conditions. Really basic research on plant genetics is conducted in only one or two companies in Asia and they are multinational companies. The small amount of basic research in the private sector is a pattern that is found worldwide.

DETERMINANTS OF PRIVATE SECTOR RESEARCH AND TECHNOLOGY TRANSFER

The second major issue of the Minnesota study was the importance of government in determining the amount and direction of private research and technology transfer. Whether private firms will invest money in the adaptation and research required to transfer technology or whether they will invest in R&D to develop new technologies depends on the profits they expect to make from these investments. The firm's expectations about profits will depend on (1) the size of the market for the new process or product that results from research; (2) the expected profit per unit of new product sold; (3) the cost of the research, development (or adaptation), production (or importation), and promotion needed to bring a new product to market; and (4) government policies that affect the other three factors.

Market size is probably the most important factor explaining the different levels of investment in research and technology transfer by the private sector. Market size is primarily determined by size of the country, its level of modernization, and the level of government invention in input and output markets.

The first line of table 16.3 shows estimates of private research expenditure from table 16.1. The rest of the table contains some indicators of (1) market size-agricultural value added, (2) agricultural modernization, (3) amount of commercialization (urbanization), and (4) government intervention. Government has affected the size of the agricultural sector and the amount of commercialization indirectly. Modernization has been due in part to government and international investments in agricultural research, extension, and input supply. The estimates of government intervention are based on both the government shares of input supply and their intervention in markets through price controls, etc. The amount of agricultural input supply and processing done by the public sector is, of course, a government decision.

The numbers in table 16.3 are consistent with the hypothesis that market size is important in determining private research expenditure. India, which has the largest agricultural sector of these countries and also the largest markets for modern inputs, has the most research. Bangladesh has the least private research and is the least developed of these countries with very little use of modern inputs except fertilizer, which until recently was distributed by the government.

TABLE 16.3
Possible Determinants of Private Agricultural Research, Seven Asian Countries

	India	Indonesia	Philippines	Malaysia	Pakistan	Bangladesh	Thailand
Private R&D (US$ millions)	16.7	2.0	4.4	10.5	0.8	.1	4.3
Private R&D as % Ag. V.A.	.03	0.1	0.6	.17	.01	.00	.05
Ag. V.A. ($ billions)	59.7	21.2	8.7	6.6	6.6	6.7	5.6
Modernization							
Fert. (kg. nutrient/ha.)	34	75	29	97	57	48	19
Tractor (per 1000 ha.)	2.7	.7	1.6	1.8	6.0	.5	5.4
Cereals yield (kg./ha.)	1486	3352	1723	2647	1637	2033	1986
Commercialization							
Urban %	25	25	39	31	29	18	18
Government Intervention	Major	Major	Minor	Minor	Major	Major	Minor

Sources: First three rows, table 16.1; Modernization and Commercialization, World Resources Institute 1986; Government Intervention, estimates of author.

However, the relationship between market size and research is not as strong as one might expect. The other countries are all modernizing their agriculture and have fairly large agricultural sectors. The main determinant of the level of private research appears to be government industrial policies. Indonesia, the next largest agricultural economy after India, has very little private research, in part because the government controls most of the input distribution and owns most of the plantations. Pakistan, which has a fairly large agricultural sector and is a large market for some modern inputs, has less private R&D than one would expect, possibly because the government supplied inputs until recently and the government has restricted private research.

The Philippines, Thailand, and Malaysia have much higher levels of private research than Indonesia and Pakistan and spend a higher percentage of agricultural value added on research than India. Their economies are smaller than Indonesia's and are about the same size as Pakistan's. Their level of modernization is about the same or less than Pakistan's and Indonesia's--they use less fertilizer than Indonesia and fewer tractors than Pakistan. The major factor that differentiates these countries from Indonesia, Pakistan, and India is that they have allowed the private sector a major role in the production and distribution of inputs. Several additional factors are the presence of large agricultural export sectors that invest in research, the regional research headquarters for a number of agricultural chemical companies in each country, and regional headquarters of large seed companies in the Philippines and Thailand.

At the industry level it is possible to examine the relationship between market size, government policies, and private research in more detail. In addition, one has to look at factors like firm size, market share, and structure of the industry. The remainder of this section presents case studies of the development of research in the seed, pesticide, farm machinery, and plantation industries.

In the seed industry, companies look at acreage under open-pollinated annual field crops--especially corn--as an indicator of potential markets. This is an explicit criterion of multinationals for entering a new market. Multinational seed companies invariably invest in seed research in the new country. In all of these countries except the Philippines, the initial investment in private seed research on field crops appears to have been by multinationals.[2] The sequence in which seed research was initiated was India, 1960; Pakistan, 1965; Thailand, early 1970s; and the Philippines, 1976. There are no private seed research programs on annual field crops in Indonesia, Bangladesh, or Malaysia.

Although the choice of country for these programs was made largely by market size, their continuation was determined largely by government action. The low level of multinational research in India and Pakistan is due largely to government restrictions on the role of multinationals. Both multinationals and local companies in Southeast Asia benefited from the Kasetsart University/Rockefeller Foundation/CIMMYT program that identified genetic resistance to downy mildew. The success of local seed companies in India and Thailand is, in part, due to the role of government research in developing improved inbreds and other technical assistance at the early stages of their development. In contrast, until recently Indonesia and Pakistan did not allow private companies to do seed research and sell new varieties to all farmers.

Pesticide technology transfer and research also followed market size. An additional factor may have been the size of other investments in chemicals that companies had made. India, which had large cotton and rice crops, a large mosquito control program, and a chemical industry, was the first place to which many companies transferred insecticide technology in the 1950s. The plantation economies of Malaysia and Indonesia were the natural places to introduce herbicides. Paraquat was introduced there in the 1950s. Subsidies and government programs to supply pesticides were important in introducing farmers to pesticides and increasing the demand for pesticides in most countries in Asia. If the government continued to supply pesticides after the point at which farmers understood their efficacy, as happened in most countries, the government became a constraint to the development of private research programs.

In countries where the private sector was allowed to play an important role in supplying pesticides, growth of local research was determined by (1) the growth of demand for pesticides in rice and plantation crops that are not grown in the West, (2) the numbers of chemicals being discovered at headquarters, (3) resistance to specific chemicals, (4) the growing regulatory requirements, and (5) publicity on the potential ecological and health hazards of pesticides.

At present, according to industry estimates, the largest consumer of pesticides is India ($300 million), followed by Indonesia ($100-140 million), Pakistan ($100-120 million), Thailand ($100 million), Philippines ($60-80 million), and Bangladesh ($10 million). This is consistent with the investments in research of India and Bangladesh, but the Philippines and Thailand conduct more research than Indonesia and Pakistan. The position of the Philippines is due to the location of regional rice experiment stations of four or five multinationals near IRRI. Indonesian investments are held back by the government policy of distributing 75 percent of pesticides through the government supply organization at highly subsidized prices. There is no simple explanation for the position of Pakistan research. It may be that since companies can test pesticides on cotton in the United States, there is less need to do applied research on cotton pesticides in Asia. Private research may simply not have caught up with the rapid growth of pesticide use brought on by denationalization of pesticide supply in the early 1980s. Another possibility is that political instability and the risk of future nationalization have reduced the incentives for companies to do research there.

The machinery story looks simple. India used 450,000 tractors in the early 1980s. The next largest market in our sample of countries was Pakistan with 120,000. With a market this large, it is hardly surprising that Indian agricultural machinery manufacturers do all of the formal R&D that is conducted in these countries. In fact, the story is probably not so simple. The government of India has had considerable influence on the amount of research done through its policies on imports, restrictions on what products large companies can sell, and policies on foreign ownership and technical agreements. A more thorough time series study is required to determine how these policies influenced research in this industry.

Informal research on simpler agricultural machinery is going on continuously in all of the sample countries. The amount has grown as agricultural mechanization has grown. Some government policies have influenced the amount of innovation that has taken place. Utility patents appear to have been an incentive for research in the Philippines. Government programs in rice mechanization that were supported by IRRI in the Philippines, Thailand, and Indonesia have induced innovative informal research in those countries.

The private plantation industry in Malaysia is the largest in Asia. The Indonesian industry is also large but much of that industry is owned and operated by government corporations. The Philippines also has some large plantations of more recent origin in bananas and pineapples. Thailand has rubber, oilpalm, and pineapples. Expenditure on research in Malaysia seems to have followed the fortunes of the industry. As yet, the purchase of some foreign-owned plantations by government-owned banks does not seem to have affected the management of research on these plantations. Plantation research in Indonesia has been determined almost entirely by government decisions. Private research stopped in Dutch plantations when they were nationalized in 1958 and in the other plantations when they were nationalized in 1965. Some of the major companies were denationalized in the late 1960s, and four companies reestablished their research programs at that time. Plantation research in the Philippines started in the late 1960s with the establishment of the banana and pineapple export industry. It grew rapidly as the industry grew but, like the industry, has stagnated or declined since 1980.

POLICY IMPLICATIONS

In countries like Bangladesh, government policies are not the main constraint to private sector research and technology transfer. The main constraint is the lack of a modern agriculture. These countries do, however, frequently have laws that restrict any private research that might occur. The major need is for public sector activities to develop infrastructure that will lead to modern agriculture. These activities include investments in physical infrastructure, human capital, agricultural research and extension, and policies to encourage private investment in modern input and processing industries.

In Indonesia, India, and perhaps Pakistan, policies appear to be a major constraint to the development of a private input industry and also to private sector research and technology transfer. Therefore, policy reform should be a major priority along with continued investment in the public infrastructure required for agricultural modernization.

Thailand, the Philippines, and Malaysia have higher investments in private agricultural research relative to the size of their agricultural sectors than do Indonesia, India, and Pakistan. This is due to the modernization of their agricultural sectors and relatively limited government regulation and government ownership of input supply, plantation, or processing industries. Their private research intensity is still low relative to the standard set by Brazil or developed market economies. There are three possible reasons for the low level of private research. First, the agriculture in these countries is still not modern; thus, the size of markets for new technology is limited. Second, there are still many government regulations in these countries and, in Malaysia, the major plantations are being purchased by government-owned banks. Third, the private sector almost always invests less than the socially optimal amount in research, so government programs are required to bring the level of private research up to the desired levels. Both Brazil and the United States have policies to encourage private research. These include government research programs that do basic research that private industry needs but cannot capture the gains from and patent policies that encourage local research.

The policy implications for Thailand, the Philippines, and Malaysia would seem to (1) continue to modernize agriculture through public investments in research, education, and infrastructure, (2) examine current policies to see if they are restricting private research and technology transfer, and (3) provide special incentives for private research.

WHY SHOULD GOVERNMENT SCIENTISTS SUPPORT PRIVATE RESEARCH?

There is both a general and a specific answer to the question, Why should government scientists support private research? The general answer is that in most Asian countries more agricultural research by the private sector would help agricultural productivity grow more rapidly. Studies available on rates of return to public sector research (Hayami and Ruttan 1985) indicate that developing countries would grow more rapidly if their governments invested more money in agricultural research. Table 16.1 indicates that the private sector invests much less in research than the public sector. Given the high rates of return to research, the total public and private investment would not be enough to bring research expenditure up to the optimal level. Evenson argues (Evenson 1986) that developing countries might aim for private research intensity of about 0.75 percent of the value added in agriculture. Private sector agricultural research in South and Southeast Asia varies from close to zero to 0.16 percent of the value of product in Malaysia (table 16.1).

Scientists and research administrators concerned about the growth of their country should therefore work for policies that will encourage more private sector research.

There are also some very specific benefits to government researchers and administrators that can arise from private sector research and development. More private sector research and development can lead to more effective government research programs. Research itself can be more efficient and productive if the private sector--farmers, input suppliers, marketing people, and processors--works with the government to plan research. The government can concentrate research on areas where the private sector will transfer the results to farmers most rapidly. The public sector can work on the areas that are not profitable to the private sector--areas such as farm management, IPM, and self-pollinated crop breeding. The results of government research will reach farmers more rapidly if private input supply companies get early access to new technology.

Private companies that do applied research and sell technology can provide valuable financial and political support for public sector research. Many government research systems are missing important opportunities for generating political and financial support by not cooperating more closely with private agribusiness. There are many examples of private financial support for public research in Asia. Exxon provides support to the government for fertilizer research in Pakistan. The poultry industry supports research seminars on corn in Bangladesh. Indian industry provides general support to some research universities and also supports specific projects that they think will benefit them. The private sector can provide important political support for government investments in research and training, a source of support that few government research systems in Asia have taken advantage of. Such support has its political perils, though, because it may appear that the government is helping big business or the multinationals.

An indirect benefit of a larger private sector research program is that government scientists' incomes could be increased in the following way. The demand for agricultural scientists will increase. This will have several effects. There will be more consulting opportunities; scientists who are hired away from the government to work in private research will receive higher salaries; as the government loses scientists, it may be forced to increase salaries to hold its remaining scientists. Private pressure does not necessarily lead to higher salaries but without this pressure higher salaries are much harder to get. The effect will be a larger and better agricultural research community.

WHAT CAN SCIENTISTS AND RESEARCH ADMINISTRATORS DO TO ENCOURAGE PRIVATE RESEARCH?

Government scientists and research administrators can encourage private research through both their research programs and their role as advisors to policymakers on science and technology policy. What they can do depends on the level of development of the country and size and sophistication of the private input and processing industries.

Government scientists can plan their research program so that it stimulates private technology transfer and research programs that are useful to farmers. The successful private corn, sorghum, and pearl millet research programs in Asia are based on public research programs that developed inbred lines needed by private research. Examples of this include the Kasetsart University corn research program in Thailand; the corn, sorghum, pearl millet, and cotton research in India; and the Philippines corn program. The local power tiller and rice harvester industry in the Philippines is an example of an industry that was induced by the Philippine Ministry of Agriculture and IRRI to do more research that fit their needs.

The government can do research on science and technology policy that might lead to better policies in the future. For example, it appears that the main beneficiary of import protection on agricultural inputs has been local industry and local subsidiaries of multinationals. The losers are poor farmers who pay high prices for seeds, tractors, and pesticides but do not get the best technology available. There is little evidence that the benefits from protection, like more local research and growth of infant industries, are greater than the cost to farmers and consumers. Research that quantifies the size of costs and benefits of various policies can lead to more rational government policies.

More cooperation and communication by government scientists with private sector colleagues can also help stimulate more effective research and technology diffusion by the private sector. The joint research and testing program of the Pakistan Tobacco Board is an example. The program is jointly financed and controlled by the Pakistan government and the tobacco industry. The board carries out joint research programs with the two companies that have their own research and conducts trials with all four major tobacco companies. The companies are able to test their varieties under a broader range of conditions than in the past. This speeds the development and spread of improved varieties. The companies are also able to get the Tobacco Board to do research that individual companies will not finance, such as research on reforestation and better fuel efficiency in flue curing.

Government scientists can also provide important scientific and technical assistance to companies' research programs. This can take the form of consulting or simply providing free technical assistance. ICAR and Rockefeller staff provided unpaid research assistance to some Indian seed companies. These companies report that this assistance was crucial in starting their plant breeding programs. These companies now interact regularly with government and ICRISAT breeders. In the Philippines many university scientists work as consultants to private research programs. Scientists and research administrators must establish policies that encourage scientists to provide technical assistance to consult with the private sector without disrupting the performance of their primary job.

Government scientists are important advisors on science and technology policies like seed laws, pesticide regulation, patents, mechanization policies, and certain import and taxation laws. Government scientists have an opportunity to promote private R&D instead of presenting a major barrier to private sector research as they have in the past. Some government scientists who sit on seed certification boards have not allowed private hybrids to be approved. In some countries, government scientists determine who can import scientific equipment. Companies complain that they hold up approval or turn down permission without justification. In contrast, government scientists could push for reduced but more effective regulation. They could push for patent policies that really do provide an incentive for productive R&D.

NOTES

*This study was financed by a grant from USAID (Contract no. OTR-0091-G-S-4195-00).

1. These industries are in the same category because many processing companies, like oilpalm mills or sugarmills, also have large plantations to supply these mills.

2. In the Philippines, San Miguel conducted research in the early 1950s but in 1964 it was prevented by the government from continuing. In 1976, Pioneer started its research program in the Philippines and soon afterward, San Miguel revived its program.

REFERENCES

Agricultural Research Institute (ARI). "A Survey of U.S. Agricultural Research by Private Industry III." Bethesda, Maryland: ARI, July 1985.

Evenson, Robert, Jonathan Putnam, and Donald Evenson. "Agricultural Inventions and Legal Systems in Developing Countries." Economic Growth Center, Yale University, 1983. Mimeo.

Hayami, Y., and Vernon W. Ruttan. Agricultural Development: An International Perspective. rev. ed. Baltimore: Johns Hopkins University Press, 1985.

Mikkelsen, Kent W. "Inventive Activity in Philippine Industry." Ph.D. diss., Yale University, 1984.

Paitoon Wiboonchutikula. "Total Factor Productivity Growth of the Manufacturing Industries in Thailand, 1963-1976." Ph.D. diss., University of Minnesota, 1982.

Thomas, T. Managing a Business in India. Bombay: Allied Publishers, 1981.

World Resources Institute. World Resources 1986. New York: Basic Books, Inc., 1986.

17 Food Transnational Corporations and Developing Countries: The Case of the Improved Seed Industry in Mexico*

Louis W. Goodman

The three basic material agricultural inputs critical to increasing the productivity of the world's farmers are improved seed, fertilizers, and pesticides. To ensure high crop yields, seeds must be tailored to the agro-environmental conditions of each planting area. Therefore, the output of the world seed industry is more differentiated, agricultural zone-by-zone and country-by-country, than any other agricultural input industry. Mexico's improved seed industry is the oldest and most advanced among developing countries, rivaled only by Brazil and India. The premise of this paper is that an examination of how the improved seed industry operates in Mexico and the options that exist for its improvement will be instructive for understanding policy choices for this crucial aspect of the agricultural programs of many developing countries.

Mexico has several options to assure appropriate development of its national seed industry:

1. Allow open competition between foreign and domestic firms. The probable result would be early domination of local markets by foreign firms.
2. Cut the ties of the Mexican seed system to the international system. The probable result would be persistent industry inefficiency and an increase in the likelihood of genetic disaster, since Mexico would lose access to the full range of plant germplasm.
3. Create a mixed system involving both Mexican and international seed companies. The probable result would be rapid incorporation of technological developments from the international industry into the national industry, and increased efficiency of the domestic industry through more effective competition.

This paper will argue that the third alternative is most appropriate for Mexico.

The present condition of Mexico's improved seed industry can perhaps be captured by two statistics. From 1979 to 1981, the Productora Nacional de Semillas (PRONASE) increased seed production of seventeen basic crops by 167 percent to 224,000 metric tons. In the same period, Mexico was the world's largest importer of seed with more than 160,000 metric tons imported between 1979 and 1981.

This contradiction forces the question, "Does Mexico's improved seed industry, as currently organized, best serve national needs?" Anticipating that the existing system can be improved, the next question is, "What are the best ways of improving the national seed industry?"

Answers to both questions must take account of Mexico's relationship to the international improved seed industry. Mexico's economy has become an increasingly important part of the international economic system and its seed industry already has close and important links to the international seed industry. Leading seed companies use highly specialized techniques to anticipate improvements in seeds needed to adapt to an ever-changing natural environment. Since changes in environments frequently occur in unanticipated ways, it is important that seed companies operating in any given nation have genetic resources available from the widest possible base in order to plan their product strategies.

In the next four sections, this paper addresses the following questions: (1) How does Mexico fit into the international improved seed industry? (2) What is the structure of Mexico's domestic industry? (3) What are the obstacles to improved utilization of Mexico's seed resources and seed industry? (4) How can these obstacles be overcome?

CHARACTERISTICS OF THE INTERNATIONAL SEED INDUSTRY

The retail market value of seeds planted in 1980 was estimated at $42 billion worldwide, with commercial improved seeds accounting for approximately 30 percent of this total. The United States is the largest market for commercial seed with sales of around $4 billion. A dozen large transnational seed companies based in the United States and Western Europe dominate the international seed trade. In 1979, United States-based corporations were the leaders, with foreign seed sales of over 200,000 metric tons valued at nearly $200 million.

Although improved seeds have been used in North America and Europe for more than 200 years, the market remained relatively small until the technology for hybrid corn seed became commercialized in the 1930s. This event radically altered perceptions of seed technology and set in motion an industry that has evolved rapidly. The development of hybrid seed corresponded in time with large migrations from rural to urban areas in industrialized countries and concomitant needs to increase farm productivity to feed growing urban populations. In the United States, public agencies took the lead in carrying out seed research and development, with private firms commercializing the seed and marketing it to farmers. This pattern continued through the 1960s with public agencies carrying out most basic research, while marketing was conducted by a large number of small, private regional seed companies and by state universities who sold their own "public varieties."

At the international level, a network of seed research agencies was set up under the sponsorship of the World Bank, the United Nations Food and Agriculture Organization (UN/FAO), and the United Nations Development Program (UNDP) and governed by the Consultative Group on International Agricultural Research (CGIAR) to carry out research on seeds suited to the needs of developing nations. These centers were instrumental in sparking the green revolution and have continued to improve the quality of seeds and provide the material for national seed programs and private seed companies marketing improved varieties throughout the world.

In the 1970s, political and economic factors combined to reduce the resources provided by public agencies in the United States and Europe to maintain the pace of basic seed research. In the United States, as federal support for genetic research declined during that decade, the largest seed companies were able to utilize economies of scale to expand their proprietary research programs and gain competitive advantage over smaller rivals through in-house production of superior varieties. At the same time, profits in seeds and most food-related industries skyrocketed as a result of the commodities boom that followed the 1973 petroleum price increases.

The result of these forces was a fundamental structural change in the international seed industry that was quickened by the acquisition of numerous regional seed companies by other larger firms. The resulting concentration of the industry linked the operations of many seed companies to transnational conglomerates who saw complementarities between seed research and their other ongoing research activities (e.g., pharmaceuticals) or with other food-production-related activities (e.g., fertilizers). Seed companies tied to large conglomerate firms acquired new resources for seed research independent of revenues from seed sales. Increased research funding further intensified industry competitiveness to meet the need for continual development and marketing of new seed varieties adapted to the constantly changing farm environment and to the changing preferences of their farmer customers.

Parallel to the developments in the organization of seed research were changes in laws governing the use of the results of this research effort. The International Union for the Protection of New Varieties of Plants (UPOV) was created by the leading seed companies and gained support of the World International Proprietary Organization (WIPO) to promote the worldwide enactment of Plant Variety Protection (PVP) legislation. The PVP legislation granted patent-like rights to individuals and corporations developing unique varieties of plants and prohibited farmers from independently multiplying and selling plant seed from varieties first purchased from seed breeders. This legal protection allowed seed companies to intensify their research and broaden their markets for both hybrid seeds and open-pollinated seeds that had patent protection.

Another result of the changes in the international seed industry was the emergence of new genetic engineering techniques that opened the possibility of quicker development of seeds and plants that perhaps are superior to genetic stock produced by traditional plant breeding methods.

The highly concentrated United States and European-dominated international industry is now expanding rapidly in developing nations. Transnational seed companies are creating new subsidiaries and licensing agreements, particularly in Latin America and Asia, with at least three aims: (1) to set up international facilities to cover shortfalls in seed production in one area through compensatory production in another area; (2) through head-to-head competition to penetrate new, growing markets outside their home countries to add to their relatively mature home country markets; and (3) to take advantage of new markets that are increasingly available in developing countries.

THE IMPROVED SEED INDUSTRY IN MEXICO

The three major groups of actors in Mexico's improved seed industry are the Mexican state, private commercial companies, and private cooperative associations of producers.

At least seven state agencies are directly involved in the industry (see figure 17.1). The Instituto Nacional de Investigaciones Agricolas (INIA) conducts basic seed research. The Productora Nacional de Semillas (PRONASE) carries out seed production, multiplication, and distribution. The Banco de Fomento Rural (BANRURAL) provides credit to PRONASE for operation and to farmers for purchase of PRONASE seed. The Servicio Nacional de Inspección y Certificación de Semillas (SNICS) is the national seed quality control agency. The Comité Calificador de Variedades de Plantas (CCVP) authorizes new varieties and grants import licenses. The Registro Nacional de Variedades de Plantas (RNVP) maintains the official registry of plant varieties. The Fondo de Garantia y Fomento para la Agricultura, Ganaderia, y Avicultura y Fideicomisos Agrícolas (FIRA), an agency of the Banco de Mexico, provides credit for seed purchases to more than 300,000 Mexican farmers and operates its own seed breeding and distribution programs.

More than thirty private companies are in Mexico's seed industry, the largest of which are joint ventures between foreign-based and domestic seed companies.

Most seed producer associations are located in Mexico's Northwest and are particularly active in the production and export of wheat seed.

Mexico is also the headquarters of the Centro Internacional de Mejoramiento de Maíz y Trigo (CIMMYT), a CGIAR-sponsored seed research organization specializing in corn and wheat improvement that has a special relationship with the Mexican seed system through INIA.

The relative importance of each of these actors in the Mexican market is not clear from available data. There is conflicting evidence on the percentage of seed supplied by state and by private sources and on the reasons for the higher prices of private sector seeds as compared with those from the public sector. There is also mixed evidence

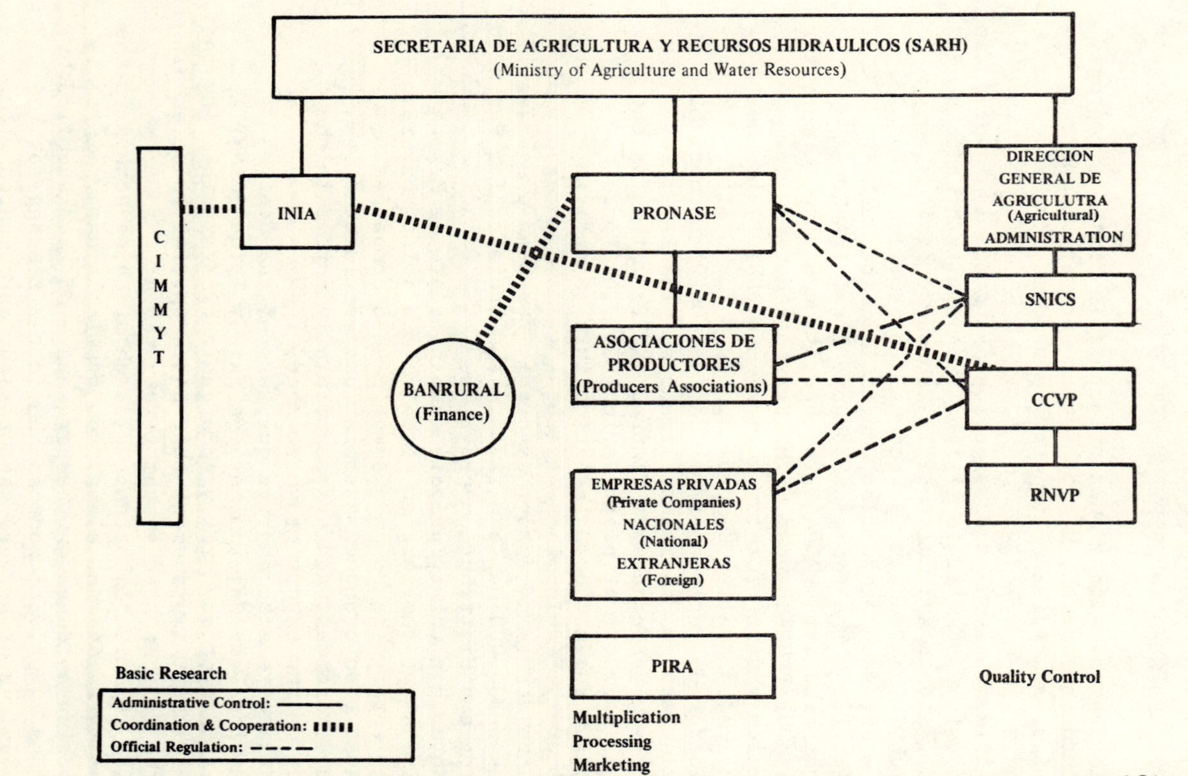

Figure 17.1. Organization of the Mexican system of seed research, production, certification, and distribution

on yield results from the seeds from both the private and public sectors.

Mexico's state seed operations are the largest and the most successful of those in any developing country. On the other hand, Mexican seed imports continue to be the largest of any developing nation. This contradiction suggests the critical need to adapt the seed industry to the changed conditions in worldwide agriculture and in the international seed industry. These changes are needed both for the future of Mexican agriculture and for the guidance of the future development of national seed institutions in developing nations throughout the world.

LIMITATIONS OF THE EXISTING MEXICAN SEED SYSTEM

A number of serious deficiencies in the Mexican improved seed industry exist. These include:

1. Disorganized and wasteful competition between major actors in the national industry
2. Unsystematic development and implementation of government priorities for the industry
3. Inefficient functioning of state institutions originating from inadequate integration of the state research organization (INIA), the state production and distribution agency (PRONASE), and official sources of finance
4. Restrictions on private sector activity that prevent full utilization of non-governmental ability to create seeds suited to Mexico and to adapt available seed technologies to Mexican conditions
5. Lack of policy direction for smallholder and _ejido_ involvement in the improved seed industry, and for the promotion and growth of seed producer associations
6. Discriminatory application of quality control regulation by SNICS and ties with other government agencies that limit the possibility of arms-length regulation of public agency activities, to the persistent disadvantage of the private sector
7. Inadequate coordination and priority setting for basic seed research between INIA, universities, and CIMMYT
8. Shortages of scientists and other professionals needed for the development and expansion of a domestic seed industry
9. Inadequate public sector ties and scientific exchanges with international seed research, professionals, and institutions

10. Counterproductive practices in international seed trade operations by public sector agencies

SEED INDUSTRY OPTIONS FOR MEXICO

Mexico's seed industry is passing through a critical period. A vast number of important changes have taken place in the thirty-five years since the Mexican state took its first important steps to control and guide the industry: rapid technological developments, huge increases in the size and diversity of the Mexican market for commercial seeds, expanded participation of foreign-based companies in local markets, plus the high priority given to national self-sufficiency in basic foods and technological capabilities. Fundamental changes occurring in the industry and in Mexico have created the need to reexamine and restructure the institutions that evolved in an earlier period. The proposals offered in the following pages are given in awareness of the need to rationalize public agency involvement in this important industry.

Three broad options exist for the organization of Mexico's seed industry:

1. A state seed industry monopoly could be created.
2. A seed industry could be fostered that is dominated by private firms and only loosely regulated, modeled after the United States pattern.
3. A mixed public-private system could evolve in which private firms (both domestic and foreign), producer associations, and an integrated state firm actively compete in a federally regulated industry.

It will be argued here that the mixed system is the most appropriate in light of the nature of the industry and present circumstances. Since the commercial seed industry is heavily dependent on genetic research and technology, a policy that isolates Mexico's industry from international technological developments will doom the seed industry--and much of Mexico's agriculture. This establishment of an autarkic state seed monopoly would reduce or eliminate Mexico's access to important sources of germplasm produced elsewhere. This would make current and future varieties of Mexican crops more susceptible to new pests and diseases than varieties constructed on the broadest possible germplasm base.

The establishment of a national seed industry based exclusively on private industry, paralleling that of the United States, would expose Mexico to foreign dominance of this key sector. The large foreign-based firms are well situated to increase considerably their market shares at the expense of locally based suppliers. If the foreign companies' market share were provided by seed imports, it would cost Mexico jobs, technological independence, and foreign exchange. With the current organization of research and development in international seed companies, Mexican agriculture could easily become largely dependent on seed technology developed outside its borders. Such dependence is in direct contradiction to the national objective of self-sufficiency in areas of national priority (particularly food and agriculture).

The mixed industry alternative would include private domestic and foreign-owned seed companies, producer associations (with _ejido_ participation) with capacity for producing their own seeds, and a state enterprise with substantial research capacity, market share, and seed stocks. The state enterprise would have an important place in the market if it were to produce satisfactory supplies of high quality seeds at fair market prices in competition with producer associations and private firms.

A NATIONAL PLANT GENETICS INDUSTRY BOARD

Under this mixed industry option, Mexico's state seed industry would be regulated by a National Plant Genetics Industry Board composed of representatives of the state seed enterprise, producer associations, private sector companies, seed consumers, and a representative of the Ministry of Agriculture (figure 17.2). The purpose of the board would be to oversee the operation of the five essential components of the Mexican national seed industry: (1) seed production and distribution, (2) seed imports and exports, (3) scientific seed research and training of seed scientists, (4) quality control of seed, and (5) the collection of seed industry information.

SEED PRODUCTION AND DISTRIBUTION

Central to the proposed restructuring of the industry would be the creation of a fully integrated state seed enterprise. This institution would combine the applied research of INIA with the production, processing, and distribution operations of PRONASE and the finance capacity

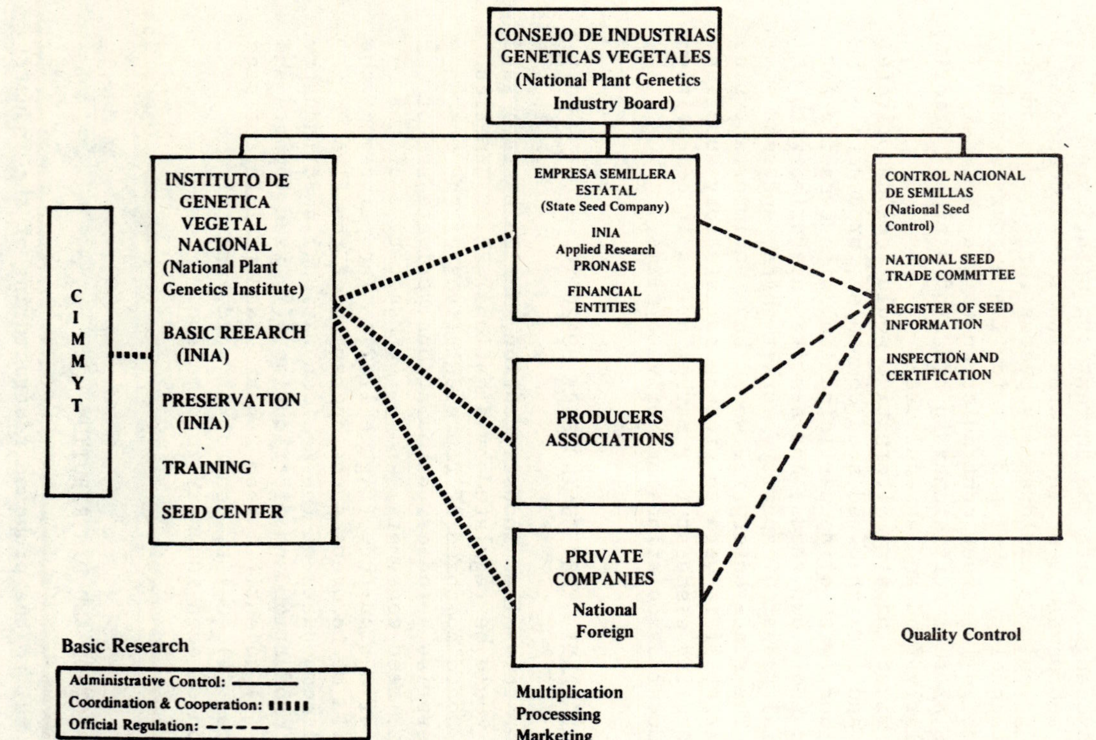

Figure 17.2. Proposed organization of the Mexican system of seed research, production, certification and distribution

of BANRURAL and/or some other Mexican financial institution. The result would be a state-owned Mexican firm capable of producing internationally competitive varieties of basic crop seeds, and of producing and distributing seeds to service national needs. It would possess an in-house financial capacity that would overcome the delays and planning problems that have traditionally plagued the public sector of the Mexican seed industry.

The purpose of this integrated state company would not be simply to compete with the private sector and other companies. Its objective would be to provide a standard and a reserve capacity to assure that Mexican farmers have the best possible seed at the lowest possible price. There would be less risk for the state than it currently takes, and proper operation of the state company would assure that high quality seed be available at low prices.

Consistent with national objectives to promote a more equitable distribution of income and to eliminate malnutrition in the countryside, this state enterprise would facilitate the increase in the number of producer associations with strong _ejido_ participation. For example, the current PRONASE practice of providing technical assistance to associations of farmers in the form of seed processing and distribution would be extended and expanded. This would be achieved by generating a national network of producer associations with the capacity for producing a substantial portion of locally adapted seeds for indigenous grower needs. Of equal importance would be the seed needs of individual smallholders. Programs such as PRONASE's Maiz Seleccionado program would receive special emphasis so that small farmers' productivity and economic well-being would be increased.

Although this state enterprise would have the capacity to carry out research on any problem of importance to Mexico's seed industry, like other large seed companies, it would specialize in crops most central to its interests. In this case the major focus would be on basic grain seeds--corn, beans, rice, and wheat--and other crops important to the Mexican economy.

Private firms would be encouraged to compete with the state seed company and producer associations on a nondiscriminatory basis. SNICS would inspect and regulate private sector seed research, certification, processing, and distribution on the same basis as it regulates the activities of the state enterprise and the producer associations. Standards governing seed quality and price would be at least as high in Mexico as those prevailing in the international market. Under supervision of the National Genetics Industry Board, the State Seed Trading Committee would monitor international market conditions.

The national seed enterprise would be organized on a regional basis, with a maximum of six regions. Such a regionally based organization has been proposed by Ernesto Badillo to increase local input into the operations of INIA and PRONASE (Badillo Navarrete 1981, 352). These two institutions now operate on a national level while local delegations do so on a state level. Badillo found that when these operations were separated into state-level subdivisions, priorities in planning in the seeds industry were imposed from above. A regional structure, sensitive to the variety of microclimates characterizing Mexico's ecology, would assure that policy initiatives are carried out through a unified national structure.

INTERNATIONAL AND INTERSTATE SEED TRADE

The National Genetics Industry Board would supervise the operation of a National Seed Trade Committee. The purpose of the seed committee would be fourfold:

1. To harmonize interstate seed trade with national policy objectives. The committee would allow producers in given states to provide seed in other states, where appropriate, and to enforce quarantines to prevent the spread of disease.

2. To supervise the importation of seeds. The committee would authorize the issuance of seed import permits to cover shortfalls in national seed production. It would also allow the sale, in Mexico, of seed whose characteristics were superior to seed grown nationally.

3. To promote the exportation of Mexican seed. The export of Mexican seed to other nations would generate foreign exchange for Mexico and would promote intra-Third World seed trade.

4. To rationalize the purchasing behavior of the Mexican State Seed Enterprise. By preserving international purchases, the National Seed Trade Committee would guide the State Seed Company and private sector firms to cover efficiently shortages in seed production and could continually monitor conditions in world seed markets. This would be accomplished by a program of limited continuous purchases on major spot seed markets and contracts for larger long-term obligations where needed.

SCIENTIFIC SEED RESEARCH: A NATIONAL PLANT GENETICS INSTITUTE

The National Genetics Industry Seed Board would also supervise the operation of a National Plant Genetics Institute. This institute would link the basic research carried out by INIA with university-based plant genetic research. This basic research capacity would be separated from INIA's applied research capacities, which would be housed in the new State Seed Company. Institute-sponsored research would pay special attention to Mexican comparative advantages, i.e., the unique germplasm reserves and the variety of ecological conditions. The operations of the National Plant Genetics Institute would be linked to basic research programs at universities and international centers throughout the world. The seed board would also establish ties with CIMMYT, with the objective of mutually coordinating and complementing research activities.

The National Plant Genetics Institute would have four operative components, discussed in more detail below: (1) basic seed research, (2) germplasm preservation, (3) manpower training, and (4) an international seeds center.

Basic Seed Research

Basic seed research would focus on areas consistent with national objectives, e.g., investigation of basic crops such as beans and corn; improvement of open-pollinated varieties, permitting isolated and micro-climate farmers to save their own seeds; and emphasis of frontier genetic engineering technologies on Mexican national objectives.

The National Plant Genetics Institute would be especially alert to advances in genetic engineering and would combine capacities for both traditional operations and research in this field. As a national center, it would have a substantial number of Mexican scientists active in genetic engineering research who could monitor the developments relevant to Mexico in the industry worldwide. The institute programs would be able to adapt developments in genetic technology to Mexican needs and to combine innovative genetic engineering techniques with traditional methods as they became scientifically feasible.

Germplasm Preservation

A systematic program of germplasm preservation would be basic to the national research effort. Mexico is a "Vavilov Center"--the locale of origin of an unusually large number of important plant species, including corn, beans, tomatoes, chili peppers, and squash. Although Mexico's germplasm preservation program is more advanced than that of any other Third World nation, even greater efforts are needed to:

1. Build a program of systematic artificial seed variety storage and growing out. This would permit the full cataloging and preservation of germplasm of special interest to Mexico.
2. Extend existing natural preserves for germplasm to harbor additional wild varieties and to accommodate genetic variety to natural drift brought about by changing ecological conditions.
3. Develop a program of germplasm preservation that is regionally based and fully responsive to the diverse conditions in Mexico's different agricultural areas.
4. Encourage "seed savers" and indigenous seed ("semilla criolla") programs among local farmers who retain and process homegrown seeds particularly well adapted to local ecological conditions.

Scientific Manpower Training

The training component of the National Plant Genetics Institute would provide facilities where agronomists, biologists, botanists, microbiologists, agricultural economists, and other professionals needed for seed development would receive specialized training. This manpower training program would be carried out both through the training operations of the National Plant Genetics Institute within Mexico and through cooperative programs with training centers in foreign universities and institutes.

International Seeds Center

An International Seeds Center would be set up at the National Plant Genetics Institute. This center would be oriented particularly to the needs and problems of seed industries of other developing nations. As such, it would facilitate scientific cooperation and exchange with other Third World nations and make direct contact with their national seed companies. It would be particularly attentive to the development of new seed varieties, the preservation of unique germplasm, and the training of seed scientists appropriate to a wide variety of agricultural conditions. Mexico's unique germplasm reserves and variety of ecological conditions make it a natural locale for such a center.

In addition to its research and training functions, the international center would serve as a worldwide monitor for germplasm mapping, plant diseases, and plant pests. At present, no center exists that pays special attention to developing country agricultural needs. Mexico's unique germplasm resources place it in an unusually favorable position to set up a worldwide germplasm center. Its long-established work on Third World plant diseases and pests also gives Mexico a comparative advantage for the establishment of a center to monitor these phenomena on a worldwide basis.

SEED QUALITY CONTROL

It is critical that the national seed industry regulatory agency operate on an "arms length" basis with all important actors involved in the production and trade of seed in order to control quality in the industry. SNICS, therefore, would be maintained as an independent agency charged with inspection, certification, and other control functions. The State Seed Enterprise, the producer associations, and private firms would all be evaluated on an equal basis according to the norms and priorities set by the National Genetics Industry Board.

SNICS' identity and functioning should be separate from INIA; it would not use INIA facilities or personnel to carry out its own evaluations. SNICS would be charged with authorizing and certifying seeds for national distribution, and with regulating the procedures employed in processing and distributing seeds to consumers. As such, SNICS would independently test yield, disease resistance, pest resistance, and other characteristics of seeds. It would also retain its capacity for evaluating seed production and seed processing techniques. It would similarly monitor seed pricing and advertising to ensure fair industry practices.

It would be appropriate to attach the National Register of Plant Varieties (RNVP) to SNICS. In doing so, it would be necessary to update the procedures of the RNVP, correctly identifying obsolete varieties and identifying owners of current varieties.

THE COLLECTION OF SEED INDUSTRY INFORMATION

The Direccion General de Economia Agricola of the SARE would be charged with collecting information on the seed industry needed for national industrial planning and national food security. With regard to national industrial planning, it would be necessary to document a census of the industry, including public agencies, producer associations, and private firms. The National Genetics Industry Board could supervise this census, which is needed to estimate both industry capacity and the ability of Mexican growers to provide seed for national needs, as well as to plan for future industry growth.

An initial census would be supplemented by periodic disclosures of information by seed companies indicating to the national board the size and location of facilities, as well as the ownership of the seed-producing firms and plans for industry expansion or contraction.

Collection of this data would also be necessary for national food security purposes. Reports of increases or decreases in capacity would be essential to maintain adequate supplies of food.

Finally, all commercial and research germplasm would need to be registered with the National Genetics Industry Board, with samples made available at cost to any party authorized by the board. Such a policy would promote the efficient and widespread utilization of germplasm resources to create new varieties to ensure both satisfactory Mexican food production and national food security.

NOTE

*This paper is drawn from a longer document, "The Improved Seed Industry: Issues and Options for Mexico," prepared in 1982 by the author with Arthur L. Domike and Charles Sands as part of a joint research project between the Sistema Nacional de Evaluación de la Presidencia de la República Mexicana, and American University, Washington, D.C.

REFERENCES

Badillo Navarrete, Ernesto. "El Sistema de Semillas Certificadas en Mexico." Chapingo, Mexico: Colegio de Postgraduados, 1981.

18
Support Structures for Technology Transfer in Agriculture

Francis W. Wolek

"Be Prepared" is more than the Boy Scout motto. It is a fundamental precept for managers in all fields. The more we capitalize on previous experience, the greater our chances for having sufficient resources, avoiding error, and anticipating competition. One of the uses of previous experience is building mental maps of situations we expect to encounter again. For technology transfer such maps tell us things like: (1) what actions need to be targeted to which groups to obtain awareness, experimentation, and adoption; (2) what skills are needed for communication, field trials, and adaptative engineering; (3) what resources in time, capital, and management support will be required; and (4) what warning signals predict conflict and wasted effort.

In short, our mental maps or models of technology transfer have much to do with our success as managers and policymakers.

A STUDY OF TECHNOLOGY TRANSFER CASES IN AGRICULTURE

Few of us are fortunate enough to have the personal experience to build valid maps of all the situations we will influence as policymakers. If we are especially lucky, we will have the opportunity to listen and learn from others. However, it is more likely that, like most of our brethren, we will be forced to generalize from limited situations seasoned generously with media reports. This is the way in which most models of technology transfer are born and this is the way we are easily led to expectations such as described by the well-known proverb "Invent a better mousetrap and the world will beat a path to your door." If only it were that simple!

Methodology

The study on which this paper is based investigated the extent to which technology transfers from the U.S. Agricultural Research Service (ARS) to suppliers and farmers conformed to the principles of technology transfer appearing in the literature (e.g., from studies of transfer involving aerospace technology, energy, and universities). To understand the full process of transfer, it was necessary to understand the experiences of both those who provided and those who accepted technologies. Therefore, equal numbers of ARS and industry sources were interviewed. The ARS sample focused on its larger laboratories and was a stratified selection representative of size, relative productivity, and type of laboratory. All interviews with ARS personnel began with the laboratory director and proceeded to interviews with the scientists in charge of the nominated case.

The industry sample sought a match with the laboratories studied and focused on trade associations active in agriculture. Wherever possible, the chairman of the association's research committee was interviewed (typically a farmer or industry executive). In a few cases, it was necessary to interview either the association's director of technical affairs (e.g., the chairman was relatively new and inexperienced) or industrial executives from firms with a strong reputation for innovation in agriculture (e.g., there was no comparable association).

Each interview began by asking respondents to: "Please nominate an actual project, preferably one ARS would be proud of, whose results have been transferred from ARS to industry within the last 3-5 years." In other words, rather than seeking a random or representative sample of ARS transfers, we *purposely biased our sample toward recent cases and cases that represented the best of ARS practice*.

Purpose of the Present Paper

This paper discusses a subset of case histories that illustrate characteristics of the social system supporting innovation in agriculture. Despite the fact that most of the cases discussed were successful, the arguments presented here are based on a limited number of anecdotes from the larger study. Therefore, further study is necessary and is encouraged to provide adequate sample sizes and measures to assure the validity of these exploratory ideas.

SCREENING MECHANISMS FOR AGRICULTURAL INNOVATION

Technology transfer is a time- and energy-consuming process for all involved. Scientists must communicate results to diverse audiences, technologists must adapt findings and technologies to fit existing markets, companies must promote products in the market, and innovative farmers must conduct realistic field trials. In short, policymakers must support an often long and trying process it if is to be successful. Such energy cannot be committed without some mechanism for evaluating the merits of a project compared with other opportunities.

The Task of Screening Inventions

Innovators in all sectors of agriculture need some mechanism for evaluating ideas and technologies and narrowing down to the very few that fit available resources. That is, they must be able to assess:

- technological feasibility of an invention (does it violate laws of nature, will it work, are materials and components available, etc.?);
- commercial feasibility of the products based on the invention (is there sufficient advantage over competitive products, are manufacturers available with the requisite resources, do potential buyers possess the resources and talent needed, etc.?); and

- investment priority for programs to promote innovations (how much resources are available, is necessary expertise at hand and will this invention contribute to a cumulative growth of market leadership, how much risk is entailed in defending proprietary position, etc.?).

The magnitude of this task of evaluating innovations is illustrated by two programs seeking to screen inventions for possible government awards. In one, the government of Sweden used a panel of technologists and industrialists to screen inventions responding to an announcement of a prize for inventions utilizing Sweden's natural resources (Ottosson 1983). Out of 2,710 submitted inventions, only 54 (2%) were selected as patentable and only 10 (.4%) were accepted for commercialization by Swedish companies.

The second program is one used by the U.S. government (Office of Energy Related Inventions (OERIP) of the National Bureau of Standards) to screen energy inventions (conservation, power-generating technologies, etc.). Over the first ten years of OERIP's existence, 22,000 inventions and ideas were submitted and 320 (1.5%) were evaluated as warranting investigation by funding authorities (e.g., Department of Energy). George Lewett, director of OERIP, summarized his experience by noting that for every 1,000 inventions submitted only 3 (.3%) will "end up as being, to some extent or the other, 'successful'" in yielding profits and energy savings (Lewett 1982).

The difficulty and complexity of the screening process are evident in the present study. In total, sufficient data were available to judge the value of the 47 technologies. The simplest test of value was used: was the technology actually put into practice and did it have an impact in use? Despite the fact that all the technologies included were ones that some authority felt were a matter of pride, 30 (64% of the total) passed this minimum test of value.

The Informal Screens of Agriculture

People familiar with industrial innovation might be led to believe that the screening mechanisms encountered were ones that formally involved top management in ARS and the companies involved. For example, a laboratory director might see that a project had potential and give it the needed backing, or a company president might recognize an opportunity and encourage aggressive action. In contrast with this expectation, the most impressive characteristic of the successful transfers was that the screening mechanisms were quite informal.

Probably the best examples of these informal mechanisms concern the technologies of new plant varieties. For example, NC82 is a new variety of tobacco that resists bacterial wilt, is easily cured, and is the most popular flue-cured tobacco developed by a public agency (ARS Tobacco Research Laboratory) in thirty years. In its clearest form, the informal nature of the screening process for NC82 is illustrated by the work of the New Variety Advisory Group of the Tobacco Workers Conference. The group and the conference that it advises take great pains to retain their informal status. The conference has no charter, central administration, or formally recognized authority. However, the state agencies that do have authority will not promote (e.g., formally register, stock breeding seed, disseminate information, etc.) a variety of tobacco that has not been "approved" by the advisory group.

The mechanism by which the advisory group gives its approval is a model of informal collegiality. New varieties are discussed at the group's annual convention. While a vote is taken, it is delayed until a consensus is evident. This primary criterion of consensus depends on successful consideration of such issues as:

- Do field and laboratory tests of the variety show it to have the characteristics (nicotine, sugar content, appearance, etc.) to pass standards that document the minimums needed for buyer interest in a variety?
- Have experimental plantings generated sufficient interest among farmers and buyers?
- Have sufficient data been obtained to understand its advantages, possible problems, and management requirements (e.g., sucker control and early flowering in cold weather)?
- Is the variety necessary given the availability of others and the pressures upon farmers for other investments?

Sometimes the consensus is to wait, sometimes to drop further work (something ARS takes seriously), and sometimes to release. However, the emphasis is on a consensus that the variety serves the needs of the market and producers, not on criteria common in formal organizations such as formal objectives, deadlines, and management pressure.

Certainly the activities of the Tobacco Workers Conference are interesting, and equally certain, it is good to hear of a field that has developed an effective means for assuring the widespread input and action necessary to achieve a market presence of sufficient scope to assure adequate prices. However, the case is introduced, not as a unique example, but as an especially clear illustration of informal mechanisms encountered in many of the successful cases in our survey. These characteristics are summarized in table 18.1 together with examples from other cases concerning technologies other than new plant varieties.

In summary, many of the successful cases of technology transfer showed reliance on informal groups that had shifting membership and leadership, reliance on both empirical data and social influence, and a willingness to support a technology with the group's status and practiced means for disseminating information. When the field was well established and the importance of consensus reflected in the market for all farmers (tobacco), informal mechanisms were as well established as the Workers Conference. When the field was less structured around one commodity or when concerted action was not necessary (aflatoxin testing for grain elevators), the informal mechanisms were likely to be ad hoc and emergent as crises or common problems arose.

SUPPORT SERVICES AND INNOVATION

The previous section began by noting that technology transfer requires careful screening by all concerned. Investments of time and energy are best made when we are united with others in appreciating the opportunities of a new technology and being committed to favorable action. This section presents a further extension of the idea that innovation requires concerted action. However, since this discussion will be breaking new ground for the generic subject of technology transfer as well as for agriculture, it will be introduced via two case examples.

A Services Infrastructure for Irrigation Equipment

The previous section pointed out how the agricultural community organizes to obtain widespread action when the market requires a significant consensus. In this case organizations also evolve, but around another motivation: a common crisis and threat of widespread loss.

TABLE 18.1
Generic Characteristics of Informal Screening in Agriculture

Characteristic	Function	Example
Flexible Leadership develops and changes as the situation requires (e.g., test genetic stability vs. assess buyer reactions).	Innovations require a long time, have high uncertainty, and require shifting expertise as new problems and opportunities arise.	Alfa Toxin Procedures resulted from an informal network of elevator operators, food companies, regulators, and scientists who found the problem, developed and tested a new method.
Results-Based Decisions rely on empirical proof of the technology's performance relative to the group's goal (e.g., assure sufficient supply of quality product to obtain a fair price).	There is no authority over the actions of independent growers and companies and voting plans of one farm, one vote or number acres = votes would not test commitment.	Cherry Fumigation Methods were refined and tested until Japanese officials relaxed bans on American imports.
Managed Risk in that no individual risks status or capital until a consensus on action is reached (e.g., field data motivate commitments to plant and buy).	The status of the group is behind the technology and lends confidence to adoptees.	Hybrid Sunflowers were pioneered by farmers and breeders working as a group to develop the approach before establishing farms, seed companies, and associations.
Network Dissemination results from group member status as opinion leaders in their constituencies (e.g., approved varieties are listed in the growing reports of experiment stations).	An established reputation with media, agencies, and associations places news and reports on meeting agendas and in publications.	Brucellosis Testing was improved only after a consensus of farmers and animal health officials created pressure for companies to modify existing products.

Only a few accidents were needed to make the writing on the wall clear: suppliers and users of irrigation equipment faced the possibility of expensive legal actions concerning liability for electrocution of workers handling irrigation equipment. The fact that people were making simple mistakes did not alleviate the need for clear standards and a way of getting information to the field on how to install wiring for irrigation systems.

The response was the organization of a diverse group of scientists, equipment producers, and farmers (organized as a new committee of the Irrigation Equipment Manufacturers Association), who developed a system of standards and information dissemination to manufacturers, equipment contractors, and farm organizations. The success of the response was attested by an industry award to the ARS scientists and a decrease in the number of reported accidents.

This case of wiring standards highlights the need for supporting services when new technologies are introduced into the field. It is not enough to sell a wonderful black box and leave the users to their own resources. Safe and successful use requires standards, contractors who know how to install and service equipment, and consultants who can respond to special problems and inspect system installations. In short, a whole host of support services is needed.

A Political and Services Infrastructure for Boll Weevil Eradication

The ambitious ARS program to eradicate the boll weevil is a total systems effort that involves the use of multiple technologies and modern field management in a carefully monitored approach to eliminating boll weevil populations. The approach is especially demanding of consensus, for all farmers in a targeted area must be bound to conformance or the project has no hope of success (i.e., the weevil must not have sanctuaries from which it can easily invade cleared territory).

When the project was first conceived, it was approached as a task of educating farmers and public organizations. If they could be shown the advantages of an integrated attack (controlled sprayings, monitoring populations, use of parasitic organisms, etc.), they could be convinced to give the area-wide cooperation necessary. A program of education in areas naturally bounded by barriers to weevil migration (e.g., mountains) was a massive task, but one that federal and state agencies undertook. Technical publications, field days, news and farm media, seminars, and presentations at fairs were all used to educate farmers, local agencies, and the public.

The extent and organization of this massive effort was an admirable exercise in technology utilization and undoubtedly has many lessons for those interested in technology transfer. However, the point made here concerns a problem encountered by the program. The eradication program was strongly, and for a time successfully, resisted by insecticide service firms (contractors and consultants). The eradication program would specify the nature, timing, and location of insecticide applications as well as promote non-insecticide methods. The program's ultimate goal was the eradication of a pest whose existence was the service companies' justification for being.

Service Firms and Innovations in Agricultural Methods

The primary message of this section is that the success of many technologies in agriculture requires the active support of service firms that provide aid in critical activities (figure 18.1). These activities include:

- consulting - designing technology applications for the specific conditions found at farms and regions;
- contracting and training - installing and/or applying the technology so as to minimize problems not evident before application and to maximize the technology's effectiveness, considering its synergy with the farm's characteristics;
- testing and setting standards - testing for stability, effectiveness, and side effects and participating in programs to generalize the lessons learned into standards and principles of good practice; and
- documenting and disseminating information and field data - documenting practice, underlying principles, and field data and disseminating these for use by farmers, local agencies, and industry.

The support of service industries was especially important in those cases in which a technology entailed innovation in the methods or processes used by farmers and agricultural firms. Such process innovation is contrasted with the adoption of new products that are substituted in existing methods (e.g., an improved vaccine, a new chicken cage, a forage protein tester, etc.).

THE INVENTIVE COMMUNITY

The above anecdotes about technology transfer excited our interest because they suggest important modifications to our mental maps about the process of innovation in agriculture. That is, they emphasize the need for concerted action by many people. When a support system exists and consensus is built, more than transfer is obtained. Consensus and service build markets, expertise, and improvements in technology and its use.

The dominant model of technology transfer already emphasizes action by several people: scientists, agents, and early users (figure 18.2). The cases discussed here modify this posture by emphasizing the contributions of other parties to transfer (screening networks and service systems). In addition, and maybe more important, these cases highlight the interactions between these parties, thus implying that the structure is a social system for innovation in agriculture (see figure 18.3). A system works best when all parties appreciate the contributions of others and they are personally linked to each other. Scientists cannot develop practical findings without input on realistic field conditions. Agents cannot substitute for service organizations who provide routine work in installing, testing, and maintaining field systems. Service organizations cannot be assured that a technology has sufficient commitment to justify start-up costs without a consensus of farmers and industry. In short, the system provides inputs of information, confidence, and support for each party.

The targets of this system are twofold: existing manufacturers of agricultural technology (industrial firms) and producers of commodities (farmers). Established organizations are often thought to possess an NIH (Not Invented Here) mentality that explains their resistance to innovation. However, the cases studied in the present survey identified many practical concerns that cause executives and farmers to take the time to be sure of a new technology before making expensive commitments that take resources from other opportunities. Such practical considerations include determining:

TESTING
(Testing for Adulteration in Syrup)

INFORMATION
(Data Base on Crossbred Cattle)

STANDARDS
(Aflatoxin in Grain Elevators)

CONTRACTING
(Design of Flow-Control Weirs)

Figure 18.1. Service organizations in agriculture

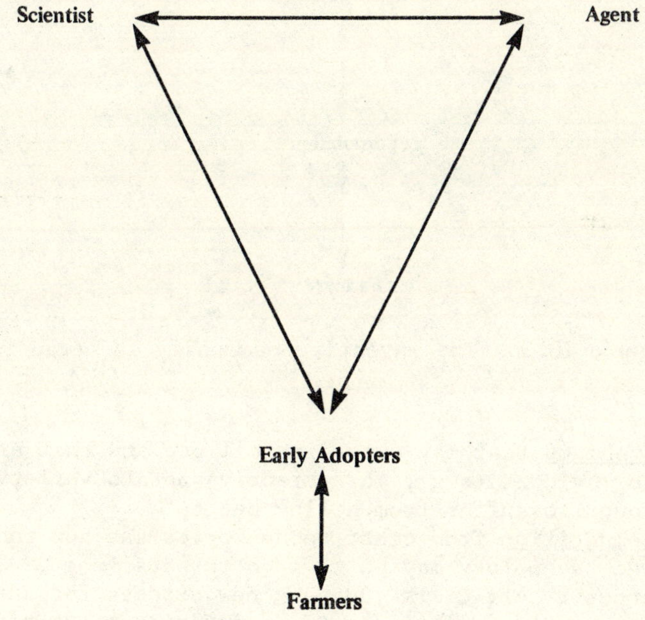

Figure 18.2. The Extension process of transfer

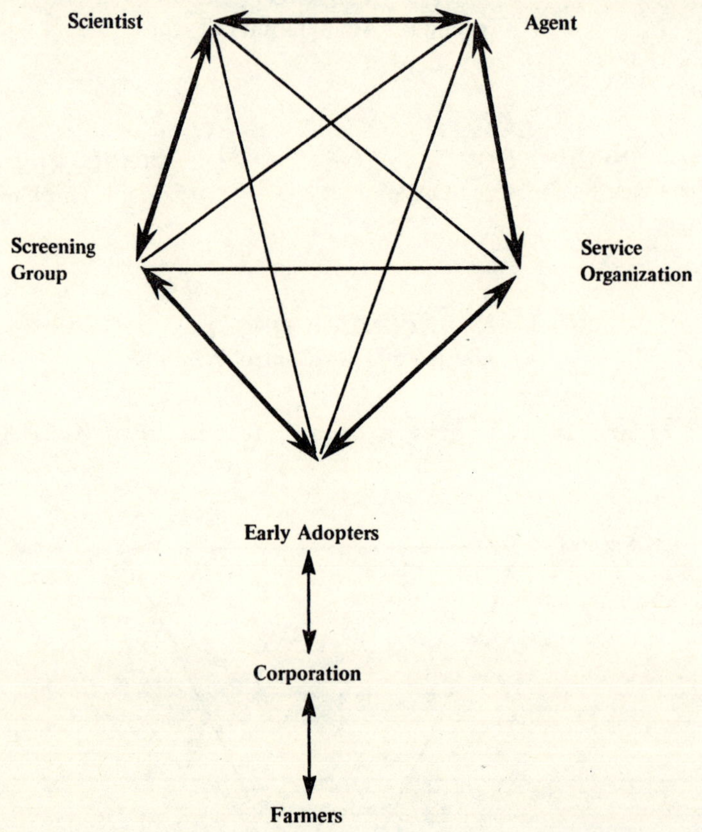

Figure 18.3. The inventive community and transfer

- problems that may arise in full use and that are not now evident (e.g., an attractive potato variety was found to suffer from hollow heart);
- competition from other technologies and how robust the technology may be to such threats (e.g., industry officials judged a new process for anerobic fermentation of wastes to be inferior to competitive technologies);
- returns that are likely to be produced and whether they provide the market presence to interest buyers (e.g., processors saw no need for a high-tech method of processing vegetable oil; therefore equipment

producers have not responded to a scientifically exciting technology);

- <u>suppliers</u> who are strong enough to perfect, produce, and market a technology (e.g., an innovative approach to sucker control is not being addressed because equipment firms are fighting for mere survival); and
- <u>changes</u> in established practice that may be necessary for a technology to realize its full advantage (e.g., a new fruit is slow to be adopted because buyers taste products before they are ripe, which is before this variety's advantage is evident).

POLICY IMPLICATIONS

The concept of an inventive community has important implications for policymakers in agricultural research. Specifically, the framework relates to resource allocation, the role of the public sector in R&D, and to the design and management of technology transfer programs.

<u>"Target to Market" Planning</u>

The last thing a policymaker needs to be told is that the process of technology transfer is more complicated than is commonly thought. To be told, in addition, that effective policy encourages long-time commitment to the activities of invisible networks that make intangible contributions is rubbing salt in an already open wound. Policymakers already know that the process is complicated and that cooperation with private organizations is necessary. The problems are that resources are not available for any but the most tangible actions and that pressure exists for clear results in technologies transferred and producing visible results.

The point of this paper is not an intellectual nicety that has no practical significance to resource-constrained agencies. While it is true that policymakers must encourage reliance on the limited technology transfer mechanisms that budgets and constraints allow, there are also times when agencies are committed to the full process of transfer in all its complexity. Clearly there are missions of national importance in which top policies and/or emergencies call for timely application of research from national laboratories (e.g., crises concerned with epidemic diseases). In such cases, policymakers will need to promote the cooperation of screening networks and the growth of supporting services outlined here. In short, there are times when national program managers must plan from initiation to use and must target programs to market.

It is also important for policymakers to remind themselves that they are not the only arbiters of program priorities. An agency's scientists have their own sense of when history demands rapid and full use of research. In other words, agencies do find their best people leaving the bench and spending all the resources allocated to them (and which they can bootleg) on both work with informal networks in active field tests and on promotion of private service businesses. The message to policymakers is that the best people will target programs to market with or without the policymakers' active support.

The framework presented in this paper offers the kind of mental map needed for policymakers to understand when their staffs are targeting markets. They may then initiate the necessary support when a national priority is involved or take the needed actions to deflect energies back to the bench. In short, in today's resource-constrained agencies, targeting to market can be a valid goal for only a small minority of programs; all others must be maintained to provide the infrastructure of knowledge that will be necessary if and when private action documents the resolve needed for commercial innovation.

The Obstreperous Community

The material presented to this point is strongly colored by the successful transfers of the cases investigated. Thus, the screening by informal groups and building of service infrastructure are presented as desirable activities for government agencies seeking to maximize the public good. This is not always the case.

While informal screens and service structures are necessary for complete transfer, these structures may also become resistant to innovations that threaten their present status and allegiance. Their loose structure and informal screens may make them even more intolerant of apparently threatening ideas. This issue was at the heart of the case involving boll weevil eradication: the ARS program was threatening the livelihood of existing service industries. However, the service industries had to change if a promising new program was to be given a fair trial. It might have been helpful if ARS had understood this potential threat and planned an approach that offered advantages to existing services, but it might also have been unable to avoid a fight with those who insisted on the entrenched technology.

The informality of much of the structure of the inventive community is both a strength and a potential weakness. The strength of informal arrangements is the force of socially sanctioned agreement and the peer pressure that supports action once consensus is reached. The weakness of such agreement is that once made, it is difficult to change and evolves into social tradition. For example, one of the screening groups studied essentially controlled the release of new plant varieties to the market. Over the years, the group had learned that farmers and buyers would be unlikely to respond unless a plant tested out on specific characteristics. So strong was this conviction that despite all desire to remain informal, these characteristics were documented as "required standards" for approval. The ARS scientists who had worked carefully with this group over many years noted that the standards had become "very strict" to the point of retarding important new introductions. However, these same scientists confessed to being impotent in the face of the tradition that had emerged.

Federal Leadership of Agricultural Innovation

What makes obstreperousness particularly dangerous from the perspective of public policy is that the impotency of federal officials can extend beyond specific cases to more general positions about agricultural policy. That is, in building relationships of trust and mutual understanding in an informal structure, federal scientists may all too easily become preempted. Instead of representing the public welfare, scientists may find themselves siding with parochial interests to retain their positions in networks and their right to be heard on other projects. The quandary is that without informal interaction, the necessary framework of network commitment and service support will not be built.

In resolving this quandary, it is important to emphasize that the public does not have the right to ask federal scientists and managers to minimize contact with private users of technology. This is the path to sterile research, frustrated professionals, and ultimate competition with private organizations. What the public does have a right to is dedication by the federal professionals to manage the interaction--in particular, to take leadership positions in targeting technologies to market that are in the broad public interest and assure the presence of voices representing the public interest in network deliberations (e.g., university faculty appeared to play this role in several of our cases). The ability to assume such leadership will not develop from passive management of federal programs. It must be actively promoted and nurtured through such steps as recruitment, sponsorship of high-visibility conferences, training of program directors, and incentive systems to reward professionals whose projects provide role models for others.

REFERENCES

Lewett, George P. "Evaluation of New Technology." Paper presented to the International Congress on Technology and Technology Exchange, Pittsburgh, May 1982.

Ottosson, Stig. "Guided Product Idea Generation." Omega: The International Journal of Management Science 11(1983): 547-57.

Wolek, Francis W. "Transferring Federal Technology in Agriculture." Journal of Technology Transfer 9(1985): 57-70.

________. Technology Transfer and ARS. A Report to the U.S. Department of Agriculture, Villanova University, May 1984.

19 Private Sector Agricultural Invention in Developing Countries

Robert E. Evenson, Donald D. Evenson, and Jonathan D. Putnam

Three important characteristics distinguish the investment in improved agricultural production technology from investment in other industrial inventions. First, compared to other major industries, the firms (farms) are generally small; even a relatively large farm is too small to capture significant rents from an invention. Although they imitate, innovate, and invent, all in some degree, farms do not conduct much formal research and development designed to earn rents. Second, much agricultural technology is easily replicated. Should a farm invest in research and gain a competitive advantage thereby, this advantage would dissipate quickly as other farms "borrow" its invention. Third, it is often hard to identify the specific invention embodied in a plant or animal. An agricultural inventor has difficulty proving that a given biological specimen, such as a sample of seed, represents the particular group of traits that constitutes his invention. Current technology is making it easier to check/trace genetic traits. In sum, a typical agricultural technology discovery is costly to make, hard to keep secret, and also hard to identify with its originator. Together these three characteristics produce weak incentives for the private sector to invest in many types of agricultural inventions.

Legal systems, notably the standard patent systems, create incentives for small firms like farms to invent, by granting a limited monopoly to the inventor in exchange for his disclosing how to make the invention. To function well, however, these systems must meet several conditions, which correspond to the above-named obstacles to inventing. First, they must require a clear disclosure of the invention. Second, their laws must prevent "borrowing," or infringement, by other would-be producers. Third, the system must describe precisely the invention and certify those properties that distinguish it from others. Finally, all these services must be available to inventors at relatively low cost. In agriculture, however, invention patent systems have not always functioned well. Until recently, for example, the U.S. patent system, generally considered the most comprehensive in the world, did not provide protection for most crop improvements (i.e., new varieties); for animal improvements (new breeds and strains), it still does not.

In response to this basic limitation in the incentives for private invention, most countries have established public sector research institutions (agricultural experiment stations). As a result, the public sector has outshone private institutions in most fields of agricultural research, especially in the biological sciences. Nevertheless, private firms that develop and manufacture certain agricultural inputs, particularly agricultural chemicals, pharmaceuticals, and farm implements, have produced a significant flow of inventions, as have firms that process agricultural outputs ("post-harvest technology"). In those fields of invention in which patent systems have been effective, farmers themselves have produced a significant number of inventions. The key in both cases is the capture of rents for invention.

This chapter assesses quantitatively the magnitude of private sector agricultural inventions throughout the world. We are especially concerned with inventive activity in developing countries. Developing countries, we will argue, possess quite specialized environments for invention that differ markedly from those in developed countries. Consequently, legal systems that work effectively in developed countries will not necessarily work effectively in developing countries. We will conclude that the institutional development of legal systems in the developing countries has been very slow and lacking in imagination. We will also conclude that agricultural invention, despite this limited institutional development, is indeed an important impetus to the growth process of developing countries.

In the first part of this paper we review several types of legal instruments used to protect inventions in both developed and developing countries. The second part summarizes the principal international agreements that govern the protection of intellectual property between countries. The third part tabulates the protection offered to representative agricultural inventions by the individual legal systems of a large number of developing countries. The fourth part presents the available data on private sector research and development and patenting in agriculture.

In the past, biogenetic inventions, which tend to dominate the improvements in agricultural technology, have had very limited "patentability." The fifth part of the paper discusses two recent developments that have changed this situation. The first development is an alternative to the classic patent system, viz., the plant patent. The second and possibly more significant development is the protection that has been extended recently to genetically engineered microorganisms.

The final part of the paper offers a brief discussion of alternatives available to developing countries to encourage more invention. We also discuss the implications of expanded private sector invention for public sector research systems.

LEGAL SYSTEMS FOR STIMULATING INVENTIONS

Types of Invention. The term invention broadly covers any new process, device, chemical composition, or thing that has been developed, devised, or discovered. Although every invention has different economic effects and origins, inventions stemming from agricultural research can be generally grouped as follows: (1) mechanical/electrical; (2) chemical; (3) biogenetic; and (4) other.

Mechanical/electrical inventions primarily include new machinery for planting, cultivating, and harvesting agricultural crops, as well as new machinery and equipment used in animal husbandry. Limited geographically in their application by varying factor price ratios and by geoclimatic conditions, these inventions frequently require modification in order to be transferred from one region to another. Mechanical/electrical inventions also include processes for making and/or using such machinery.

Chemical inventions include fertilizers, herbicides, insecticides, and pharmaceuticals for animals. These inventions are usually more adaptable than are mechanical/electrical inventions. For example, most fertilizers have the same basic chemical ingredients. Current research focuses on improving the processes of production. On the other hand, there is a great deal of research still conducted to find new pesticides and herbicides that meet local conditions and regulations and, ironically, that replace older chemicals to which the pests themselves have adapted.

Biogenetic inventions include improved plant varieties and animal breeds, the products of the new bioengineering processes such as recombinant DNA technology, and such processes themselves. The products of biogenetic research, especially plants, are usually quite limited in their geographic adaptability due to varying soil and climatic conditions, as the large and increasing number of corn and soybean varieties planted in the United States attests. The processes for carrying out biogenetic research may, however, have much broader adaptability.

"Other" inventions include computer programs and general business systems for improving the conduct of agricultural research or farm management. These other inventions are usually not protected by patent and related legal systems and are mentioned here only to illustrate the breadth of activities within the scope of agricultural research.[1]

Legal Systems for Protecting Inventions. Legal systems for securing private rights to inventions implement diverse types of protection: (1) seed and breed certification, (2) copyrights, (3) trade secret enforcement, (4) invention patents, (5) utility models or "petty patents," (6) inventor's certificates, (7) industrial design patents, and (8) plant patent and variety protection. All of these systems provide some type of legally enforceable right to restrict the use of inventions by someone other than the inventor and his licensees/assignees.

The seed and breed certification systems normally require that seed and animals be marketed with sufficient labeling to identify the origin of the seed or animal and give its genetic heritage. Such certification operates like a trademark to prevent others from trading on the reputation that a breeder establishes with a new plant or animal variety. These systems usually permit others to use and sell the same varieties as long as they identify the source. Thus, private rights depend largely on the capability of the breeder to advertise and market his products. For small breeders the laws may be of little assistance unless they prohibit large-scale seed replication for resale.

A copyright prevents unlicensed copying of works of art or an author's writings. The "copyright" is quite literally limited to "copying" the publication and does not preclude the use of the information contained therein. For example, an article describing new plant breeding techniques could be copyrighted, but this would only restrict the "copying" of the article and would not prevent anyone from using the technique.[2]

For inventions that can be maintained in secrecy, such as manufacturing processes that are not readily apparent in the marketed product, trade secrets contracts prevent anyone (primarily ex-employees and collaborators) from disclosing secrets of manufacture and the like to competitors. The standard example is the recipe for Coca-Cola. For the most part, trade secrets weakly protect mechanical/electrical inventions but may be quite potent for chemical and biogenetic inventions. In the biogenetic area, an inventor may maintain certain strains of microorganisms in secrecy indefinitely as "parent" stock from which to develop commercial stocks.

An invention patent system, which governs the usual type of patent, gives the inventor the right to exclude others from practicing the invention for a certain time period (usually 15-20 years). Under such systems, an application for a patent must include an "enabling disclosure" that sufficiently describes the invention so that others skilled in the same technical field can reproduce it successfully. Patent laws thus encourage early publication of the invention in return for granting a limited monopoly, the limiting parameters being the length of time the patent remains in force and the scope of the right to practice the invention.

An invention patent is not an exclusive right to practice a particular invention, but rather a right to exclude others from practicing the invention. The claims that describe the novel contributions made by the inventor define the scope of the exclusion. A classic hypothetical example is the invention of the telephone and invention of the dial system some years later during the pendency of the telephone patent. The patentee for the telephone can prevent the patentee for the dial system from effectively commercializing the dial system, since the dial system requires a telephone in order to be used. On the other hand, the patentee for the dial system can prevent the patentee for the telephone from using his telephones with any kind of dial system, thereby effectively limiting the commercial advantages of the telephones. The usual consequence of such a situation is that the patentees will reach some kind of an accommodation that will give them both the commercial advantages of their respective contributions to the technology.

To be valid, an invention patent must disclose an invention that is novel, useful, and an improvement over the prior art. An invention must be novel in the sense that it has not previously been published, exhibited, or otherwise described. As to its utility, the invention must be capable of industrial or agricultural application, and not be purely ornamental. It need not, however, be profitable.

The degree of "improvement over prior art" that an invention must exhibit defines the single most important attribute of a patent system. Also called the "inventive step" or "level of invention" requirement, this increment must be greater than what would be obvious to the average person skilled in the art. The height of this step varies from country to country. For instance, because of its high standards for an inventive step, among other things, Germany grants only 35 percent of the patent applications it receives; in the United States, the figure is over 70 percent and in France it is about 90 percent. On the other hand, some countries do not require explicitly an inventive step at all.[3] Countries that do not require an inventive step in their patents generally do not have a separate system to govern utility models. We will return to this point later in our discussion of policy alternatives available to developing countries.

<u>Utility models</u> or "petty patents" are similar to invention patents in that they give the inventor the right to exclude others from practicing the invention for some period of time. They differ from the invention patents in requiring only novelty and utility, without any "inventive step" above the prior art. Thus, petty patents preserve rights to minor variations of known devices rather than to major technical innovations having broad adaptability. Countries usually grant petty patent protection for a much more limited time than is the case for invention patents, and then only to their own citizens.

The purpose of these systems is, to quote the Korean Utility Model Act,

> to encourage, protect and foster pragmatic devices, and thereby seeking technological progress, and contributing to the development of national industry.

Since "the development of national industry" was at the heart of these laws, utility model patents, unlike invention patents, are not granted generally to foreigners.[4] As such petty patent systems tend to stimulate domestic inventions to accommodate local conditions, they may help the local economy but may not be valuable abroad. Nationals of these countries are more likely, therefore, to utilize petty patents rather than the more costly and difficult-to-obtain invention patents.

An _inventor's certificate_ is a notice given in socialist countries that entitles an inventor to receive compensation for his invention, which as as matter of property belongs to the state. This non-market alternative to the standard patent aims to reward the inventor while removing his monopolistic control over the invention.

Industrial design patents provide protection to _ornamental designs_ as opposed to inventions _per se_. They thus do not play a large role in stimulating technical change, since they protect a shape or form that is simply _different_, rather than a product or process that _improves_ the state of the art.

Analogous to invention patent systems, _plant patent_ and _variety protection_ systems provide breeders and developers with limited rights to exclude others from commercializing new plant varieties they have developed. A plant variety must establish novelty by exhibiting uniformity, stability, and distinctness from all known varieties. Typically, some type of control depository is provided to preserve seed samples for the plant varieties being protected, in addition to the requirements for verbal and pictorial description. To encourage the use of the protected varieties in further breeding/research programs, most plant protection laws exempt experimental breeding from infringement claims.

THE ROLE OF INTERNATIONAL CONVENTIONS

In keeping with the treatment of inventions as "intellectual property," most countries of the world are party to at least one international agreement, the intent of which is to protect an inventor's rights to his invention from country to country, much in the same way that free trade agreements protect commerce from tariffs and other unilateral restrictions. The most widely held of these agreements is the International Convention for the Protection of Industrial Property, usually called the

"Paris Convention" for the seat of its first formulation in 1883. This agreement, as subsequently amended at The Hague (1925), London (1934), Lisbon (1958), and Stockholm (1967), provides that any country belonging to the convention should grant to citizens of another convention country the same rights as those belonging to its own citizens. This means simply that a German should be able to obtain a patent in France as easily as a Frenchman. It does not mean that a German patent is equivalent to a French patent.

This "laissez-faire" approach to intellectual property protection has its adherents and critics for many of the same reasons as free trade agreements do. Among developing countries, arguments against participating in such an agreement tend to focus on the allegedly inhibiting effects on domestic research and development of allowing more technologically advanced countries to compete with fledgling domestic industries. Because of the relatively long-term and monopolistic nature of patent protection, the disagreements over the benefits and costs of multilateral treaties tend to be even sharper than those over free trade. Furthermore, unlike typical examples in international trade, it is unusual for a developing country to have a comparative advantage in some field of research and development.

Nevertheless, most countries of the world belong to the Paris Convention. Among agricultural inventions, this uniformity of protection may not be as significant, for in agriculture the utility of an invention may often be confined to particular subregions that share the same geo-climatic characteristics. Inventions can be adapted and modified, however. Thus, to the extent that such conventions encourage the importation of foreign technology as a source of "raw material" for local modification, they may also stimulate some domestic innovation.

The next logical step in providing uniform protection for citizens of different countries is to allow uniform application procedures among countries, to reduce the application burden on inventors, and to provide for the exchange of information gleaned from searches of the prior art. Under this arrangement, an inventor can apply in one country and subsequently designate in which other countries he intends his application to be examined. An agreement of this type called the Patent Cooperation Treaty was signed by thirty-five countries in 1978. Among its stated aims is "to foster and accelerate the economic development of developing countries through the adoption of measures to increase the efficiency of their legal systems...."

Although many of its features are noteworthy, the treaty pertains most directly to our discussion by explicitly seeking to foster economic development and by its attempt to cover a broad range of inventions, including patents, utility models, inventor's certificates, and patents of addition (these expand the protection given to an inventor's prior patent or utility model). In addition, it allows each participating country to establish different or additional criteria for the granting of patents.

About half of the parties to the Patent Cooperation Treaty are developing countries.[5] With the exception of the Philippines, North Korea, and Sri Lanka, underdeveloped Asia is missing entirely. Latin America, too, has few participants. We may ascribe the relatively strong rate of participation by African countries to the formation of the African Intellectual Property Organization (open to any country but presently comprising former French colonies), a group of countries that binds its members by a modern agreement modeled on French patent law. In particular, this agreement provides for the granting of a community patent, valid in all participating countries, thereby anticipating the substance of the Patent Cooperation Treaty. The countries of Asia have no comparable agreement; the agreements binding Latin American countries are nearly seventy-five years old and have few signers.

The most integrative among the international patent agreements in force is the European Patent Convention, which provides a limited type of international patent that is examined by the European Patent Office. Although it relies on the national offices and courts for the interpretation and enforcement of claims, it nevertheless constitutes an advantage to applicants that wish to file in four or more European countries and wish to defer expenses for the applications as long as possible. A major consequence of this European patent system is that Japanese and American multinational corporations that file extensively in Europe choose the EPC, prosecute the applications in English, and thus defer the expenses for translations, etc., until they are assured that they will receive a patent.

Two other treaties have a more direct bearing on agricultural inventions: the International Convention for the Protection of New Varieties of Plants, and the Budapest Treaty on the International Recognition of the Deposit of Microorganisms for the Purpose of Patent Procedure. The first of these, the Plant Variety Convention, was amended most recently in 1978 and provides for patent or patent-like protection to breeders of new plant varieties who belong to member countries. These plants may be sexually as well as asexually reproduced (which gives protection to hybrid varieties), but member states may exclude hybrid varieties from protection at their discretion (on the grounds that the breeder retains control over the parents, which renders protection unnecessary). At present, no developing country belongs to this treaty, and only one, Argentina, has passed a law to give protection to new varieties.

The Budapest Treaty on Microorganisms was signed in 1977. It provides for an "international depository authority" in several nations, which keeps samples of patented microorganisms. This special arrangement takes the place of the usual written and/or graphic description that regular patent documents employ. The treaty aims to lower the cost and reduce the inconvenience of trying to deposit multiple samples in each country in which the inventor desires protection.

The treaty does not grant patent protection per se, but merely commits a state to the system of recognizing deposits made in other countries as equally valid as those made in its own. Thus, the treaty leaves a considerable degree of freedom in the hands of the individual countries to decide what constitutes a patentable microorganism. Among developing countries, only Senegal and the Philippines belong to this agreement; the only other non-Western participant is the Soviet Union.

Thus, one or more international agreements govern most intellectual property for which the inventor from a developing country might seek protection. Moreover, at the domestic level, the World Intellectual Property Organization (an agency of the U.N.) has written a "model law" for patent, design, and utility model protection as the suggested format for a developing country. Many countries have employed this model in drafting their own patent laws, despite its having some questionable provisions. As an example of an instance where conformity to worldwide practice may be unsuitable, this model law recommends that the country adopt an "inventive step" requirement among its standards for issuing a patent. This effectively means that the country grants only invention patents. Two of the important features of the petty patent system, namely its effective preferential treatment of citizens and its reduced cost, remain inaccessible to countries that implement only the model law. Furthermore, this arrangement places a burden of technological sophistication on the domestic patent office that in all probability it cannot carry, since the "obviousness" of an inventive step depends on one's knowledge of the state of the art. One may fairly argue that developing countries have not shown a great deal of institutional inventiveness in implementing more optimal laws that suit their specific needs. It is also true, however, that the international status quo establishes a reasonably strong current against which it is difficult to run should a country desire to protect its inventions differently. Ironically, in the cases where developing countries have chosen to go against the international tides (most notably in India), the innovations have usually taken the form of reduced protection, by giving an exemption to food, medicine, or other human necessities. A few countries have adopted more flexible laws to encourage domestic invention; we discuss the effects of these efforts in the third and fourth parts of the chapter.

LEGAL SYSTEMS FOR PROTECTION OF INVENTIONS IN DEVELOPING COUNTRIES: INTERNATIONAL COMPARISONS

Origins of Patent Systems in Developing Countries. Article I of the Constitution of the United States declares that one of the fundamental powers and obligations of the Congress shall be

> To Promote the Progress of Science and the Useful Arts, by securing for limited Times to Authors and Inventors the exclusive Right to their respective Writings and Discoveries.

This essential philosophy was also held by the major European powers of the time. Patent systems were developed to promote the progress of science and the useful arts in all industrializing countries during the late eighteenth and early nineteenth centuries. By the end of the nineteenth century the industrialized countries had joined the Paris Convention.

Most colonial powers, particularly Great Britain and France, established patent systems in their colonies. For the most part, these systems served mother country interests rather than domestic interests. In Latin America, however, a number of legal systems were developed without strong colonial pressures in the late nineteenth century.

In the past thirty years or so, a good deal of discussion about, and dissatisfaction with, legal systems have emerged from the developing world. As the next section of this paper shows, most invention patent systems in developing countries grant a very high proportion of invention patents to foreign inventors. Political activists often see this as evidence of "dependency" and support the curtailment or elimination of patent systems. Planners and bureaucrats, on the other hand, argue that patent systems facilitate inflows of foreign capital and associated technology. Yet, even in relation to the level of domestic invention, many developing countries have patent offices with minimal staffing and administrative support. Others lack the efficient judicial environments in which patent systems work most effectively.

As an example of the situation facing developing countries and the attitudes they have expressed, we quote extensively from an Explanatory Memorandum, which the government of Kuwait attached to its patent law in 1965. The memorandum is revealing in that it openly acknowledges Kuwait's status as a developing country, and explains its eclectic use of the legislation of industrialized countries in light of that status. (One must remember that at the time the memorandum was written--prior to the enormous influx of foreign currency following the OPEC oil embargo--Kuwait was perhaps a more representative developing country than it is now considered to be.)

> The law has...adopted the principles established by modern legislation, particularly the Swiss law of 1907, the Italian law of 1934, the German law of 1936, the Finnish law of 1943 and the Egyptian law of 1949, without disregarding the requirements of the industrial development which Kuwait is bound to undergo sooner or later.
>
> There are two main systems regarding invention patents: the French system, which is based on the freedom of the grant of patent on the strength of mere deposit without examination or comparison, and the English system which is based on the grant after thorough examination to ascertain the existence of the objective elements which the law requires the invention to possess.... Between these two, there are intermediate systems ranging from the French to the English....
>
> The French system involves the essential vice of granting patent on the strength of mere deposit, which

> leads to the granting of patent in respect of matters which are not regarded as inventions in the sense which makes them worthy of legal protection. This vice does not exist in the English system since it makes the grant of patent subject to prior search and thorough investigation undertaken by a government organization which possesses the necessary technical means and has its own traditions and experience in the various arts and sciences.
>
> Thus legislative thinking in various countries has started to drift away from the simple deposit system and to lean towards the prior examination system.
>
> Since it is not appropriate for Kuwait at this stage of its development to adopt the French system which the nations have started to abandon, and since it is not practically feasible for it to start by adopting the English system as such, it has been deemed appropriate to follow a middle course....
>
> In this way it is possible to achieve some of the results of the full examination system. The law leads to the training of the new government administration and to the formation of a nucleus of technicians permitting the future adoption of the English system which is internationally considered to be a model system.

Apart from having a "model system," the United Kingdom exerts an influence on patenting in many developing countries in a more direct manner. As is noted in table 19.1, several countries, principally in Africa, do not have independent patent systems, but merely provide for the registration of British patents. This saves the country the cost of a patent system, but also preserves vestiges of colonial influence on the flow of technology.

Types of Protection Offered in Developing Countries. In this part of the chapter we review the current legal systems of a number of developing countries in terms of the protection written into current statutes. The reader should bear in mind that the administration and adjudication of these statutes often determine whether these systems encourage invention.

We will focus our attention in this section on the scope of protectable "subject matter" in the various countries. (We will not provide detail on examination requirements, renewal fees, and so forth, although they alter significantly the character of each country's

TABLE 19.1
Availability of Patent/Variety Protection for "AARC" Inventions in Selected Developing Countries

	Mechanical/ Electrical		Chemical				Biogenic						Other	
Country	Plough I Major	Plough II Minor	Fertilizer	Insecticide	Herbicide	Chemical Vaccine	Soybean	Corn Hybrid	Rose	Beef Cattle	Nitrogen-Fixing Bacteria	Live Virus Vaccine	Computer Program	Accounting System
U.S.A.	yes	no	yes	yes	yes	yes	yes	no	yes	no	yes	yes	no	no
Asia														
Bangladesh 1,2	no	no	no	no	no	no	no*	no*	no*	no*	?	?	no*	no*
India	no*	no	no**	no**	no**	no**	no*	no*	no*	no*	no*	no*	no	no
Indonesia 2	--	--	--	--	--	--	--	--	--	--	--	--	--	--
Korea (South)	yes	yes	no**	no**	no**	no*	no*	no*	yes	no*	yes	no*	no	no
Malaysia 1,2	yes	no	yes	yes	yes	yes	no*	no*	no*	no*	no	no	no*	no*
Nepal 4	yes	no	yes	yes	yes	yes				see footnote			no	no
Pakistan 3	yes	no	yes	yes	yes	yes	no*	no*	no*	no*	yes	yes	yes	no**
Philippines 5	yes	yes	yes?	yes?	yes?	no	no	no	no	no	yes	yes	no	no
Singapore 1,3	yes	no	yes	yes	yes	yes	no*	no*	no*	no*	no	no	no**	no**
Sri Lanka 3	yes	no	yes	yes	yes	yes	no*	no*	no*	no*	no	no	no	no**
Taiwan	yes	yes	no**	no**	no**	no**	no*	no*	no	no*	no	no*	no	no
Thailand	no*	no	yes	yes	yes	no**	no*	no*	no*	no*	no*	no*	no*	no
Latin America														
Argentina	yes	no	yes	yes	yes	no*	yes	yes	yes	no	yes	no*	no*	no*
Bolivia	yes	no	no**	no**	no**	no**	no	no	no	no	no	no*	no	no*
Brazil	yes	yes	no**	no**	no**	no*	no*	no*	no	no*	no*	no*	no*	no*
Chile	yes	no	no*	no*	no*	no*	no*	no*	no	no*	no	no*	no	no*
Colombia	yes	no	no**	no**	no**	no**	no*	no*	no	no*	no*	no*	no*	no*
Costa Rica	yes	no	yes	yes	yes	no*	no	no	no	no	no	no*	no	no
Dominican Rep.	yes	no	yes	yes	yes	yes	no	no	no	no	no	no	no	no*
Ecuador 7	yes	no	yes	yes	yes	no*	no*	no*	no	no*	no	no*	yes	no*
Guyana	yes	no	yes	yes	yes	yes	no	no	no	no	no	no	no	no
Haiti 5	yes	no	yes	yes	yes	yes	no	no	no	no	no	no	no	no
Jamaica	yes	no	yes	yes	yes	yes	no	no	no	no	no	no	no	no
Mexico	yes	no	no*	no*	no*	no*	no*	no*	no*	no*	no*	no*	no*	no*
Nicaragua	yes	no	no**	no**	no**	no**	no	no	no	no	no	no	no	no
Panama 5	yes	no	yes	yes	yes	no*	no*	no*	no	no	no	no*	no	no
Paraguay	yes	no	yes	yes	yes	no**	no	no	no	no	no	no*	no	no
Peru	yes	no	no**	no**	no**	no**	no*	no*	no*	no*	no*	no*	no*	no*

Surinam	yes	no	no**	no**	no**	no**	no*	no*	no*	no*	no*	no	no	no
Uruguay	yes	yes	no**	no**	no**	no**	no	no	no	no	no	no*	yes	no*
Venzuela	yes	no	no**	no**	no**	no**	no	no	no	no	no	no*	no	no*
Near and Middle East														
Aden 1	yes	no	yes	yes	yes	yes	no*	no*	no*	no*	no	no	no*	no*
Bahrain 1	yes	no	yes	yes	yes	yes	no*	no*	no*	no*	no	no	no*	no*
Egypt	yes	no	yes	yes	yes	no**	no	no	no	no	no*	no*	no	no
Iran	yes	no	yes	yes	yes	no**	no	no	no	no	no	no**	yes	no
Iraq	yes	no	yes	yes	yes	no**	no	no	no	no	no	no*	no	no*
Israel	yes	no	yes	yes	yes	yes	no*	no*	no*	no*	yes	yes	no	no
Jordan	yes	no	yes	yes	yes	yes	no	no	no	no	yes	yes	yes	no
Kuwait	yes	no	yes	yes	yes	no**	no	no	no	no	no	no*	yes	no
Lebanon	yes	no	yes	yes	yes	no**	no	no	no	no	no	no*	no	no*
Syria	yes	no	yes	yes	yes	no**	no	no	no	no	yes	no*	yes	no*
Africa														
African Intellectual Property Organization 3,6	yes	no	yes	yes	yes	yes	no*	no*	no*	no*	no	no	no*	no*
Algeria 3,7	yes	no	yes	yes	yes	yes	no*	no*	no*	no*	no	no	no	no
Botswana, Lesotho 3	yes	no	no	no	no	no	no*	no*	no*	no*	no	no	no*	no*
Burundi Rwanda, Zaire 5	yes	no	yes	yes	yes	yes	no	no	no	no	no	no	no	no
Ghana 3	yes	no	yes	yes	yes	no**	no*	no*	no*	no*	no	no*	no*	no*
Kenya 3	yes	no	yes	yes	yes	yes	no*	no*	no*	no*	no	no	no*	no*
Liberia	yes	no	yes	yes	yes	yes	no	no	no	no	no	no	no	no
Libya	yes	no	yes	yes	yes	no**	no	no	no	no	no	no*	no*	no
Malawi, Zambia, Zimbabwe 8	yes	no	yes	yes	yes	yes	no	no	no	no	yes	no*	yes	no

(continued)

Table 19.1 (Cont.)

Country	Mechanical/ Electrical		Chemical				Biogenic						Other	
	Plough I Major	Plough II Minor	Fertilizer	Insecticide	Herbicide	Chemical Vaccine	Soybean	Corn Hybrid	Rose	Beef Cattle	Nitrogen-Fixing Bacteria	Live Virus Vaccine	Computer Program	Accounting System
Nigeria 3	yes	no	yes	yes	yes	yes	no*	no*	no*	no*	no	no	no	no
Sierra Leone 3	yes	no	yes	yes	yes	yes	no*	no*	no*	no*	no	no	no*	no*
S.W. Africa 5	yes	no	yes	yes	yes	yes	no	no	no	no	no	no	no	no
Sudan	yes	no	yes	yes	yes	yes	no	no	no	no	no	no	no	no
Tanzania	yes	no	yes	yes	yes	yes	no	no	no	no	no	no	no	no
Tunisia	yes	no	yes	yes	yes	no**	no**	no**	no	no	no	no	no	no

*This invention is specifically excluded from patent protection by national law.
**Although this chemical substance is specifically excluded from patent protection by national law, the process used to produce the substance is not excluded.

NOTES: This table summarizes the data that record the principal constitutents of each country's patent laws.
1. British patent law is assumed to hold in this country, owing to the provisions in its laws. British patent applications (whether or not by British citizens) have priority. In practice, a prior British patent is routinely granted approval in this country at the applicant's request. We refer the reader to chapter 37 of the Patents Act of 1977 of Great Britain. The U.K. prohibits the patenting of microbial processes or products for use on humans or animals. Ghana independently prohibits patents on pharmaceutical and medical substances.
2. This country has no patent act of its own.
3. "Microbiological processes and the products of such processes" are patentable. Whether this protection extends to microorganisms per se is not known and will depend on the interpretations of the domestic courts. In the absence of specific indications to the contrary, we have assumed that the nitrogen-fixing bacteria and the live virus vaccine are not patentable under these circumstances.
4. A patent is granted to a foreign inventor if he has obtained a patent in his own country and any three other countries. Presumably, patentability standards in those countries apply.
5. Other than meeting public standards of heath and morality, no other criteria for patentability are cited. In general. we take mechnical. chemical, and electrical inventions to be patentable, and others to be unpatentable. In the Philippines, U.S. law is assumed.
6. The following countries are signatories to the Libreville Agreement of 1962, which establishes a regional Intellectual Property Office: Benin (Dahomey), Cameroon, Central African Republic, Chad, Congo, Gabon, Ivory Coast, Madagascar, Malagasy Republic, Mauritania, Niger, Senegal, Togo, and Burkina Fasso (Upper Volta). The revised agreement of 1977 has been signed by Cameroon, the CAF, Gabon, Ivory Coast, Mauritania, Niger, Senegal, and Togo. In the absence of laws to the contrary, we apply the revised standards to the other countries as well.
7. The inventor is entitled to state indemnification for the rights to some or all of these inventions. In this case, he does not own the rights himself.
8. Food and chemical patents require mandatory licensing.

system.) In order to summarize concisely the scope of protectable subject matter in the various countries we will assume a hypothetical Amalgamated Agricultural Research Corporation (AARC). AARC's agricultural research program covers the entire spectrum of research and development activities. Below we list several AARC inventions. For the purpose of this discussion, we assume these inventions satisfy novelty, utility, and inventive step requirements as they may exist in each country.

Mechanical/Electrical Inventions

1. AARC Plough I: a basic innovation in ploughs that is adaptable to a wide range of soil and climatic conditions.

2. AARC Plough II: a minor modification for adapting a plough to be utilized in a specific localized soil condition.

Chemical Inventions

3. AARC Chemical Fertilizer: a new fertilizer chemical compound that optimizes tolerance to extremes in soil moisture conditions.

4. AARC Insecticide: a new and useful chemical compound insecticide.

5. AARC Herbicide: a new and useful chemical compound herbicide.

6. AARC Pharmaceutical for Animals: a chemical compound pharmaceutical for controlling disease in farm animals.

Biogentic Inventions

7. AARC Soybean: a new improved soybean variety developed in a plant breeding program.

8. AARC Corn: a new hybrid corn seed variety developed in a plant breeding program, with AARC retaining control over the hybrid parents.

9. AARC Rose: a new variety of asexually reproducible ornamental rose.

10. AARC Beef Cattle: a new pure breed of beef cattle developed in a selective breeding program.

11. AARC Bacterium: A new and improved nitrogen-fixing strain of bacteria developed using recombinant DNA techniques.

12. AARC Live Virus Vaccine: A new strain of virus to be used as a vaccine for animals, developed using recombinant DNA techniques.

Other Inventions

13. AARC Computer Program: a new and improved computer program that determines the optimal mix of chemicals in the AARC Chemical Fertilizer (invention #3).

14. AARC Accounting System: a new and improved accounting system used for optimally allocating the research personnel and facilities in the various research projects, adaptable to any large agricultural research organization.

From table 19.1, one can see immediately that the degree of protection given to agricultural intellectual property varies considerably among countries. Of particular interest are those entries that denote that the device in question is specifically prohibited from being patented. Agriculture is somewhat peculiar among traditional production industries in that many agricultural inventions are either ineligible for protection or have an indeterminate status under most national patent acts. In this respect, many countries group agriculture with medicine rather than with manufacturing; effectively, they view improvements in the production of food, like improvements in the provision of health care, as public, not private, property. Since these countries treat many agricultural inventions as public goods, the return to them cannot be legally appropriated by the inventor. We will return to the effects of this policy after we discuss the pertinent points of table 19.1.

As one can see from table 19.1, several countries ban the patenting of chemical substances and the processes for making them; others ban only the substances; still others permit chemicals in general to be patented but do not allow chemical vaccines to be patented because they are medicines. Among Asian countries, India, Korea, and Taiwan do not allow chemical patents of any kind; Thailand excludes only chemical medicines. As a group, Asia falls midway between Latin America and Africa in allowing chemical patents. In Latin America, half of all countries ban patents on chemicals of any type, and three-fourths of them ban patents on chemical medicines. By contrast, all the African countries permit some kind of chemical patent, and only two exclude chemical medicines from protection.

In the Middle and Near East, all countries allow non-medicinal chemical patents, but most forbid patents on medicinal chemicals.

For Plough II-type inventions (those that do not meet the "inventive step" requirement of a regular patent, but contain some adaptation or modification of existing technology), five semi-industrialized countries provide protection via a utility model system: Brazil, Uruguay, Korea, the Philippines, and Taiwan, as well as the parties to the African Intellectual Property Organization.

As noted above, these patents are particularly relevant to agriculture, because of the common problem of adapting known technology to local conditions. We examine the results of the Philippine utility model system in table 19.6 below.

As to biogenetic inventions, most countries specifically exclude plant varieties and animal species from protection, either as such or by excluding foodstuffs. Among developing countries, only Argentina and Korea make provision for plants of any kind to be patented, and only Argentina permits sexually reproduced plants to be patented. The United States does not provide plant variety protection for the hybrid corn--presumably because such protection is not needed, since corn breeders can effectively maintain control over the hybrid parents, and thus they already have so-called "genetic" protection. Although there are legal systems in some countries (e.g., Hungary and Romania) that provide breeder's rights to new animal breeds, none of the countries listed in table 19.1 has such a provision, so none of them would protect AARC's new strain of beef cattle.

As to microorganisms, the United States, Argentina, Korea, and Israel specifically allow the patenting of microorganisms not occurring in nature; Argentina and Korea disallow the live-virus vaccine on the grounds that it is a medicine. We may divide the other countries into four groups: those where the status of microorganisms is indeterminate; those that ban the patenting of plants or animals, which ban presumably includes microorganisms; those, like Brazil, that specifically forbid the patenting of microorganisms; and those that while banning plant, animal, and other biological-process patents, provide for the patenting of "microbiological processes and the products thereof." Whether this provision extends specifically to microorganisms appears to depend on the ruling of domestic courts. We note this group in the table. As we explain in more detail below, the U.S

Supreme Court ruled only quite recently on the possibility of patent protection for the AARC's nitrogen-fixing bacteria and the live-virus vaccine. Like most countries, the United States had no guidelines as to whether one could actually obtain an invention patent on a living organism.

The exemplary "other inventions," namely the computer program and the accounting system, do not generally receive patent protection or the like under any of the legal systems. A copyright system that included computer programs might, however, provide some limited protection. In the United States, a computer program is not patentable, but processes, and equipment for carrying out processes, especially manufacturing processes, wherein the novelty is really the specific application of a generally programmable computer, can be patented.

[Recently, this pattern has been broken. In a 1983 survey the following countries indicated that they gave patent protection to computer programs: Ecuador, Iran, Japan, Jordan, Kuwait, Pakistan, Syria, and Uruguay. Curiously, developed countries appear to lag behind developing countries in adopting this broader view of intellectual property protection (see Baxter and Sinnott, vol. 2A, pp. 232:22-232.25).]

The rationale for banning patents on certain agriculture-related inventions is set forth in the Kuwaiti Explanatory Memorandum of 1965:

> Article 2 refers to certain specific types of inventions which, for reasons connect with morals, public policy or public health, are not patentable. This is in line with the general legislative trend (Article 2 of the U.A.R. law). Thus no patent is granted in respect of inventions the utilisation of which causes breach of morals or public policy, nor in respect of inventions connected with foodstuffs, medical preparations or pharmaceutical compounds because the monopoly of such products is harmful to public health. Nevertheless, the prohibition concerning chemical inventions provided for in Article 2 (2) applies only to the products themselves, not to their method of manufacture. The adoption of the system of patenting the method of manufacture of chemicals and not the chemical products themselves is calculated to encourage the chemical industries and to enable them to flourish. This is the system followed by the German and Egyptian legislators.

India's patent law, adopted in 1970, is the most restrictive in this sense; we will consider it in some detail here as an example. India forbids the patenting of "a method of agriculture or horticulture"; "any process for the medicinal, surgical, curative, prophylactic, or other treatment of human beings or any process for a similar treatment of animals or plants to render them free of disease or to increase their economic value or that of their products"; or any substance "intended for use, or capable of being used, as food or medicine or drug" [Article 3(h), (i); Article 5(a)]. Furthermore, "medicine or drug" includes "insecticides, germicides, fungicides, weedicides and all other substances intended to be used for the protection or preservation of plants" [Article 2(l)(iv)]. It is readily apparent that these categories cover most agricultural inventions. While it is true, as one can see from table 19.4, that a certain number of agriculture-related devices have been patented in India since 1970, one can see from table 19.7 that the level of patenting in agricultural implements has dropped sharply. It is doubtful whether any given agricultural implement is patentable--and, as table 19.1 shows, the other typical agricultural inventions are clearly unpatentable.

There are two general effects of this type of legislation. The first effect deprives foreign inventors of economic rights to their invention in India; this may make the borrowing of foreign agricultural technology less costly.[6] The second effect deprives domestic inventors of incentives either to invent on their own or to modify foreign agricultural technology. Without offering protection to domestic inventors, the investment required to adapt foreign inventions to local climate and soil conditions may not be forthcoming. This adaptation process is crucial to the diffusion of agricultural technology. The public sector, in general, has neither the resources nor the flexibility to engage in this kind of broad-based adaptation. Thus, by not providing domestic inventors with incentives to modify the inventions they borrow, the Indian legislation may actually increase dependence on foreign agricultural technology.

INVENTION IN THE PRIVATE SECTOR: A QUANTITATIVE SUMMARY

In this section we discuss data that summarize private sector agricultural invention in developing countries. For our purposes, only pre-harvest and harvest research activities qualify as agriculturally related. The only available data on private sector expenditures are reported in Boyce and Evenson (1975). Their data show that the private sector probably spends less than 10 percent of the total agricultural research and development budget in developing countries. (This includes not only agricultural experiment station research in this total but university research as well.) In the United States and Canada, this share is about 35 to 40 percent, and it is growing rapidly.

We use data on total patents granted to draw inferences about inventive outputs and, indirectly, about inventive inputs. Table 19.2 reports a tabulation of invention patents granted in all fields for a large number of countries. Our primary interest is in patenting related to agriculture, but these data illustrate several important patterns of international invention. First, for most developing and semi-industrialized countries, the ratio of imported technology (1),

$$\frac{\text{patents granted to foreigners}}{\text{patents granted to nationals}} \tag{1}$$

is quite high. Symmetrically, the ratio of exported technology (2),

$$\frac{\text{patents granted abroad to nationals}}{\text{patents granted at home to nationals}} \tag{2}$$

is quite low.

These data indicate that invention patent systems in the developing countries do not stimulate high levels of domestic invention in the face of foreign competition. Although the levels of foreign patenting in developing countries have declined from 1967 to 1980, this decline is due mainly to a general drop in patenting by some of the major industrialized countries (especially the U.S., U.K., and France). One does not see strong evidence, however, for aggressive expansion of invention patenting by nationals in the semi-industrialized and developing economies. The low ratio of patents granted abroad to patents granted at home indicates that most inventions patented in the third world are of an "adaptive" nature, rather than being highly original and suitable for export. They often derive from inventions originating in the developed countries.

We have stated earlier that the utility model system best promotes the adaptive and relatively minor inventions of third world countries. Table 19.3 reports data for this legal instrument. Comparatively few third world countries include this instrument among their intellectual property rights. In each case, we observe that they grant relatively more utility models to nationals than is the case for invention patents. Foreigners (generally, multinational firms) do not make much use of this form of protection. Utility models turn out to be quite important in agriculturally related patenting as we will demonstrate in our discussion of Philippine invention below.

We now turn to agricultural inventions per se. Three legal instruments cover most of these inventions: the invention patent, the utility model, and the plant patent. Table 19.4 presents the available data for invention patents related to agriculture. These data show the number of patents granted annually in International Patent Class (IPC) A01, which covers most agricultural implements and related animal and crop husbandry inventions. It does not cover food and fiber processing, or agricultural chemicals and pharmaceuticals.

Table 19.4 shows that the level of annual patent flows represents a significant amount of agricultural invention in many countries. Some of this invention is undertaken in public sector research institutions, but the data available on this point suggest that their share of total patenting is very low. In the United States, for example, fewer than 5 percent of patented agricultural inventions are produced by the state agricultural experiment stations or the USDA. Predictably, the table shows that patenting is highest in the industrialized economies; among developing countries, only a handful grant more than a dozen patents in agriculture each year.

Table 19.5 provides more detail for invention patents granted in several subclasses of agricultural implements from 1973 to 1978. Only five third world countries grant significant numbers of invention patents in these fields. For comparison we provide data for the United States and four developed countries. The data show the number of patents of domestic origin and of LDC origin. Of the third world countries, only Brazil has a reasonably large number of invention patents. Patents originating in other LDCs are not significant except perhaps in grain harvesting inventions, the most active of the subfields.

We have also included a measure of the "trade balance" for these inventions in the form of a ratio (3).

TABLE 19.2
Invention Patents Granted by Country: Selected Years

	Patents Granted to Nationals				Patents Granted to Foreigners				Patents Granted to Nationals in Foreign Countries			
	1967	1971	1976	1980	1967	1971	1976	1980	1967	1971	1976	1980
I. Industrialized Market Economies												
A. Rapid Growth												
Japan	13,877	24,795	32,465	38,032	6,896	11,652	7,582	8,074	6,843	15,832	20,246	20,663
Austria	1,188	1,230	1,177	1,277	6,896	7,460	5,235	4,745	1,913	2,399	1,065	1,669
France	15,246	13,696	8,420	8,433	31,749	37,760	21,334	19,622	14,393	17,150	12,677	12,511
Denmark	338	252	208	192	2,002	2,212	2,068	1,453	1,165	1,650	1,217	1,103
Germany	5,126	8,295	10,395	9,826	8,300	9,854	10,570	10,362	41,775	44,862	37,316	33,708
Belgium	1,586	1,345	1,034	837	15,041	15,004	12,110	5,081	2,701	2,894	1,905	1,720
Norway	225	386	210	276	1,831	2,343	1,883	1,843	618	658	617	549
Netherlands	322	318	370	417	1,913	2,396	3,219	2,907	7,283	8,745	5,901	5,964
B. Slow Growth												
Canada	1,263	1,587	1,301	1,503	24,573	27,655	20,449	22,392	2,789	3,201	2,661	2,200
Italy	9,076	4,320	--	1,810	26,180	13,180	--	6,190	5,621	6,749	5,416	5,877
Ireland	28	16	27	24	635	788	1,055	1,407	113	151	146	106
Switzerland	5,388	4,165	3,482	1,475	16,462	11,914	8,818	4,486	12,452	15,409	10,954	9,827
Sweden	1,776	2,245	1,888	1,394	7,532	7,748	6,956	3,604	5,031	6,327	5,719	4,769
U.S.A.	51,274	55,988	44,162	37,152	14,378	22,328	26,074	24,675	73,960	87,589	90,273	54,360
Australia	752	979	910	620	10,371	9,662	10,074	7,805	905	986	1,065	2,690
U.K.	9,807	10,376	8,855	5,158	28,983	31,178	30,942	18,646	17,579	21,179	14,072	11,140
Finland	231	350	291	439	739	1,312	921	1,467	345	559	650	923
New Zealand	--	--	211	137	--	--	1,314	1,122	135	1,420	91	235

II. Semi-Industrialized Market Economies												
A. Rapid Growth												
Spain	2,758	2,042	2,000	1,485	6,827	7,764	7,500	7,739	627	933	766	1,180
Israel	178	202	200	305	935	1,225	1,200	1,419	219	231	146	316
Greece	975	1,227	1,343	1,114	2,302	698	1,285	942	61	70	81	691
Singapore	5	2	--	1	26	334	--	548	--	--	5	5
Portugal	84	214	46	95	1,045	3,238	1,319	2,200	53	57	50	50
Brazil	262	429	450	349	684	1,543	1,500	3,494	63	85	88	113
Korea (5)	207	200	1,593	258	152	117	1,727	1,161	20	20	50	50
B. Moderate to Slow Growth												
Chile	80	58	60	60	1,237	1,115	514	514	--	--	--	--
Venezuela	41	237	50	55	954	1,599	514	408	--	--	--	--
Argentina	1,244	1,346	1,300	1,264	4,488	3,484	2,800	2,343	81	152	102	133
Mexico	1,981	412	300	174	7,922	5,199	3,000	1,831	149	148	181	141
Turkey	30	52	35	34	438	357	588	424	--	--	--	--
Uruguay	165	88	46	41	351	161	110	236	--	--	--	--
III. Developing Economies												
Ecuador	5	8	7	7	126	180	103	103	--	--	--	--
Iraq	22	5	12	14	146	67	150	24	--	--	--	--
Morocco	28	24	23	21	391	313	334	330	--	--	--	--
U.A.R.	48	13	16	10	873	236	511	317	--	--	--	--
Colombia	49	62	30	36	851	651	600	808	--	--	--	--
Philippines	16	46	108	82	498	946	767	755	--	--	--	--
Kenya	0	1	5	--	104	121	98	97	--	--	--	--
India	428	661	433	500	3,343	3,256	2,062	2,000	72	70	73	57
Sri Lanka	1	10	4	5	4	148	156	36	--	--	--	--
O.A.P.I.	1	15	3	26	573	455	545	545	--	--	--	--
IV. Planned Economies												
Germany E.	11,520	8,295	3,755	4,455	8,351	9,854	2,735	1,371	976	2,240	1,652	992
Czechoslovakia	3,613	2,824	4,880	6,763	787	1,276	2,220	1,854	1,718	1,735	927	515
U.S.S.R.	24,008	33,534	40,259	92,897	662	2,098	1,883	7,852	1,379	2,973	3,309	2,601
Hungary	414	559	594	760	663	1,054	1,155	1,018	596	1,020	1,116	1,294
Poland	1,564	2,331	5,619	5,736	485	543	2,380	1,962	447	538	347	629
Bulgaria	423	674	750	1,271	90	240	393	102	78	164	167	242
Yogoslavia	173	143	58	58	650	706	355	355	95	90	87	110
Romania	2,955	1,075	1,123	1,194	1,283	1,246	572	314	224	313	106	103

Source: Industrial Property Statistics, World Intellectual Property Organization, Geneva.

TABLE 19.3
Utility Models (Petty Patents) Granted

	Applications						Utility Models Granted					
	Nationals			Foreigners			Nationals			Foreigners		
	1967	1975	1980	1967	1975	1980	1967	1975	1980	1967	1975	1980
Germany (FR)	42,214	30,114	26,094	11,344	11,938	8,153	20,948	12,099	10,252	2,400	2,181	1,879
Italy	4,418	--		778			3,935			702		
Japan	109,154	178,992	190,388	1,906	1,668	1,397	20,601	47,449	49,468	721	957	533
Philippines	141	565	762	2	7	24	94	331	465	--	9	3
Poland	1,647	1,896	2,523	22	31	36	411	1,775	1,680	4	25	20
Portugal	139	78	118	25	13	15	77	153	159	9	25	6
Spain	7,601	7,650	5,830	710	1,353	1,162	6,177	4,128	3,845	600	2,041	1,131
Brazil	--	--	1,657	--	--	89	--	--	131	--	--	13
Korea	--	7,052	7,936	--	238	622	--	1,032	1,315	--	14	438

Source: Industrial Property annual statistical reports.

(3) $\dfrac{\text{total patents granted at home to national and to foreigners}}{\text{patents obtained by nationals abroad}}$

This provides an index of international "leadership" in these invention fields.

Table 19.6 provides a more detailed picture of a relationship between invention patents and utility models in one important developing country, the Philippines. In this table, patents are assigned to industries using a concordance developed by Mikkelsen, Medalla, and Evenson (1982). The industries related to pre-harvest and harvest agricultural production include: industry 0, general agriculture; 3119, other industrial chemicals and fertilizers; and 362, agricultural machinery. Agricultural processing (post-harvest) industries include: 20, food manufacturing; 21, beverages; 22, tobacco products; and 29, leather.

The agricultural production industries produced 53 invention patents of domestic origin from 1971-1980 (or 9 percent of all domestically originated patents). They produced 247 utility models (8 percent of the total). The processing industries obtained 54 invention patents (9 percent of the total) and 71 utility models (2 percent of the total). The utility model option was used particularly heavily in the agricultural machinery industry. Mikkelsen, Medalla, and Evenson (1983) report that of a sample of 54 small agricultural implement firms in the Philippines (average size, 26.8 employees), 25 had obtained or applied for one or more utility models while only one had obtained or applied for an invention patent.

Finally, table 19.7 provides data for India, showing invention in agricultural machinery. Two conclusions are immediately evident from this table. First, foreign patenting was, until 1970, the dominant source of agricultural technology. Second, with the passage of the Patents Act of 1970, the flow of patents in agricultural machinery slowed to a trickle.

On the whole, the data on patenting show the following:

1. The poorest developing countries grant few patents in agriculture. They generally have not devised legal systems to encourage indigenous invention of any kind.

2. The medium-income developing countries grant a significant number of patents but, with few exceptions, they grant them to foreigners protecting their products in growing national markets. Many of these patents are in

TABLE 19.4
Average Annual Patenting in Agriculture,* Selected Countries

	Average Annual Agricultural Patents			Breakdown by Country of Origin of Patents Issued, 1977-80							
	1965-68	1969-72	1977-80	Indigenous	U.S.	U.K.	West Germany	Other Western Europe	Eastern Europe & USSR	Japan	Other
I. Industrialized Market Economies											
A. Rapid Growth											
Australia	304.0	327.0	242.5	40.3	20.5	7.0	104.8	57.8	6.8	3.5	2.0
Belgium	415.3	362.7	150.3	31.8	20.3	7.8	24.5	57.8	3.3	2.0	3.0
Denmark	186.5	179.5	38.5	12.5	4.0	1.5	6.3	12.0	.8	.5	1.0
France	1178.5	836.0	509.0	243.8	48.5	16.3	70.3	98.8	11.8	8.8	11.0
West Germany	473.7	299.2	269.0	150.8	30.3	3.5	150.8	62.8	9.8	8.3	3.8
Netherlands	76.5	65.0	40.3	20.3	3.3	.8	6.5	5.0	1.0	.3	3.3
Norway	n/a	70.8	46.3	12.5	9.3	1.3	4.8	15.0	.3	1.3	1.8
B. Moderate to Slow Growth											
Australia	116.5	133.5	151.5	28.0	60.3	10.0	4.0	23.5	0.0	6.0	19.8
Canada	576.7	444.7	n/a	--	--	--	--	--	--	--	--
Finland	44.2	49.7	42.3	18.8	3.3	.3	1.3	16.8	0.0	1.8	.5
Italy [c]	n/a	792.5	145.0	n/a	n/a	n/a	n/a	n/a	n/a	n/a	n/a
Ireland	48.7	45.5	26.3	**	3.3	5.0	2.5	8.8	.3	0.0	6.5
Sweden [a]	206.2	185.2	81.3	n/a	n/a	n/a	n/a	n/a	n/a	n/a	n/a
Switzerland	497.0	417.5	79.3	n/a	n/a	n/a	n/a	n/a	n/a	n/a	n/a
U.K.	476.7	427.0	368.3	n/a	n/a	n/a	n/a	n/a	n/a	n/a	n/a
U.S.A.	785.0	830.0	n/a	--	--	--	--	--	--	--	--
II. Semi-Industrialized Market Economies											
A. Rapid Growth											
Hong Kong	n/a	n/a	5.5	**	2.0	.8	.3	.5	0.0	1.3	.8
Israel	87.2	118.7	n/a	--	--	--	--	--	--	--	--
Portugal [d]	35.0	n/a	0.0	--	--	--	--	--	--	--	--
Spain	448.0	393.7	197.0	53.9	22.0	7.3	19.3	70.8	3.5	2.8	7.3

B. Moderate to Slow Growth											
Chile	53.0	79.7	5.7	**	1.7	0.0	0.0	2.0	0.0	0.0	2.0
Uruguay	32.0	24.0	16.0	**	2.5	.8	5.5	3.0	0.0	.3	4.0
Venezuela	25.5	111.7	n/a	--	--	--	--	--	--	--	--
III. Developing Economies											
A. Asia and Pacific											
Bangladesh	n/a	n/a	0.0	--	--	--	--	--	--	--	--
India[d]	32.5	33.7	14.5	**	2.5	0.0	.5	.5	1.0	1.5	8.5
Iran[e]	n/a	n/a	370.0	**	121.0	25.0	49.0	139.0	2.0	9.0	25.0
Philippines	19.2	43.7	67.3	**	23.5	3.5	7.0	10.0	0.0	8.5	15.0
Sri Lanka	8.0	n/a	n/a	--	--	--	--	--	--	--	--
Thailand[d]	n/a	n/a	0.0	--	--	--	--	--	--	--	--
B. Latin America											
Barbados[f]	n/a	n/a	2.0	0.0	0.0	1.0	0.0	1.0	0.0	0.0	0.0
Colombia	26.0	30.5	0.0	--	--	--	--	--	--	--	--
Costa Rica[d]	22.0	10.0	.5	0.0	.5	0.0	0.0	0.0	0.0	0.0	0.0
Haiti[b]	n/a	n/a	.5	0.0	0.0	0.0	0.0	.5	0.0	0.0	0.0
Jamaica[h]	n/a	n/a	.3	0.0	.3	0.0	0.0	0.0	0.0	0.0	0.0
Peru[d]	n/a	n/a	5.5	**	1.5	0.0	.5	2.0	0.0	0.0	1.5
Trinidad & Tobago	7.2	5.5	0.0	--	--	--	--	--	--	--	--
C. Africa											
Algeria[d]	n/a	n/a	0.0	--	--	--	--	--	--	--	--
Burundi	n/a	n/a	.8	**	0.0	0.0	0.0	0.0	0.0	0.0	.8
Egypt	n/a	n/a	3.3	**	.3	.3	.5	1.3	.3	0.0	.8
Ghana[d]	n/a	n/a	4.0	**	.5	.5	.5	2.0	0.0	.5	0.0
Kenya	8.0	10.2	3.3	0.0	.5	0.0	2.0	.5	0.0	.3	0.0
Malawi	22.0	15.0	2.5	**	.8	.3	.5	.3	0.0	0.0	1.0
Mali[d]	n/a	n/a	0.0	--	--	--	--	--	--	--	--
Mauritius	n/a	6.5	7.5	**	5.0	.5	0.0	.8	0.0	0.0	.3
Morocco	29.3	19.0	11.8	**	1.0	.3	1.8	7.8	.5	0.0	.5
O.A.P.I[f]	n/a	n/a	21.0	**	2.0	2.0	3.0	10.0	0.0	3.0	1.0
Rwanda	0.0	0.0	0.0	--	--	--	--	--	--	--	--
Sierra Leone[b]	0.0	.5	0.0	--	--	--	--	--	--	--	--
Tanzania[b]	n/a	n/a	.7	0.0	0.0	0.0	0.0	0.0	0.0	.7	0.0
Uganda	7.0	6.7	3.0	**	.5	.5	1.0	.5	0.0	.3	.3
Zaire	n/a	6.0	2.8	**	.5	0.0	.3	1.0	0.0	0.0	1.0
Zambia	26.7	8.0	4.3	**	1.5	.5	.3	.5	0.0	0.0	1.3
Zimbabwe[d]	n/a	n/a	1.5	**	0.0	0.0	.5	.5	0.0	0.0	.5

(continued)

TABLE 19.4 (Cont.)

	Average Annual Agricultural Patents			Breakdown by Country of Origin of Patents Issued, 1977-80							
	1965-68	1969-72	1977-80	Indigenous	U.S.	U.K.	West Germany	Other Western Europe	Eastern Europe & USSR	Japan	Other
IV. Planned Economies											
Bulgaria	48.0	61.7	36.8	**	8.8	2.3	19.0	8.5	2.0	1.0	.3
- Utility Models	n/a	n/a	89.0	86.3	0.0	0.0	0.0	0.0	2.8	0.0	0.0
Cuba[b]	11.0	4.0	4.3	**	0.0	0.0	.3	1.0	0.0	0.0	3.0
Czechoslovakia[d]	95.7	162.5	234.5	**	0.0	0.0	0.0	0.0	0.0	0.0	234.5
- Utility Models	n/a	n/a	210.0	191.0	0.0	0.0	0.0	0.0	0.0	0.0	19.0
Hungary	34.5	115.5	64.3	37.5	5.0	1.0	3.8	7.5	9.0	0.0	.5
Korea, North[i]	n/a	n/a	10.0	n/a	n/a	n/a	n/a	n/a	n/a	n/a	n/a
Poland	n/a	49.5	91.0	**	.8	.5	5.3	7.0	4.3	1.0	72.3
Romania	n/a	n/a	53.3	**	4.0	1.0	3.8	4.0	3.3	0.0	37.3
USSR	n/a	n/a	16.0	0.0	4.0	.3	3.5	5.8	1.5	.5	.5
- Utility Models	828.0	1231.3	1300.8	1294.8	0.0	0.0	0.0	.3	5.8	0.0	0.0

Sources: Industrial Property Statistics, 1977, 1978, 1979, 1980, chart III, World Intellectual Property Organization, Geneva; for 1965-72, the source is the annual supplement to Industrial Property.

Notes: *The figures for 1965-72 and those for 1977-80 are not directly comparable because of differing definitions of "agricultural" patenting used in the two periods.

**Indigenous patents are included in the "other" category.

[a]1978, 1979, 1980 only.

[b]1977, 1979, 1980 only.

[c]1980 only.

[d]1979, 1980 only.

[e]1978 only.

[f]1977 only.

[g]1977, 1978 only.

[h]1977, 1978, 1979 only.

[i]1977, 1979 only.

TABLE 19.5
Origin of Agricultural Patents Granted by Selected Countries and Ratio of Total Patents Granted to Total Patents Granted to Nationals, 1973–78

	Plows				Harrows				Tractor-Related Implements			
	Total Patents Granted	Domestic Origin	LDC Origin	Trade Balance Ratio	Total Patents Granted	Domestic Origin	LDC Origin	Ratio	Total Patents Granted	Domestic Origin	LDC Origin	Ratio
Developing Countries												
Argentina	3	1	0	3/1	2	2	0	2/2	1	1	0	1/3
Brazil	47	39	0	47/39	55	42	1	55/42	46	28	0	46/28
Cuba	0	0	0	0	1	1	0	1/1	1	1	0	1/3
India	2	2	0	2/2	1	1	0	1/1	0	0	0	0
Philippines	1	1	0	1/1	9	5	0	9/5	3	0	0	3/0
Developed Countries												
U.S.	232	200	0	232/285	291	182	0	291/274	347	304	0	347/786
U.K.	80	37	2	80/98	249	73	4	249/108	229	52	0	229/151
France	330	172	0	330/211	642	240	1	642/265	837	203	0	837/266
Japan	77	70	0	77/74	302	261	0	302/264	408	371	0	408/384
USSR	322	322	0	332/322	301	298	0	301/298	462	452	0	462/456

TABLE 19.6
Invention Patents and Utility Models Obtained in the Philippines by Philippine and Foreign Inventors, 1951-1980

SIC Code	Industry	1951-60	1961-70	1971-80	1951-60	1961-70	1971-80	1951-60	1961-70	1971-80
Agriculture-Related Industries: Pre-Harvest										
0	Agriculture, hunting, etc.	4	7	36	2	15	30	1	23	87
3119	Other industrial chemcials and fertilizers	2	2	6	18	80	150	0	1	7
362	Agricultural machinery	3	3	11	7	16	33	1	19	153
Agriculture-Related Industries: Post-Harvest										
20	Food manufacturing	1	2	44	11	41	230	0	1	1
21	Beverages	0	0	8	1	48	66	0	3	35
22	Tobacco products	4	12	2	7	48	73	0	3	10
29	Leather	0	1	0	0	0	3	0	1	24
Resource-Based Industries										
1	Mining and quarrying	0	8	15	4	131	98	3	14	99
25	Wood and cork, exc. furniture	0	0	3	7	17	14	0	6	50
27	Paper and paper products	0	3	9	3	63	52	0	11	33
30	Rubber products	0	0	0	4	47	23	0	1	5
32	Petroleum and coal products	2	2	12	3	22	17	0	1	11
33	Other non-metal mineral prod.	0	1	6	2	45	68	3	20	74
34	Basic metal industries	2	0	5	2	40	65	1	26	41
Chemicals										
315	Drugs, medicines, cosmetics	0	1	19	54	447	1079	5	6	87
31	Other chemicals, not 315/3119	5	16	100	42	532	2784	6	37	105

<u>Manufacturing Industries</u>

4	Electricity, gas, steam	1	0	35	2	60	107	0	23	118
5	Construction	1	3	6	16	88	192	1	5	45
23	Textiles	3	3	35	11	108	158	1	13	21
24	Clothing and footwear	0	2	12	7	31	20	3	14	90
26	Furniture and fixtures	3	3	19	10	38	35	4	46	256
28	Printing and publishing	2	0	9	2	16	12	0	5	49
35	Fabricated metal products	2	7	30	6	102	121	15	46	239
369	Machinery, not agr. or elec.	7	27	118	69	424	673	15	19	395
37	Electrical machinery and appl.	2	8	62	93	208	303	4	30	292
38	Transportation	1	5	24	3	23	42	3	21	146
39	Instruments and misc. mfg.	6	11	108	50	414	578	22	82	702

Table 19.7
Schedule of Patenting Activity in Agricultural Machinery in India (Indian Classes 5A, 5C, 5D, 5E), 1912-1980

Years	Domestic Patents	Foreign Patents	Total Patents
1912-1920	9	49	58
1921-1925	9	32	41
1926-1930	4	36	40
1931-1935	8	14	22
1936-1940	7	14	21
1941-1945	0	26	26
1946-1950	6	46	52
1951-1955	18	66	84
1956-1960	25	63	88
1961-1965	37	141	178
1966-1970	51	116	167
1971-1975	12	25	37
1976-1980	3	8	11

Source: Indian patent specifications lodged at the Patent Offices in Delhi and Calcutta.

chemical fields. Countries in this group (notably the Philippines) that have attempted to devise petty patent systems to encourage local adaptive invention appear to have succeeded in this effort.

3. The advanced developing countries do more patenting. Again, unless some imagination is employed in designing legal systems, many of these countries grant most patents to foreigners. Where utility models or other devices are used to stimulate domestic invention, they appear to have an extensive impact. Only the more advanced countries in this semi-industrialized category appear to be generating inventions that, in turn, are patented in other countries.

4. Among developed countries, the patenting pattern has changed quite rapidly in recent years as Japan and Europe (especially Denmark and France) have expanded their foreign patenting. The United States, which once dominated many of these technology fields, is now relegated to a lesser role.

BIOGENETIC INVENTIONS: SPECIAL CONSIDERATIONS

Historically, inventions and discoveries involving living organisms have contributed greatly to increases in agricultural productivity.[7] Until recently, however, most of these advances stemmed from the domestication and selective breeding of plants and animals. Recent discoveries regarding the genetic structure of living things and new techniques for manipulating this structure have brought new strains of microorganisms, new plant varieties, and new animal breeds. In addition, technological innovations have speeded up the process of selective breeding by employing sperm storage, artificial insemination, estrus synchronization, superovulation, embryo transfer, embryo storage, sex selection, twinning, *in vitro* fertilization, and the like.

One may reasonably expect that more and more agricultural research will seek to employ this new biogenetic technology in specific agricultural inventions. Thus, it is particularly appropriate at this time to examine the history of patent/variety protection for living organisms in conjunction with the history of this technology, to suggest what may, or should, be forthcoming in the development of these legal systems.

Historically, patent/variety protection for living organisms has been much more restrictive than patent protection for devices and chemical compositions. Even in the United States, it was not until 1930 that a very limited Plant Patent Act was enacted, the current version of which reads:

> Whoever invents or discovers and asexually reproduces any distinct and new variety of plant, including cultivated sports, mutants, hybrids, and newly found seedlings, other than a tuber-propagated plant or a plant found in an uncultivated state, may obtain a patent therefor, subject to the conditions and requirements of title.
>
> In the case of a plant patent the grant shall be of the right to exclude others from asexually reproducing the plant or selling or using the plant so reproduced. (Emphasis added.)

To this day, the 1930 U.S. Plant Patent Act in practice restricts patenting to orchard fruits and ornamental flowers (primary roses); most of the patents cover roses, apples, peaches, and chrysanthemums. The U.S. Plant Patent Act provides no meaningful protection for new varieties of major farm crops such as corn, wheat, soybeans, oats, etc., since they cannot be commercially exploited without utilizing seeds, thereby falling outside of the "asexual reproduction" class.[8]

It was not until 1970 that the United States enacted a Plant Variety Protection Act (PVPA), which covered new varieties of sexually reproduced plants. The passage of this act coincided with the emerging consensus that plant breeding technology had advanced to the point where new, stable, uniform varieties could be sexually reproduced. Fungi, bacteria, first-generation hybrids, and six soup vegetables (okra, celery, peppers, tomatoes, carrots and cucumbers) did not qualify. In 1980 Congress amended the U.S. PVPA to include the soup vegetables.

In table 19.8, we present data from seventeen countries on the number of plant patents granted in 1975 and in 1980 (sometimes called "registrations" in those countries, like the United States, where separate legislation governs plant variety protection). The "registrations in force" column gives the cumulative total of registrations that had not expired in 1980.

In the United States, the PVPA has encouraged a significant expansion in private sector breeding activity. The USDA had issued over 1100 PVPA certificates by the end of 1983. Only 14 percent of these were awarded to public sector breeders in the USDA and state agricultural

TABLE 19.8
Plant Patents Granted 1975, 1980

	Registrations Granted				Registration in Force
	Nations		Foreigners		
	1975	1980	1975	1980	1980
Argentina		--		--	
Belgium		19		82	284
Bulgaria		16		--	
Denmark	19	19	35	90	428
France	114	187	29	19	963
Germany	151	260	52	143	2,210
Hungary	4	--	--	4	31
Netherlands	187	282	90	91	1,912
New Zealand	--	43	--	--	98
Romania		6		--	62
South Africa	12	10	0	22	91
Soviet Union	--	94		--	2,826
Spain		2		--	2
Sweden	9	22	8	14	172
Switzerland		--		3	20
U.K.	73	75	66	139	1,013
U.S.A.	124	201	26	49	3,064
U.S.A. (PPA)	97	150	1	3	700

Source: Industrial Property.

experiment stations. Some 56 private companies now constitute the National Council of Commercial Plant Breeders and many new biogenetic firms are entering the field as well. Private varieties are growing in importance in several commodities (including wheat and soybeans), in addition to hybrid corn, sorghum, and certain vegetables where the private sector has long had a major role.

As to living organisms other than plants, namely microorganisms, it was not until the Chakrabarty decision in 1980 that the U.S. Supreme Court construed U.S. patent law to cover genetically engineered bacteria. Previously the Court had held that processes for the development of new living organisms were properly patentable subject matter.[9]

Advances in biotechnology have now spawned several dozen "venture" firms with the objective of developing agricultural biotechnology. Millions of dollars have been invested and projections of the potential agricultural market in the year 2000 are in the $50-100 billion range.

While it is still difficult to predict with much certainty the actual impact of the modern private biotechnology industry, it is reasonable to conclude that it will have a major impact. At this point, third world countries by and large have not given consideration to effective legal systems designed to deal with its peculiarities.

POLICY ISSUES

This survey of legal systems and agricultural invention, while incomplete in some respects, does indicate that existing legal systems in most third world countries are not producing optimal results. The invention patent appears to be stimulating significant numbers of domestic inventions in only a handful of countries. This may be partly due to inefficient administration. Many countries also expressly limit protection for agriculture-related inventions.

Many developing countries have joined international patent conventions without full consideration of the consequences. Membership typically has two effects. The first is that members freely grant domestic protection to foreign inventors. The second is that they adhere to an international standard regarding the granting of invention patents. This second effect is probably more detrimental than the first because it effectively means that many national legal systems do not stimulate domestic invention. They only provide a means for the purchase of technology from abroad. An effective and useful system should do both.

Legal systems that apply international standards do not stimulate national invention because they serve only one part of a differentiated international demand for technology and most developing countries are at a serious disadvantage in serving that particular demand.

The large "primary" markets for most types of technology lie in the developed high-wage economies serving high-income consumers. The demand for new technology in developing countries, on the other hand, is for technology suited to the lower wage conditions and for the commodities produced by these technologies. Firms and individuals in developing countries are at a comparative disadvantage in competing with firms from Japan, Germany, France, and the United States to produce new primary inventions. Not only do they not have the financing, or the access to skills, but more importantly, they do not have the economic "laboratories"--e.g., the capacity to "test market"--that firms in the high-wage countries have. On the other hand, they have a comparative advantage in the production of "adapted" technology because they can modify primary market technology to suit it to the smaller, secondary markets in their own economies. The standard legal systems of many third world countries do not stimulate this secondary adaptive invention. A good deal takes place, but if anything, the legal systems discourage it by providing easy patenting for a broad range of "blocking" patents to foreign and multinational firms.[10]

Most developing countries would be wise to review their general industrial strategy and to question whether they are not placing domestic inventors at a serious disadvantage vis-á-vis foreigners. During import substitution phases of development, many governments have pursued policies that favor large national and multinational firms over smaller firms. The standard legal systems may reinforce this bias by providing mechanisms for the large firms to purchase foreign technology, but not small firms.

An effective patent system should enable small firms and individuals to receive rewards in competition with larger firms. The experience of the Philippines, Brazil, Japan, and other countries with petty patent systems seems generally to have been good. In these countries, the invention patent and petty patent systems operate side by side, providing the advantages of both, with few concomitant disadvantages. By providing a weaker and narrower standard of patentability and by administering the system efficiently, these countries have stimulated local adaptive invention.

As for extending patent protection to plants and animals, the type of research that produces these inventions is generally so location-specific that it has to be done _in situ_, so the issue of international patent recognition may not be too important. We suggest that a system that adequately safeguards public breeding and research programs and stimulates private biotechnical product development makes good sense.

In many third world countries, private sector research and development is now quite significant, and problems of conflict and adjustment between public and private sector systems have arisen. It is instructive to note that public sector agricultural researchers, including those at the International Agricultural Research Centers (IARCs), have generally not attempted to foster more private sector activity. They have worked with some private sector researchers in the hybrid corn and hybrid sorghum seed industries, and in the broiler industries, but have not provided leadership to expand private sector activity to a broader range of technology. They have not, for example, encouraged more effective plant variety protection or petty patent systems.

In view of the rather clear evidence that most countries spend less than is socially optimal on agricultural research (Evenson, Waggoner, and Ruttan 1979) and that the level of scientific training and sophistication of their research programs is too low, it is somewhat puzzling that development strategies (including the agricultural research establishments themselves) are not doing more to encourage private research. We suspect that development economists themselves are largely responsible for this. The welfare and efficiency implications of patent laws are complex. The skills of the well-trained economist are required to lay a solid analytical foundation on which effective policy can be built. By and large, the fraternity of development economists generally (and of agricultural economists in particular) has not laid such a foundation in this field.

Indeed, development economists may well have contributed to the confused policy base that has hampered the effective development of legal systems. Most non-economists see the granting of the "monopoly power" inherent in a patent as improper, particularly if granted to a foreign firm. Economists have failed generally to point out that with competition the monopoly rents captured by an inventor with patent rights accrue only to an invention that otherwise would not have existed. Furthermore, they fail to point out that trade secrecy permits the capture of monopoly rents in a much more concentrated fashion.

Agricultural scientists often argue that giving patent protection to private breeders allows the private breeder to capture monopoly rents produced at public expense. This would be true if public sector varieties of germplasm were provided to private breeders with an exclusive license for the production and sale of seed. The situation is very different if several private firms <u>and the public sector</u> are competing suppliers in the market. Indeed, if the public sector research institutions have a modestly competent seed supply enterprise, they themselves can prevent private breeders, or even a single private breeder, from collecting revenues or rents in excess of the economic value of his improvement over the public variety.

Agricultural scientists and development planners and administrators often fear the growth of the private sector on the grounds that public sector enterprises should be protected from competition. Many development planners, including most economists, also have a bias against the private sector, particularly as it affects the agricultural sector. The growth of large private agricultural supply firms raises the fear that traditional rural systems will be destroyed and that welfare losses will follow.

Many agricultural research administrators in the third world are in fact attempting to staff and develop research institutions in a protected environment. They attempt to pay low salaries and to compensate scientists in the form of housing and graduate study fellowships. Unfortunately, one of the most pervasive problems throughout the third world today is that the economic environment for such institutions is actually unprotected and that staff erosion to the private sector and to international institutions is a severe problem. These problems may be exacerbated by an expansion of private sector research. Discouraging private sector research will not, however, solve the problem.

Our purpose in this paper has been to attempt to identify the major outline of private sector agricultural research activity. We conclude that the general patterns emerging from our limited data reflect legal and institutional systems that are far from optimal.

NOTES

1. Within the last two years, however, the courts in the United States have rendered decisions that reveal their increasing sympathy with the view that computer programs constitute protectable intellectual property like any other invention. As we show in table 19.1, several developing countries now extend formal patent or patent-like protection to computer programs.

2. There is one limited area of invention where copyright protection may be meaningful to an agricultural researcher, namely computer software. Since in practice the use of computer software many times requires actually

copying the software itself, copyright protection may be significant.

3. As a result, there seems not to be any meaningful distinction between an invention patent and a utility model, which is the other type of patent that countries commonly grant, and which does not require an inventive step.

4. Since the existence of an inventive step need not be determined, such systems also cost less to administer than do most invention patent systems.

5. Algeria, Argentina, Brazil, Egypt, Iran, Israel, Madagascar, Malawi, North Korea, the Philippines, Sri Lanka, Syria, along with the following members of the African Intellectual Property Organization: Cameroon, Central African Republic, Chad, Congo, Ivory Coast, Senegal, and Togo. The remainder comprises Western Europe, North America, Japan, and half of Eastern Europe.

6. It may also result in the invention's being withheld altogether, if the transfer thereof requires cooperation or other participation by the inventing firm.

7. The most noteworthy are the high-yielding rice, wheat, and maize varieties that together constituted the "Green Revolution."

8. Congress apparently based this requirement for asexual reproduction on the belief that sexually reproduced varieties could not be reproduced true to type and that it would be senseless to try to protect a variety that would change in the next generation.

9. In view of the way in which the bacteria would have to be exploited commercially, this decision significantly expanded the scope of patent protection. Once the bacterium strain was introduced to the public, even by a single sale, it would be difficult, if not impossible, to prove infringements even of the process patent claims. For this reason, prior to the court's decision, such process claims for many of these inventions had very limited value.

10. Of course, as noted previously, the failure to grant protection to foreign technology may result in the withholding of the technology altogether.

REFERENCES

Baxter, J. W., and J. P. Sinnott. World Patent Law and Practice. New York: Matthew Bender, 1985.

Boyce, James, and R. E. Evenson. National and International Agricultural Research and Extension. New York: Agricultural Development Council, 1975.

Evenson, R. E. "International Inventions: Implications for Technology Market Analysis." In R&D, Patents and Productivity, edited by Z. Griliches. Chicago: University of Chicago Press, 1984.

Evenson, R.E., P. Waggoner, and V. Ruttan. "Economic Benefits from Research: An Example from Agriculture." Science 205 (Sept. 14, 1979): 1101-7.

Greene, Anne Marie. Designs and Utility Models Throughout the World. New York: Trade Activities Division of Clark Boardman Company, Ltd., 1983.

________. Patents Throughout the World. New York: Trade Activities Division of Clark Boardman Company, Ltd., 1985.

Mikkelsen, K., E. Medalla, and R. E. Evenson. "Invention in Philippine Industry." Journal of Philippine Development. 9(1-2)(1982): 16.

APPENDIX A

APPENDIX A

AGRICULTURE: RESEARCH PLANNING PARALYZED BY PORK-BARREL POLITICS*

Nicholas Wade

The agricultural research system is an organizational leviathan that employs some 10,000 scientists and consumes about half a billion dollars a year. The system is replete with paradoxes. It is credited with a major share in the remarkable productivity of American agriculture yet, in the opinion of some economists, the present burgeoning of food prices may be the delayed result of a decline in research quality. It has caused over the years a revolution in every aspect of agriculture, yet it has itself changed hardly at all. Its pattern of growth has been determined by history rather than the needs of the time, and when its growth is curtailed, as at present, it finds adaptation painful or impossible. Its business is innovation, but it innovates conservatively, choosing to achieve numerous small advances rather than revolutionary breakthroughs. Its productive workers are its scientists, but it is ruled by a higher caste that consists chiefly of administrators. It is finely attuned to the immediate needs of its clients, but its central nervous system has only vestigial control over its working parts. It is said to be governed by the rational dictates of planning, priority-setting, and coordination. In fact, it obeys a quite different logic. The agricultural research system is politicized from crown to grass roots. Its operation needs to be understood not in terms of the administrators' organization charts, but as the behavior of a highly political animal.

The animal that the agricultural research system most nearly resembles, in elegance, coordination, and singleness of purpose, is without doubt Siamese twins. One twin is the system of state agricultural experiment stations (SAES), of which there is one in each state, usually located on the campus of a land-grant college or university. The SAES are designed to serve needs down to a quite local level, and to this end each station may have up to 53 branch stations. Under the terms of the Hatch Act of 1887, the SAES receive a significant proportion of their funds, virtually without strings attached, from the federal government. (Federal Hatch funds, totaling $83 million in the current fiscal year, account for 22 percent of the average station's budget, varying within wide limits according to a formula based on the size of the state's rural population and other factors.) Of the rest of the SAES budget, about half is derived from state legislatures and the remainder from industry, foundations, and other federal agencies such as the National Science Foundation and National Institutes of Health.

The counterpart to the SAES is the federal research system of the U.S. Department of Agriculture (USDA), comprising the Agricultural Research Service (ARS), the research arm of the Forest Service, and the Economic Research Service (ERS). The present budgets of these three organizations are $200 million, $57 million, and $17 million, respectively. The total USDA effort amounts to about 40 percent of the public investment in agricultural research, the rest being performed by the SAES. The research conducted by industry is roughly equal to that of the USDA and SAES combined.

It might be logical to suppose that, if the 53 state stations and their numerous branches cater to regional needs, the USDA would elect to look after national needs and hence would concentrate its resources in a few laboratories of more than critical size. The ARS and Forest Service do support some large installations, notably the ARS laboratory complex in Beltsville, Maryland, the large utilization research labs in each of the ARS's four regions, and the forest products laboratory in Madison, Wisconsin. But the resources of the two agencies are in fact spread over a total of some 280 separate laboratories, stations, and work sites. Many of these laboratories are evidence of an extreme fragmentation of effort. The Southwest Cotton Insects Laboratory in Waco, Texas, for example, is manned by just two professional scientists. And the Boll Weevil Research Laboratory in Tallulah, Louisiana, is home to just a single researcher.

Both the state and federal research systems are thus doing essentially the same kind of work and are following the same fragmented approach. There are, of course, some particular differences--the ARS accepts research responsibility for certain national problems such as the threat of foot-and-mouth disease--but by and large the general character of the research undertaken by the two systems is indistinguishable.

The reason why the ARS developed into a mirror image of the state station system goes back to before 1954, when the agency was created out of a number of different bureaus devoted to entomology, dairy science, and other disciplines. Following the customary laws of bureaucratic survival, each bureau chief built up ties with the particular farm industry his bureau served and with the congressmen who shared this constituency. The congressmen liked to place new facilities where the voters could see them, and each bureau chief tended his own satrapy without caring what the others did. Federal research stations were thus distributed piecemeal over the country on a strictly pork-barrel basis.

Power of Officials Broken

When the ARS was created, the bureaus were renamed divisions, but the power structure remained the same. The old alliance between the division chiefs and Congress continued, and the administrator of ARS was effectively powerless to coordinate or shift resources from one division to another without the consent of the division chiefs and Congressman Jamie L. Whitten (D-Miss.), the chairman of the House agriculture appropriations subcommittee. George W. Irving, ARS administrator from 1965 until last year, tried and failed three times to assert control over the divisions. Meanwhile, ARS laboratories continued to be created in accordance with political realities. The ARS was reorganized last June in order to simplify administration, as the official explanation has it, which is true as far as it goes. The real purpose was to strip the division chiefs of their power and to channel more authority through the ARS administrator's office. "It looks to me like the whole thing may be an effort to get away from responsibility to the Congress," growled Whitten when shown the reorganization plan. It was, but he could not stop it.

The same forces that fragment the federal research effort are also operative in the states. There are economic data to suggest that large stations give almost twice as much value for money invested as smaller stations. But SAES directors who would like to close out some of their less productive, small branch stations know that their state legislature will soon hear from the farmers whom the branch stations serve. The SAES directors have considerable lobbying power with the federal government, both individually through their own congressmen and collectively through their Washington-based lobbying organization, the National Association of State Universities and Land-Grant Colleges (NASULGC). The NASULGC discusses the agricultural research budget with officials of the USDA, the Office of Management and Budget, and Congress but has generally been more effective at warding off threatened cuts than at wheedling more money from the federal coffers. Some believe that the SAES lobbyists have held back the ARS by insisting that the two organizations grow in step. Joint SAES-ARS statements speak of the necessity that "each of these two partners be funded in such a manner as to maintain a reasonable balance"--which suggests that the two partners see themselves as rivals. Certainly the SAES tend to complain bitterly when the ARS receives a larger budget increase. Sometimes this is of their own doing: SAES directors often lobby individually with their own congressmen in favor of ARS facilities proposed for their own state, and the effect of 53 SAES directors lobbying for the ARS is not negligible. This year, for the first time in recent memory, the federal allocation to the SAES was cut back, and the SAES are uncertain whether to blame the Office of Management and Budget (OMB) or the report of the Pound committee (see below), or both. The reason for the cutback is that the Secretary of Agriculture chose to let the budget ax fall on the SAES rather than elsewhere in his domain.

How is research policy made in the agricultural research system? Outward signs of fervent policy-making are not too evident. Comparison of the total SAES budget allocated to 15 different research areas in the periods 1951 to 1954 and 1961 to 1964 shows that only three of these areas changed in their relative share of support by more than a single percentage point. A similar conservatism is evident in the research allocations of the ARS. The formula funding system under which the states receive their federal subventions does not assist the allocation of resources according to merit or need. In the case of the ARS, policy-makers have been hampered by a number of constraints of varying degrees of severity.

Many of the constraints spring from Congress. The autonomy accorded to the division chiefs was a major obstacle to research planning. So is the circumstance that Whitten, who has been chairman of the House agriculture appropriations subcommittee since 1949, knows the USDA like the back of his hand and is liable to query item changes as small as $2000. Congress also earmarks funds for particular commodities, such as cotton, often to an extent disproportionate to their relative economic worth. Earmarks, however, tend to remain in the budget after their designated problem has been solved and thus lose some of their restrictiveness.

Another source of inflexibility is Congress's possessiveness toward even minor research installations. "It's often less trouble to close down a military base than a two-man agricultural research station," says an OMB official, not entirely in jest. The ARS has succeeded in closing down only 30 or so stations in the last 10 years. Construction of facilities at Congress's behest has probably been the major determinant of the ARS research program. Over the last 100 years Congress has followed a cyclical pattern of erecting buildings and then neglecting to fund them properly. At present the ARS is in the trough of a cycle, and its latest batch of research laboratories are only 60 percent staffed.

ARS Budget a Christmas Tree

A different sort of problem is that the ARS budget is sometimes used as a Christmas tree whereon to hang the goodies that will secure the appropriation for the whole USDA an easy passage through Congress. USDA officials have also been able to play Congress against the executive branch, warding off cuts threatened by the OMB by citing the threat of congressional opposition.

A major constraint internal to the ARS--and SAES--is that almost all research is conducted by tenured staff whose specialized training makes it difficult to switch them to new research fields. New work is most easily undertaken with new resources, but in the last few years the ARS has lost both funds and personnel, making major shifts of emphasis almost impossible.

The attention of the ARS administrator has often been diverted from research planning by day-to-day crises, whether regulatory duties (until these were transferred in 1971) or outbreaks of epidemics such as southern corn blight, citrus blackfly, or exotic Newcastle fowl disease. These crises have been met successfully, although sometimes not without effort. "The ARS did respond to southern corn blight, and they did it well, but there was some screaming and gnashing of teeth," says a close observer of the USDA, who adds that the department "was not always anticipating as well as they could have done" the regulatory decisions coming out of the Food and Drug Administration and the Environmental Protection Agency.

Maybe the most serious obstacle to formulating a research policy in the ARS is the apparent caste barrier between the top and bottom of the organization, or between the scientists and administrators. The higher echelons of the ARS are filled with people who long ago left the laboratory and have worked their way up through the ranks. Nothing wrong with that, but so few working scientists rise to the highest salary grades in ARS that research is clearly the harder path to promotion, and active scientists seem to be seriously underrepresented in the agency's top counsels.

Scientists in the ARS are usually referred to by administrators as "bench scientists," as if the administrators conceived of themselves as a higher order of desk scientists. "We are the bench scientists, the foot soldiers of science, at the same level as the plumbers and technicians," says an ARS scientist of national reputation. As a graphic example of the scientists' place at the bottom of the totem pole, he cites the case of a power failure at the Beltsville laboratories that lasted for several days. "Quickly they ran emergency lines to the administrative office so that the paperwork could go on, but no one thought about our freezers. I had to go grabbing around for dry ice to save my specimens." Another well-known scientist says of the period before last year's reorganization (the effects of which, he says, it is too early to judge): "There was a horrible morale problem in the whole of ARS. Things here [at Beltsville] were so bad that we were saying that what was needed was a bomb out here so that you could start all over again."

While morale is good or not so bad in many other laboratories, in some it is rather worse. The committee chaired by Glenn S. Pound, the report of which has been discussed in previous articles (Science, 5 Jan., 27 April, 4 and 18 May), chanced upon one large ARS laboratory where scientists reported that mail was censored, telephone conversations monitored, and the staff on the verge of mutiny. The director of the laboratory reportedly 'intends to rule this laboratory by calculated intimidation' and he was said to operate 'by threatening people with reassignment to more unpleasant jobs, demotion, abolition of their jobs, and dismissal. Nothing riles him as much as basic research and professional recognition.' Pound says that this laboratory was not unique. His committee learned of other laboratories where there was a "question whether the administrators had the kind of philosophy that would provide the atmosphere for an unfettered quest for truth." Other sources have said that the laboratories in question are the four utilization laboratories, which, after Beltsville, are among the largest of the ARS's installations. (It may be significant that the work on utilization has not been considered a very successful effort in total.)

Steps have been taken to assist the director of the laboratory that horrified the Pound committee, and morale at Beltsville has improved since last year's reorganization. Talcott W. Edminster, administrator of ARS since July 1972, says that he has never sensed a division between scientists and administrators in ARS. Asked if scientists are concerned about their status in the ARS, Edminster says that "If they were, they would leave. I have talked with 2000 of our scientists in the last few months and I think most of them are pretty happy."

Unlike SAES scientists, many of whom are located on campus and hold dual appointments with the university, ARS scientists are relatively isolated from academic life. Many ARS stations are located off campus. The agency performed only 3 percent of its research work extramurally last year and hired the services of only ten outside consultants. University scientists, in turn, have often made their agricultural colleagues feel like poor cousins at the academic table, and for this reason, some say, have accorded them relatively few academic honors. Does the ARS lack its fair share of outstanding researchers, as the Pound committee suggests? "I would like to see this as unmerited," says Irving, "but if I were to make a case against it, there is not a great deal of evidence I could find. One measure is the number of people anointed by the National Academy of Sciences. If you compare the ARS with the numbers that come from M.I.T. or Illinois, it makes agricultural research look pretty puny."

Academic merit, the criterion by which the Pound committee measured the USDA-SAES system, is in some ways an unfair yardstick. Unlike universities, the system is not designed to produce Nobel Prize winners, although this it does do. It is designed to solve seemingly pedestrian but economically important problems in response to the needs of its clients, which is why much of the direction comes from the grass roots rather than the top. Such apparatus for directing research policy as is visible to the outside observer appears, on closer inspection, to play a largely ceremonial role.

The principal reef on which research planning founders is the jealously guarded autonomy of the 53 state stations. The SAES directors supposedly plan research through their own organization, the Experiment Station Committee on Organization and Policy (ESCOP). In practice, ESCOP is chiefly a lobbying organization with little effective influence on individual state policies. It collects and coordinates the directors' wish list but has no power to tell the directors what to do.

A similar degree of impotence characterizes the other body supposed to coordinate state research, the Cooperative State Research Service (CSRS). The CSRS, an agency of the USDA, is charged with disbursing federal funds to the state stations and with reviewing the projects the states propose to undertake with the funds. To this end, CSRS has a staff of 111 and an administrative budget of $2.3 million. The teeth of the review process, however, have been drawn by the station directors. Few proposals are rejected. Some are deferred but, according to an internal CSRS report,[1] the common practice of state scientists is to rewrite the proposal without changing the project. The CSRS does not veto any project which a state director believes should go through. As for the coordination of SAES research strategy, the previous CSRS administrator, T. C. Byerly, attempted to play such a role but was, in the words of one observer, "crucified" by the state directors. The present administrator, Roy L. Lovvorn, is a former state director. "We don't have any direct control over the state stations," Lovvorn explains. "We cannot tell them what to do, but we can point out duplication to them." The two principal coordinating bodies for the states, ESCOP and CSRS, thus have no power to coordinate.

The fact that there is no means to ensure the coordination of state agricultural research raises certain obstacles, if not an impassable roadblock, in the path of harmonizing state research with the federal effort. The body supposed to do this is the Agricultural Policy Advisory Committee (ARPAC), which includes as members the chairmen of ESCOP, the administrators of ARS, CSRS, and ERS, and the head of the Forest Service's research arm. It is hard to see that ARPAC has exerted any more control over the agricultural research system than does ESCOP or CSRS.

ARPAC's functions are by definition advisory. Lloyd Davies, the executive secretary of ARPAC, cannot name any new line of research ever initiated by the committee.

ARPAC might be expected to resolve boundary disputes between USDA agencies. The entomology research programs of the ARS and the Forest Service, for example, are described by one source as "two separate empires--on one campus I visited they weren't even talking to each other, although they are working on the same problems." ARPAC's power of persuasion with the state stations seems to vary inversely with their degree of financial independence from the federal treasury. Stations attached to the North Central universities have access to several sources of funds and derive only a small part of their support from the federal government. They behave fairly independently of the USDA and each other. Southern universities, on the other hand, caught in financial straits, are more amenable. (Historically, however, the competitiveness of the SAES system, and its freedom from USDA bureaucracy, have been advantages that helped make it the stronger of the two rivals.)

Following the reorganization of the ARS, ARPAC has initiated a regional research planning system. The system is coordinated by Davies and a member of ESCOP, but no one has overall direction, and the system, as is usual with agricultural research planning bodies, is purely voluntary.

Yet another body said by administrators to provide coordination and direction is the National Planning Staff of the ARS. A product of last year's reorganization, the National Planning Staff has four assistant administrators and a staff of 40. Its functions are purely advisory. It does not yet seem to have found a role in life. Finally, the ARS also supports a ten-man group known as the Program Analysis and Coordination Staff, the usefulness of which it is too early to assess.

ESCOP, ARPAC, CSRS, the National Planning Staff, the Program Analysis and Coordination Staff, the regional planning system--the common feature of these bodies is that their powers are considerably less grand than their titles. They are the window dressing on the political realities, producing for both public and internal consumption a geocentric explanation of a heliocentric system. This may be one reason why the agricultural research system has never been able to come to grips with the structural faults described in the report of the Pound committee. These included pedestrian research, inadequate support of basic science, duplication of effort, inept management of scientists, and administrative philosophies repressive of the vitality of science. Perhaps the remarkable feature of the report was not the forcefulness of its conclusions, but their lack of novelty. A series of previous committees, whose reports never saw the light of day, apparently reached similar conclusions. "I am not displeased with the Pound report," says former ARS administrator Irving. "They said a good many things that have been said before, and in some ways they have said them better." (The USDA deserves credit for its sense in releasing the Pound report, although this was only bowing to the inevitable.) The Pound committee was originally asked to look simply at scientific issues. When it broadened its scope to include the management of science, a high USDA official asked Pound to desist but was ignored. Until its report was delivered, the Pound committee was intended to have a permanent existence.

Assuming that the Pound committee and its predecessors are correct in saying that agricultural research is not managed in a sensible fashion, it does not necessarily follow that the system's output has suffered, likely though this would seem. The SAES have probably maintained the research leadership but the official list of even the USDA's research achievements over the last 30 years, an impressive document by any standards, lists page after page of economically significant discoveries and improvements. As the defenders of the system are wont to say, agricultural research is good for agriculture, regardless of whether or not it meets various academic criteria. The clients whom the system serves want a more efficient dairy cow, not a substitute for milk; they need a stream of small improvements that will increase profitability, not a revolutionary discovery that will drive them out of business. This is precisely what the agricultural research system has provided. "Constant improvement in animals and plants is more important in the long run than flashy breakthroughs," says ARS administrator Edminster.

How has U.S. agriculture come to be so marvelously productive if the state of agricultural research is as poor as the Pound committee believes? This paradox, frequently posed by USDA officials, is perhaps not as tight an alibi as it might at first seem. Research is only one among other factors, notably capital investment, that have raised the crop yield per acre by some 70 percent in the last 40 years. Even the productivity gains attributable to research do not necessarily justify the present research system in its entirety. The gains could be the work of only a minority of the system's 10,000 scientists. They might also derive from work done long ago; the land-grant colleges, for example, probably found it harder to compete for good people after the 1950's, when other universities started to support large research efforts.

The pattern of productivity gains in fact suggest that the quality of agricultural research may, if anything, have dropped off in the last decade. According to Robert E. Evenson, a Yale University economist specializing in agricultural development, there has been an apparent slowdown in productivity growth since the early 1960's. "With a lag, that is showing up in the present rise in farm prices. It also suggests that the contribution of the research system is lower than it once was." The lower contribution, Evenson believes, could be caused by the type of deficiencies described in the Pound committee report.

One reason for this suggestion is that a similar drop-off in research productivity which occurred in the 1920's seems to have been related to the isolation of agricultural research from basic science. The reintegration with basic science that was effected at that time led to the improvements in the breeding and health of plants and animals that underlay the productivity gains made in subsequent decades. The agricultural sciences "have by and large neglected their ties with the basic sciences," Evenson says. The recent major advances in biology do not seem to have worked their way into agriculture, despite the potential for large gains in efficiency. Evenson doubts if the system is capable of integrating these findings. The time may have come for another reintegration with basic science, similar to that which occurred in the 1920's. Others have expressed the idea that the agricultural research system may need some new source of inspiration. Sterling B. Hendricks, an eminent researcher now retired from the ARS, suggests that the methods which have underwritten the success story of agriculture--chiefly improvements in the control of disease and breeding--may already have yielded their full return, and some other source of payoff must be looked for in the future, maybe from more fundamental kinds of research.

Is the agricultural research establishment likely to recast its endeavors in this way? The short answer is no. Congress is content with the system as it is. The OMB has only eight examiners to monitor the entire USDA and seems in any case to take the general view that further research is of questionable value while the government is doling out subsidies to farmers to keep 60 million acres idle. (The counterargument is that agricultural produce is a major American export whose importance to the balance of trade is likely to increase in the years ahead; more and better research would therefore be justified.)

Within the ARS, all energies for change are still occupied in the recent reorganization which, however, was undertaken for political reasons and was not designed to affect the conduct of research one way or another. (The effect of the reorganization is to decentralize decision-making and to place it on a geographic instead of a disciplinary basis.) The contention of some ARS administrators that the reorganization has dealt with all the problems raised by the Pound committee is unconvincing. For one thing, the reorganization was planned long before the Pound Committee reported. Peer review, one of the committee's chief recommendations, is a difficult process to apply to a largely tenured staff. Nonetheless, limited forms of peer review are being tested out in ARS, notably by the ARS deputy administrator for the northeastern region, Steven C. King.

In the years ahead, both the ARS and SAES seem likely to face small or negative growth in their budgets, the SAES especially as the rural power base in state legislatures continues to be eroded. There is already a case to be made for consolidating some of the state stations--all of the New England stations, for example, might be rolled into one, or a single mountain state station set up. There is also a case to be made for consolidating the rival ARS and SAES systems into a single organization that would cater to national, regional, and local needs on a rational instead of an historical-political basis. The integrated system might be subjected to a national peer review process of the type operated by the National Institutes of Health, in order to secure uniform judgments as to priority and scientific merit.

A radical restructuring of this nature is not at all likely to occur tomorrow. The system has in the past served its clients extremely well, and the arguments for change have so far convinced only a few. Too many powerful forces are combined in keeping the system as it is--decentralized, uncoordinated, fragmented, undirected, and easy for special interest groups to manipulate. Things will have to become a lot worse before they get any better.

---

*Reprinted from Science 180 (June 1, 1973):922-37.

1. J. J. Endean, "CSRS Administrative Procedures: An Outsider's Appraisal," mimeographed (16 August 1971).

APPENDIX B

APPENDIX B

SCIENCE FOR AGRICULTURE

Report of a Workshop on Critical Issues in American Agricultural Research

Jointly sponsored by The Rockefeller Foundation and the Office of Science and Technology Policy Executive Office of the President United States of America

June 14-15, 1982
Winrock International Conference Center
Petit Jean Mountain
Morrilton, Arkansas

Published by the Rockefeller Foundation
1133 Avenue of the Americas
New York NY 10036

PREFACE

Agriculture is vitally important to our national security, the strength of our economy, and the health and welfare of our citizens as well as those of many other countries. Agricultural research and technology development have been responsible for the remarkable productivity of American agriculture over a period of many decades. The public institutions created to carry out agricultural research and its communication were set in place between 1862 and 1914. There is concern that these institutions may no longer be able to sustain their level of past performance. Numerous studies and reports indicate a need to strengthen agricultural research in this country if it is to meet the numerous, diverse, and complex challenges facing American agriculture now and in the years ahead.

Assuring the scientific and technological advances needed to meet those challenges will require national commitment and leadership and the mobilization of all appropriate resources including those of the Federal Government, state land-grant colleges, other state and private colleges and universities, non-profit research institutes, and the business and industrial sectors. Unfortunately, much of the debate over the future of agricultural research is characterized by interinstitutional and interdiscipline tension, defensiveness, and rigidity at a time when institutional collaboration and flexibility, interdisciplinary efforts, and a focus on scientific and technological problems are essential for progress.

The critical need to provide a constructive framework for a national dialogue over the future of agricultural research led 15 individuals (Appendix I) from government, academia, and the private sector to hold two days of concentrated discussions at the Winrock Conference Center (Petit Jean Mountain, Arkansas; June 14-15, 1982). These discussions focused on identifying the problems in agricultural research and exploring the roles of the various sectors and institutions in solving those problems and strengthening this country's agricultural research enterprise.[1]

This is a report of those discussions. It is not a definitive catalogue of the problems in agricultural research, nor do we have a complete solution for the problems discussed. Rather, it is an effort to initiate a constructive dialogue intended to define those problems and explore possible solutions. It is hoped that the widespread dissemination of the report will help influence the nature of deliberations over the future of agricultural research in this country so that they focus on the future; emphasize strengthening an already productive enterprise; involve all individuals, institutions, and sectors capable of contributing to meeting the scientific and technological challenges facing American agriculture; and concentrate on substantive collaboration in solving problems in agriculture through research and development.

John A. Pino
Denis J. Prager
September 1, 1982

INTRODUCTION

Agriculture[2] is vital to the U.S. economy. In 1980, gross farm income was $150 billion; farm production assets were valued in excess of $900 billion; and agricultural exports were valued at $40 billion, accounting for 19% of all U.S. exports. The food and agricultural sector as a whole is the largest of all U.S. industries, employing 20% of the civilian labor force. Expenditures for food now account for 16% of consumer disposable income.

However, American agriculture is disturbingly fragile. Production and productivity continue to increase, but more slowly than in the past. Those remaining in farming operate on an increasingly thin margin, straining their land and financial solvency and pushing our collective knowledge and technologies to the limit. This fragile enterprise faces even greater challenges as the need for food increases worldwide, domestic and international markets compete, and commodity and food sales return proportionately less to producers, while the inputs to agriculture--land, water, energy, chemicals, labor, and capital--become more costly and less available. The seriousness of these threats to the future of American agriculture indicates the scope and magnitude of the challenges facing the research enterprise on which agriculture depends.

The agricultural research system is responsible for the scientific and technological advances which, for decades, have sustained agricultural productivity. Annual rates of return on research expenditures in agriculture range from 35 to 50%, well above the returns to other public investments and a clear indication that we are seriously underinvesting in this national resource.

It is just its value to the country which is at the root of current concern over the ability of the agricultural research system to sustain its past performance. It was the consensus of the participants at Winrock that the continued success of the agricultural research enterprise will depend upon its ability to retain the strongest elements of a system in existence for 120 years, while changing and adapting in response to: new and ever-tougher challenges in agriculture; rapidly advancing scientific frontiers; shifting social values and government responsibilities; growing fiscal and resource constraints; and increased involvement of research institutions and scientists outside the traditional system.[3]

IDENTIFICATION OF CRITICAL ISSUES

The Winrock discussions began with an attempt to identify the bases for the collective concern for the future of agricultural research. Generally, the issues identified can be grouped under three headings: issues of public policy; issues related to institutions and their relationships; and issues related to the performance of the enterprise.

Public Policy Issues

- The lack of a coherent national agriculture policy, relating productivity goals and domestic and international policies with an explicit understanding of the value of agriculture to this country, greatly hampers efforts to establish national goals and priorities for agricultural research.
- The lack of clear, implementable national goals, priorities, and plans for agricultural research results from, and accentuates, the tremendous diversity and complexity of a research system designed to meet both national and local, immediate and long-term, and basic and applied research needs.
- Inadequate public understanding of the importance of agriculture to the strength and vitality of this country and the well-being of its citizens leads public officials generally to undervalue agricultural research. As a result, agricultural research has a relatively low priority at the national level, federal appropriations for agricultural research are well below the level consistent with its payoff and value to society, and distribution of funds is largely on the basis of geopolitics rather than need or expected return.
- The changing political balance of power in Congress, from primarily rural to predominantly urban, portends even greater difficulty in rationalizing public policies related to agriculture and agricultural research.

Issues of Institutions and Their Relationships

- Institutional change--change in the government, university, and industrial sectors, in the individual institutions comprising those sectors, and in the ways in which the sectors relate and the institutions interact--is the factor most affecting agricultural research today.

- Over the last three decades, institutional roles and relationships have changed as American agriculture has evolved, research challenges and opportunities have changed, and other sectors and institutions have become increasingly involved in agriculture-related research. At least ten federal agencies, in addition to the U.S. Department of Agriculture, now fund research pertinent to agriculture; public and private universities outside the land-grant system perform agricultural-related research and train scientists who conduct such research; and industry is increasingly investing in basic and applied research and technology development leading to new agricultural products and processes.
- Nearly a decade of federal and, in some cases, state funding constraints has greatly exacerbated natural institutional resistance to change, resulting in excessive parochialism and preoccupation with institutional protection and maintenance.
- The tendency of scientists outside the agricultural research establishment to treat agricultural scientists as something other than equals (witness the relatively small number of agricultural scientists in the National Academy of Sciences) creates additional barriers to needed institutional change and accommodation and to the application of all appropriate scientific resources to the solution of problems in agriculture.
- The cumulative effect of these factors is a level of interinstitutional tension that has hindered needed change within and among the institutions which comprise the agricultural research system. As a result, institutional energies have been devoted to administrative matters and relative budget levels rather than to the identification of critical research needs or the development of interinstituional relationships to bring about the technological advances needed by agriculture.
- The seriousness of this situation highlights two major issues: 1) there is a critical need for more high quality, perceptive leaders of national stature in agricultural research; and 2) it is unclear who represents and can speak for the

various components of the agricultural research system. The resulting leadership vacuum leaves agricultural research with inadequate, confused, and often contradictory representation at the national level during a critical period for the country, for agriculture, and for agricultural research.

Issues Related to the Performance of the System

- The criticism expressed most by observers of the agricultural research system is the unevenness of the system's focus on critical scientific problems. As a result, for example, there has been a piecemeal approach to gaining crucial, fundamental knowledge about the biology of the organisms on which the future of American agriculture depends. Some experiment station and USDA laboratories have excellent basic research programs; however, many do not.
- This concern over the adequacy of basic science in experiment station and USDA laboratories has major implications. There is a reluctance at the federal level to increase funding for research perceived neither to be of the highest quality nor focused on the most critical scientific problems. Federal funding and top scientific talent are going to institutions outside the system, conducting "cutting-edge" basic science. And, industry is increasingly performing basic research to develop the knowledge base necessary for new agricultural products and processes, raising new questions about the appropriate roles for, and relationships between, the public and private sectors in agricultural research.
- Inadequacies in the quantity and quality of basic research in the agricultural research system are traceable to: 1) loss of a clear national priority for basic agricultural research; 2) increasing substitution of political goals for scientific goals in Congress and in some state legislatures; 3) the need for experiment station research to reflect the more applied priorities of the states; and 4) an overall lack of adequate public funding for agricultural research.

- Another criticism of the system is its perceived inability to address, in a coherent manner, truly national issues. This is traceable to: 1) a diversity of state experiment stations with their necessary principal focus on state and local problems; and 2) a USDA so constrained by Congressional political priorities and by Executive Branch budget, personnel, and management restrictions that it cannot exert real leadership in determining national scientific needs and priorities and in focusing the energies of its laboratory system on meeting those needs and priorities.

INSTITUTIONAL ROLES AND RELATIONSHIPS

Many of the critical issues in agricultural research identified above relate directly to change, and the degree to which institutions are successfully adapting to change. Accordingly, a major portion of the discussions at Winrock focused on defining: 1) appropriate and complementary roles for institutions involved in supporting and conducting science for agriculture; and 2) the ways in which those institutions should relate.

The Federal Roles

Since enactment of the Hatch Act in 1887, the Federal Government, acting through the USDA, has been a key participant in the agricultural research arena, providing funds to experiment stations for research, conducting research in in-house laboratories and field stations, and coordinating state and federal research activities. For much of this time, agricultural research accounted for a major portion of the entire federal R&D budget; in fact, other research agencies such as NIH and NSF did not exist. Research and education activities were the raison d'être for the USDA and accounted for most of its budget. The federal role was clear and prominent: as a member of the so-called "federal-state partnership," the Federal Government was viewed as an integral part of the powerful coalition which, for many years, set research priorities, influenced the size of budgets and the allocation of resources, and nurtured and protected the USDA and the land-grant institutions via the political process. Through its research facilities, the USDA performed much of the basic science related to agriculture, led in setting the standards for research quality and excellence in the field, and provided the scientific and intellectual leadership, both domestically and internationally, for agricultural research.

However, much of this has changed since the 1930s. The research-education coalition, which once dominated the agricultural political scene, has largely been replaced by special interest groups representing commodity, consumer, agribusiness, and specialized farm interests for whom research is a relatively low priority. The "farm bloc" which once controlled Congress has all but disappeared as the country has become urbanized. The USDA has long since lost its primary focus on research and education, as its political attention has been diverted to providing direct economic benefits to farmers, agribusinesses, and consumers. And, agricultural research has become a small part of the total federal R&D budget (only 1.25% of the total federal R&D budget, and 2% of the total USDA budget), while other federal agencies fund major research programs of significance to agriculture.

As a result of these changes, the federal role in agricultural research has become less clear. The land-grant institutions question the federal commitment to the federal-state partnership, as the flow of research funds to the states has ceased to grow in real terms. Federal support for USDA research has also ceased to grow and the USDA research capability has become diffuse and its quality uneven. Federal scientific and intellectual leadership is no longer clearly visible.

. . .

Strong federal support for, and leadership in, agricultural research must be reestablished and sustained if all relevant research and development resources in this country are to be applied efficiently and effectively to the development of new knowledge and technologies for American agriculture. That support and leadership should include:

- *sustaining a vigorous and healthy long-term national capability in science for agriculture at the state and national levels, including adequate support and conduct of basic science and a cadre of trained scientific manpower working in up-to-date facilities;*
- *providing intellectual leadership and facilitating coordination for all institutions contributing to*

science for agriculture, focusing their collective efforts on the solution of high priority scientific and technical problems;

- *analyzing and anticipating long-term agricultural problems of national concern;*
- *mobilizing the resources necessary to respond to national emergencies and crises;*
- *providing a focus for efforts to relate the specific scientific and technical concerns of agricultural research to the overall goals of agriculture and to the broad social context within which agriculture operates; and*
- *stimulating and coordinating interactions between American agricultural scientists and their international counterparts.*

. . .

THE AGRICULTURAL RESEARCH SERVICE. Research capability and quality within the USDA's Agricultural Research Service (ARS) are perceived by some to have declined in recent years as the ARS mission has become diffuse and basic food production research has been de-emphasized in favor of more applied research and development. Political interests have been responsible for the establishment and retention of a large number of field sites and major facilities, many not justifiable in terms of research need or efficient allocation of scarce resources. Personnel ceilings have largely thwarted efforts to bring in bright, young talent, reducing the influx of new ideas and methods and increasing the average age of the ARS scientific staff. Administrative complexity, inflexible constraints on research administration, and a regionally decentralized structure lessen the ability of ARS managers to control funding and human resources so as to induce constructive change.

. . .

The advantages and benefits of a federal agricultural research capability argue for greatly strengthening the ARS and enhancing its scientific excellence.

- *The principal role of the ARS is research in support of short- and long-term federal missions, including:*
  - *basic science*
  - *requirements of federal action and regulatory agencies*
  - *management and preservation of federally owned resources*

  - *assuring the security of the American food supply*
- *ARS should be restructured to take maximum advantage of the inherent strengths of a federal research capability:*
  - *a clearly defined primary role--research*
  - *a stable cadre of full-time scientists*
  - *the ability to direct and control funds, personnel, and facilities so as to meet national needs*
  - *relatively stable funding focused on coherent research programs.*
- *The principal emphasis of ARS should be basic science. ARS should strive to re-establish its excellence and leadership in basic science for agriculture.*
- *Funding for this long-term, coherent focus on basic science should comprise two distinct sources.*
  - *base funding: a stable, predictable source of funds for long-term programs and for institutional development and maintenance*
  - *competitive funding: ARS scientists should be permitted and encouraged to seek research project grant funds from competitive funding programs in USDA and other federal agencies to support state-of-the-art research in high priority areas of basic science. All new ARS funds, above and beyond inflation, would be expected to derive from various sources of competitive funding.*
- *Research centers of excellence should be established in ARS through reorganization and consolidation of existing activities and facilities. Such centers should contain a critical mass of talent and facilities necessary to conduct cutting-edge science for agriculture.*
- *Each existing ARS facility should be evaluated thoroughly and objectively to determine its utility in meeting ARS research missions. For each facility, one of the following options should be recommended for implementation:*
  - *retain as a federal ARS research facility with roles and responsibilities consistent with the missions of a strengthened ARS*
  - *turn over to the host state with federal funding continued for a specified period, on a declining basis*

   - *sell to a public or private university or to industry*
   - *close.*
- *Congress should permit ARS to determine the fate of individual research facilities and programs solely on the basis of their ability to contribute to attainment of national research goals and priorities.*
- *ARS research managers should be delegated sufficiently flexible control of funds and personnel to facilitate reinvigoration of the ARS basic research capability, then evaluated on the basis of their ability to recognize and promote high quality, imaginative research.*
- *Rigorous peer review of proposed scientific programs, ongoing research, and individual scientist performance must be an integral and routine part of ARS research management, drawing, to the degree possible, on nationally recognized scientists external to the ARS system.*
- *A variety of new and innovative ways of broadening the experience of ARS scientists and of increasing the flow of new ideas and talent through the ARS laboratories should be implemented, including:*
   - *post-doctoral fellowships for new graduates*
   - *sabbaticals for ARS scientists*
   - *exchanges with universities and industry*
   - *support of ARS research fellows in universities.*

. . .

COOPERATIVE STATE RESEARCH SERVICE. The Cooperative State Research Service (CSRS) was established to represent the Federal Government in its partnership with the State Agricultural Experiment Stations. Due to past personnel and budget constraints and to lack of real support from the states, CSRS personnel and funds are largely limited to administration and oversight of formula funds. As a result, the federal-state dialogue increasingly focuses on budgetary and administrative matters instead of on substantive research issues.

. . .

The critically important relationship between the two principal partners in the agricultural research enterprise--the USDA and the state experiment stations--would be greatly enhanced by a Cooperative State Research Service more responsive to the individual and collective scientific programmatic needs of the states.

- *Formula funds to the states for agricultural research should be provided by the Federal Government with an absolute minimum of administration and oversight. The CSRS role in the administration of formula funds should be minimized accordingly.*
- *CSRS should comprise a small staff of qualified scientists whose principal role vis-à-vis the states is to provide intellectual leadership in:*
  - *focusing attention on research needs of national and regional concern*
  - *enhancing coordination among the states and between the states and various elements of the USDA and other federal agencies*
  - *facilitating the collection and dissemination of up-to-date, accurate information on agricultural research in progress.*
- *A new, critically important role is envisioned for a restructured and reoriented CSRS:*
  - *organization and mobilization of consortia of state and other research capabilities to address specific, high priority, national and regional research needs*
  - *funding of such consortia through a variety of non-formula mechanisms designed to be highly flexible and targeted and to permit participation of the highest quality and most relevant expertise, regardless of sector or institution.*
- *State support for a greatly strengthened CSRS, providing scientific leadership, coordination, and information, is vital to efforts to enhance the federal-state partnership in agricultural research.*

. . .

OTHER FEDERAL AGENCIES. A great deal of research pertinent to agriculture is supported or conducted by a number of other federal agencies. Public and private research institutions outside the traditional agricultural research system receive significant amounts of funding from these agencies for research applicable to agriculture. To an increasing degree, the more research oriented land-grant colleges are also seeking competitive funding from these sources. However: 1) there is still a tendency among many experiment station scientists to pass up these potentially fruitful sources of research funds, 2) there is no formal mechanism for identifying research funded by these agencies which is of particular relevance to agriculture, and 3) there is virtually no coordination at the federal level between these agencies and the USDA.

. . .

It is important that all federal resources allocated for research pertinent to agriculture--science for agriculture--be identified and exploited and that the results of the research funded by those resources be applied to the solution of agricultural problems.

- *The USDA should take the lead in identifying all federal programs funding research related to agriculture and in disseminating appropriate information to experiment station and ARS scientists, as well as to scientists outside the traditional system.*
- *The USDA should make a major effort to assure that information on relevant research funded by other agencies is readily available through the Current Research Information Systems (CRIS) or other information mechanisms.*
- *The USDA should utilize the interagency committee authorized by the 1977 and 1981 farm bills to bring about meaningful collaboration and coordination among federal agencies supporting science for agriculture.*

. . .

The States

State Agricultural Experiment Stations are organized principally to conduct agricultural research of high priority to their home states. Their state-oriented activities have important spillover benefits for other states, their regions, the nation, and the world, although these are secondary priorities. In addition, the vast majority of agricultural scientists employed in government, industry, and academia today obtained advanced degrees at land-grant universities where most of the cost of their graduate education was supported by the experiment stations.

The State Agricultural Experiment Stations and the land-grant agricultural colleges receive their mandates and a large part of their resources from state legislatures. These institutions reflect the extraordinary diversity of interests and priorities of the several state agricultural systems they serve and the needs, priorities, and fiscal stringencies of the states in which they operate. They conform to no single model, varying from state to state as local problems, priorities, and pressures, and host university policies and practices vary.

State agricultural research programs have grown considerably over the last several decades and now occupy the dominant position in agricultural research once occupied by the Agricultural Research Service. However, because each experiment station must, perforce, focus its energy primarily on problems of high priority to its home state, less attention is given to the growing number of regional, national, and international issues. In addition, because the exigencies of the state appropriations process require each experiment station to emphasize research on relatively short-term problems of high priority to state agriculture interests, there is too often inadequate attention to replenishing the knowledge base through basic research.

There is the sense that substantive communication among experiment stations, and between the experiment stations and other public and private research universities, industry, and the Federal Government is inadequate, resulting in the putative "isolation" of agricultural scientists so often cited. The lack of an effective, facilitatory CSRS at the federal level greatly hampers interexperiment station and federal-state communication, and federal administrative requirements inhibit regional efforts. Other research universities and industry lie outside traditional channels and are thus harder to access.

For the states, the major problem at the national level is redundant and highly confusing representation of the so-called "land-grant community" by multiple groups (NASULGC, Directors-at-Large, ESCOP, CAHA, CARET), all claiming to speak for "the states." The resulting confusion has increasingly negative impacts on Congressional and Executive Branch officials continually exposed to competing and conflicting views from the very community they need to understand and are trying to serve.

. . .

The system of State Agricultural Experiment Stations forms the backbone of this country's agricultural research enterprise. If that system is to retain its pivotal role, it must recognize the rapidly changing world of science for agriculture and take steps to adapt and change accordingly.

- *Experiment station directors should take steps to organize in such a way as to be able to assume greater responsibility for the program coordination, oversight, and quality control efforts necessary to sustain and enhance the scientific productivity of the state experiment*

station system. Specifically, these steps would include:
- *increasing communication and coordination among the states on scientific issues*
- *assuming full responsibility for local control and expenditure of formula funds*
- *organizing peer reviews of projects, programs, and scientist performance*

- *Actions should be taken immediately to simplify and clarify "land-grant community" representation at the national level, designating, if possible, a single entity empowered to represent "state" views on important national agricultural research issues.*
- *To the degree possible, the state-federal partnership should be structured around, and focused on, research needs, priorities, and outcomes, rather than budgetary and administrative matters.*
- *Increased effort should be devoted to the development and implementation of regional research projects focused on common research interests and complementary research capabilities.*
- *Innovative steps should be taken to increase the substantive interaction between land-grant scientists and those from other public and private universities and from private industry.*
- *Steps should be taken to assure that significant and relevant research findings produced "outside the system" are expeditiously put into practice through experiment station/extension activities.*
- *Mechanisms should be designed to keep experiment station scientists aware of relevant research funding programs in federal agencies outside of USDA.*

. . .

Other Colleges, Universities, and Research Institutes

The USDA and the State Agricultural Experiment Stations are generally considered to be the principal partners in the agricultural research system. However, other research institutions with few ties to that traditional system are increasingly conducting research--particularly fundamental biological research--directly relevant to the solution of agricultural problems. These institutions, collectively, represent a resource of research, facilities, and talent capable of making major contributions to science for agriculture. However, this capability is principally focused on problems other than agriculture, because federal research funds targeted toward basic science for agriculture are so limited.

. . .

Meeting future agricultural research needs will increasingly require that the very best scientific expertise this country can offer be brought to bear on agricultural problems. Because significant expertise exists outside the traditional agricultural research enterprise, means will have to be devised to stimulate the interest of this capability in agriculture.

- *USDA and its land-grant partners should sponsor a series of workshops, seminars, symposia, etc., designed to bring together experts from all relevant research settings to discuss the state-of-the-art of various basic science areas, identify research needs, and explore collaborative arrangements for meeting those needs.*
- *ARS and the experiment stations should explore various innovative mechanisms--fellowships, sabbaticals, exchanges, training leaves, etc.--designed to stimulate the flow of talent through leading government, university, and industry labs.*
- *Research project grants, awarded on a competitive basis, should be used increasingly as an incentive to attract the best scientific talent to agricultural problems, regardless of where that talent resides.*

. . .

Industry

Food and agricultural industries play a major role in the U.S. economy. They develop and market a range of products and services for farmers and consumers. The development of these products and services is dependent upon new knowledge and technologies resulting from basic and applied research. Traditionally, industry has relied on USDA and the state experiment stations to perform basic research and to test the results of that research. Collaboration and interaction between industry and the land-grant system has been generally quite positive and productive, contributing significantly to productivity increases in the U.S. agricultural sector.

However, the role of industry with respect to basic agricultural research is currently undergoing change. Industrial research managers are increasing their investments in fundamental plant science--particularly in molecular biology, growth regulation, and physiology--reflecting their perception of the strong potential for proprietary products as well as their recognition of public underinvestment in basic agricultural research.

. . .

In the context of a broadened agricultural research system involving all relevant research capabilities, agricultural industries have a unique role: translation of new knowledge and technologies into commercial products. Major elements of this role include: applied research, technology development and commercialization. Secondary elements include basic science, extramural support for research and testing, and training.

- *Private sector expertise should be fully utilized in efforts by the public sector to identify future research needs, estimate future demand for scientific and technical manpower, and define appropriate, complementary roles and responsibilities for the various sectors and institutions involved in science for agriculture.*
- *Mechanisms should be developed for strengthening the linkages between the findings of basic and applied research performed in the public sector and their development and commercialization by industry.*
- *Public-private sector relationships should be actively promoted by including industry scientists in symposia, consultations, and research review teams, and by seeking the contributions of such professional scientific associations and organizations as the Industrial Research Institute.*

PUBLIC FUNDING FOR AGRICULTURAL RESEARCH

Funding for agricultural research in the public sector comes from a number of federal, state, and private sources. Originally, federal formula funds provided the major source of support for the establishment and maintenance of the state experiment stations. These funds (approximately $140 million today) are still critical for the continued operation of experiment stations in states with relatively small agricultural sectors, but account for a minor share of research budgets in states with large agricultural research budgets. A portion of the federal formula funds is specified for regional research projects involving the joint efforts of states sharing geographic proximity and research interests and needs. State funds appropriated by state legislatures to at least "match" federal formula funds are now the most important source of public funds for agricultural research, totaling four times the federal contribution.

"Special grants," for purposes largely specified by Congress, target particular research problems. The USDA competitive grants program was established in 1978 to provide support for basic agricultural research of high quality and significance without regard to the nature of the research institution. Competitive grants funding is currently $16.5 million. The CSRS administers all federal formula, regional project, special, and competitive grant funds. Experiment stations obtain additional funding through corporate research grants and contracts, "check-offs" on commodity market orders, and the sale of commodities produced on the stations.

Formula Funds

Federal formula funds, with required state matching, are a source of long-term, predictable support for research at state experiment stations. These funds lend stability and coherence to experiment station research programs, functioning as a kind of "glue" binding the various aspects of station operations. They also act as an interstate "glue," facilitating systematic coordination and cooperation among the several states and attention to regional, inter-regional, and national research issues. From the federal perspective, a relatively small investment in formula funds to the states has high return, as it leverages four times as much state support and forms a basis for federal-state dialogues on research priorities and directions.

There are some disadvantages to the use of formula funds to support research. Since the preponderance of funds to support work at experiment stations comes from state appropriations, research supported by these funds tends to focus on shorter-term, site-specific problems. For these reasons, formula funds, alone, do not always stimulate the level of basic science required to address the most critical and complex research issues. Control of research quality also can be a problem with formula funding, if research leadership is inadequate. Allocation of research funds on bases other than research merit has the potential for generating complacency among scientists and their administrators. The five-year program reviews organized by CSRS are intended, in part, to counteract that potential.

In theory, formula funding of research should be a highly efficient mechanism due to the very low "transaction costs" for working scientists. However, perennial conflicts over the level of formula funding and the nature of the priorities to be addressed create substantial costs in research administration, reducing efficiency and raising political questions about the usefulness of the funds and the equity of their distribution.

. . .

Federal formula funds to the states for research should be sustained to assure the long-term, stable base-funding necessary for maintenance of coherent state agricultural research programs.

- *A mechanism for periodic evaluation of the formula and the appropriateness and equity of its allocations should be established.*
- *A mechanism should be established for automatically adjusting the level of formula funding each year to eliminate the need for the intense negotiations and lobbying which now dominate federal-state interactions.*
- *Federal formula funds are block grants to the states for agricultural research and, as such, should entail a minimum of justification, federal oversight, accountability, and paperwork. Management and control should be delegated to the state agricultural experiment stations with periodic federal review of institutional management procedures.*
- *State-CSRS interaction should concern primarily scientific issues, not primarily administrative ones.*

. . .

Competitive Research Funds

The competitive grants program in USDA was established in 1978 on the recommendation of a number of advisory groups and reports. It is intended to complement and supplement formula funds and to offset some of the weaknesses in that funding mechanism. Competitive grants are designed to focus on basic research issues of high priority to the nation, emphasize scientific merit as judged by scientific peers, and bring to bear on research problems of significance to agriculture the highest quality scientific expertise both from within and without the traditional agricultural research system.

On the negative side, project grants support relatively narrow research for relatively short periods, decreasing flexibility, stability, and coherence of overall effort. Further, the "transaction cost" of each grant awarded is high, given the time and effort of working scientists involved in the proposal writing and review processes.

. . .

Formula funds for base support of long-term, coherent research programs at state experiment stations, and competitive research funds for targeted support of high priority research by scientists from many institutions, constitute a combination of research funding mechanisms ideal for assuring that the country's basic and applied and national and state research needs are met.

- *Competitive research funding programs in USDA should be greatly strengthened as a means of stimulating increased attention to high priority science for agriculture and of attracting to the problems of agriculture scientific capabilities heretofore dedicated to research in other areas.*
- *For the immediate future, new (real increases above inflation) federal funds to support science for agriculture should be made available through competitive funding programs targeted on high priority national research needs. Such competitive funding programs should grow to the point that they are able to support a critical mass of scientists in several major scientific areas.*

CONCLUSION

The American agricultural research system is a proven, productive enterprise responsible for the technological advances underlying the success of American agriculture. The critical importance of agriculture to the vitality and strength of this country, and the increasing diversity, complexity, and intractability of the problems facing American agriculture make it imperative that the agricultural research system be able to sustain its level of past performance. Sustaining that level will require that the institutions comprising the enterprise change and adapt in such a way as to retain the strongest elements of the existing system while responding to changing scientific frontiers, evolving social values and government roles, growing fiscal constraints, changing political realities, and increased involvement of institutions and individuals outside the system. These critical changes imply a change in emphasis from "agricultural science" to "science for agriculture."

National priorities for agricultural research should be established and continually updated, drawing upon all relevant expertise. The agricultural research system must be broadened to include all institutions and individuals conducting science relevant to solving agricultural problems. Accomplishing this will require institutional change, entirely new relationships and linkages, concentration on high priority research needs, and levels of public funding of agricultural research more nearly in line with its value to society and its demonstrated rate of return.

NOTES

1. Although participants recognized the integral roles of teaching and extension in American agriculture, their discussions focused primarily on research. The scope of the discussions was also limited by the nature of the participants to the biological and physical sciences in agriculture. It is recognized that social science research is critical to achievement of the changes advocated in this report.

2. In this report, the term "agriculture" refers to the supply and service, production, and processing, distribution and marketing sectors which comprise the entire system responsible for the food on our tables. However, specific discussions tend to focus on, and cite examples from, problems and research needs in production agriculture, reflecting the primary knowledge, experience, and concerns of the participants.

3. Throughout the report, italicized print denotes conclusions and recommendations.

APPENDIX I

Participants

Dr. Perry Adkisson
Deputy Chancellor for Agriculture
Texas A&M University System
College Station, Texas 77843

Dr. James T. Bonnen
Professor Agricultural Economics
Michigan State University
East Lansing, Michigan 48823

Dr. Winslow R. Briggs
Director, Department of Plant Biology
Carnegie Institution of Washington
Stanford, California 94305

The Honorable George E. Brown, Jr.
U.S. House of Representatives
Washington, D.C. 20515

Dr. Irwin Feller
Director, Institute for Policy Research and Evaluation
Pennsylvania State University
North 253 Burrowes Building
University Park, Pennsylvania 16802

Dr. Ralph Hardy
Director, Life Sciences
Central Research and Development
E. I. Dupont
Wilmington, Delaware 19898

Dr. James B. Kendrick, Jr.
Vice President, Agriculture & University Services
University of California
2200 University Avenue, Suite 317
Berkeley, California 94720

Dr. Terry B. Kinney, Jr.
Administrator, Agricultural Research Service
U.S. Department of Agriculture
Washington, D.C. 20250

Dr. Lowell Lewis
Director, Agricultural Experiment Stations
University of California
2200 University Avenue, Suite 317
Berkeley, California 94720

Dr. Judith Lyman
Visiting Research Fellow, Agricultural Sciences Division
The Rockefeller Foundation
1133 Avenue of the Americas
New York, New York 10036

Dr. James Martin
President, University of Arkansas
Room 425, Administration Building
Fayetteville, Arkansas 72701

Dr. John Marvel
General Manager, Research Division
Monsanto Agricultural Products Company
800 N. Lindbergh Boulevard
St. Louis, Missouri 63167

Dr. John A. Pino
Director for Agricultural Sciences
The Rockefeller Foundation
1133 Avenue of the Americas
New York, New York 10036

Dr. Denis J. Prager
Assistant Director, Office of Science & Technology Policy
Executive Office of the President
Washington, D.C. 20500

Dr. Peter van Schaik
Associate Area Director, Agricultural
Research Service
U.S. Department of Agriculture
P.O. Box 8143
Fresno, California 93747

ACKNOWLEDGEMENTS

The participants in the Winrock Workshop wish to express their appreciation to Dr. Judith Lyman for the excellent job she did in reporting the discussions at Winrock and in preparing the first draft of this report; to Ms. Suzanne Tudor for the considerable skills she demonstrated as Workshop Coordinator and Secretary; and to the staff of the Winrock International Conference Center for the professional assistance and warm hospitality which helped make the Workshop such an unusually productive event.

About the Contributors

Eliseu Alves is president of CODEVASF, Public Enterprise for the Development of the São Francisco Valley. The valley is located in a poverty-stricken area of Brazil. In his research work on institutional development, he seeks to understand how public support arises for agricultural institutions in Brazil. For fourteen years he was involved with the development of EMBRAPA, a public enterprise for agricultural research in Brazil.

Jock R. Anderson is professor of agricultural economics at the University of New England, Armidale, New South Wales, Australia. In 1984-1985 he directed the CGIAR Impact Study and has been a consultant to many universities and research institutions in the developing world. He has been deputy director of the Bureau of Agricultural Economics, Canberra, Australia, and dean of the Faculty of Economic Studies at the University of New England. His research focus is the economics of uncertainty, with emphasis on efficient resource allocation under risk.

Hans P. Binswanger is chief of the Agriculture Research Unit of the World Bank. His published work deals with a broad spectrum of agricultural development issues including the following: induced innovation, research resource allocation, and income distribution consequences of technical change; decision theory; the measurement of risk and risk aversion; econometric studies of supply response, factor demand and consumer demand; rural labor markets and contractual choice; rural nonfarm activities and employment; agricultural mechanization and mechanization policy; the evolution of farming systems in

Sub-Saharan Africa; and the behavioral and material determinants of production relations in agriculture. In his current position he combines research with policy functions.

James T. Bonnen is professor of agricultural economics at Michigan State University. His research and teaching are in public policy for agriculture. His most recent research interests include information systems theory, the design and management of statistically based policy decision systems, and agricultural research policy. He served as chairman of the National Academy of Sciences Panel on Statistics for Rural Development Policy in 1979-1980; as director of the President's Federal Statistical System Reorganization Project in 1978-1980; as a member of the President's National Advisory Commission on Rural Poverty in 1966-1967; and as a senior staff economist with the President's Council of Economic Advisers in 1963-1965.

James K. Boyce is an assistant professor of economics at the University of Massachusetts at Amherst. He is the author of Agrarian Impasse in Bengal: Institutional Constraints to Technological Change (New York: Oxford University Press, 1987); co-author (with Betsy Hartmann) of A Quiet Violence: View from a Bangladesh Village (London: Zed Press, 1983); and co-author (with Robert Evenson) of National and International Agricultural Research and Extension Programs (New York: Agricultural Development Council, 1975).

Edwin A. Crosby was president of the Agricultural Research Institute, Bethesda, Maryland, until his retirement in May 1986. Before joining the Institute in 1983, he served as senior vice president of the National Food Processors Association with general responsibilities for protecting and promoting member interests in the production and acquisition of raw agricultural commodities for canning, including related research and economic and regulatory matters.

Howard Elliott is deputy director general, research and training, at the International Service for National Agricultural Research (ISNAR). His research at ISNAR has focused on methodology for reviewing agricultural technology management systems, creation of a database on national agricultural research systems, and analysis of conditions of service for agricultural scientists. Before joining ISNAR, he served as Rockefeller Foundation representative in Brazil, where he also taught labor economics at the Federal University of Bahia, and in Zaire, where he taught agricultural economics and served as director general of the Faculty Institute of Agronomic Studies at Yangambi. His work in Africa also includes three years as Ford Foundation assistant representative for West Africa and two years as visiting lecturer at Makerere University, Kampala.

Donald D. Evenson is a partner in the law firm of Barnes & Thornburg, practicing primarily patent and trademark law in their Washington, D.C., office. He has degrees in aerospace engineering and law from the University of Minnesota and has worked for the U.S. Patent Office as an examiner.

Robert E. Evenson is a professor in the Department of Economics at the Economic Growth Center at Yale University. He has held appointments at the University of Minnesota, at the University of the Philippines, and with the Agricultural Development Council. He has conducted pioneering research on agricultural productivity growth, returns to agricultural research, and research resource allocation.

Irwin Feller is professor of economics, director of the Institute for Policy Research and Evaluation, and director of the M.S. Program in Policy Analysis at Pennsylvania State University, University Park. His research has been in the area of the diffusion of innovations in the public and private sectors. He is the principal investigator of the study, The Agricultural Technology Delivery System (University Park, Pa: Institute for Policy Research and Evaluation, Pennsylvania State University, 1984), and author of the book, Universities and State Governments (New York: Praeger, 1986). He is currently conducting research on the economic and political aspects of university-industry-state research and development partnerships.

Louis W. Goodman is dean and professor in the School of International Service at American University, Washington, D.C. Before coming to American University, he served on the faculty of the Yale University Department of Sociology and its School of Organization and Management, and directed the Latin American and the Caribbean program of the Social Science Research Council and the Latin American program of the Woodrow Wilson International Center for Scholars. He is the author of numerous books and articles, including Small Nations, Giant Firms (Holmes & Meier, 1987), a study of investment decision-making in transnational corporations operating in Latin America.

Richard R. Harwood is deputy director of the Technical Cooperation Division of the Winrock International Institute for Agricultural Development. He has served as farm and research director at the Organic Gardening and Farming Research Center of Rodale Press and as head of the Department of Multiple Cropping at the International Rice Research Institute (Philippines). He received his B.S. degree from Cornell University (1964) and his M.S. and Ph.D. in vegetable breeding from Michigan State University (1966 and 1967).

Gary H. Heichel is a plant physiologist, USDA-ARS, and adjunct professor of agronomy at the University of Minnesota, St. Paul. He is co-leader of a multidisciplinary team of USDA scientists that is seeking to improve the symbiotic nitrogen fixation capability of alfalfa, and to understand nitrogen cycling in cropping systems. He has served in managerial or panelist roles in grants programs of the USDA Competitive Research Grants Office, and the U.S.-Israel Binational Agricultural Research and Development Fund.

Robert W. Herdt is a senior economist of the Rockefeller Foundation assigned to the International Food Policy Research Institute in Washington, D.C. Before joining the Rockefeller Foundation in 1986, he was scientific advisor to the CGIAR, where he had responsibility for the CGIAR Impact Study. Between 1973 and 1978 he was economist and from 1978 to 1983 head of the Economics Department at the International Rice Research Institute, Los Baños, Philippines. He is co-author, with Randolph Barker, of The Rice Economy of Asia, which received the 1986 AAEA award for Outstanding Quality of Communication in research.

Francis S. Idachaba is a professor of agricultural economics at the University of Ibadan, Nigeria. During 1984-1986, he was on leave from Ibadan to serve as head of the Federal Agricultural Coordinating Unit (FACU). His recent publications include Agricultural Research Policy for Nigeria (International Food Policy Research Institute, 1980) and Rural Infrastructures in Nigeria (Ibadan: Ibadan University Press, 1985). His current research interests are rural development and the political economy of food and agricultural policies.

M. Ann Judd is a research associate in the Economic Growth Center at Yale University. She studied at Swarthmore. She has worked with Robert Evenson on a number of studies dealing with agricultural research investment and productivity.

Jean Lipman-Blumen is the Thornton F. Bradshaw Professor of Public Policy and Organizational Behavior at the Claremont Graduate School, Claremont, California. Her research spans several fields including agricultural research policy, Third World policy, organizational behavior, crisis management, leadership and power theory, gender roles, and achieving styles theory. She is the senior author of The Paradox of Success: The Impact of Priority Setting in Agricultural Research and Extension, a two-year strategic planning study commissioned by the USDA assistant secretary of science and education. She is the author of more than fifty papers and several books.

Prabhu L. Pingali is an economist with the International Rice Research Institute. He has been associated with Hans P. Binswanger, in the Agricultural and Rural Development Department at the World Bank, in a series of studies dealing with agricultural research priorities in Africa.

Carl E. Pray is an associate professor of agricultural economics, Department of Agricultural and Food Economics, Rutgers University, New Brunswick, New Jersey. He directed a major series of studies of the performance of Asian agricultural research systems while a staff member at the University of Minnesota. His current research is on the emerging role of private sector research in developing countries.

Jonathan D. Putnam is a graduate student in the Department of Economics, Yale University.

Vernon W. Ruttan is a Regents Professor of Economics and Agricultural Economics and an adjunct professor in the Hubert H. Humphrey Institute of Public Affairs at the

University of Minnesota. His research has been in the fields of agricultural development, resource economics, and research policy. He is the author of Agricultural Research Policy (Minneapolis: University of Minnesota, 1982) and (with Yujiro Hayami) Agricultural Development: An International Perspective (Baltimore: Johns Hopkins University Press, rev. ed. 1985).

Richard J. Sauer is vice president for agriculture, forestry and home economics and director of the Minnesota Agricultural Experiment Station, University of Minnesota, St.Paul. As vice president he coordinates the teaching, research, and extension programs of the seven units of the Institute of Agriculture, Forestry, and Home Economics. As director he administers agricultural, forestry, home economics, and some related research programs. He has published in the areas of entomology, pest control, pesticides in the environment, interdisciplinary research, and agricultural research policy and priorities.

Eduardo J. Trigo is director of the Technology Generation and Transfer Program with the Interamerican Institute for Cooperation on Agriculture (IICA) in San José, Costa Rica, and formerly a senior research officer with the International Service for National Agricultural Research (ISNAR) in the Hague, Holland. His research has been on national agricultural research policy and organization mainly in Latin America and the Caribbean. He is the author (with Martin E. Piñeiro and Jorge Ardila) of Organización de la Investigación Agropecuaria en América Latina (San José, Costa Rica, IICA, 1982) and (with Martin E. Piñeiro) of Technical Change and Social Conflict in Agriculture (Boulder, Colo.: Westview Press, 1983).

Nicholas Wade is science editor with the New York Times. At the time he wrote this article he was on the Research News staff of Science.

Francis W. Wolek is a professor of management in the Department of Management, College of Commerce and Finance, at Villanova University, Villanova, Pennsylvania. His research has been on the transfer of agricultural technologies from public laboratories to industry and on the commercialization process for agricultural innovations. His past activities include service as deputy assistant secretary for science and technology at the U.S. Department of Commerce.